T5-AGM-980

MECHANICS OF SKELETAL AND CARDIAC MUSCLE

MECHANICS OF SKELETAL AND CARDIAC MUSCLE

Edited by

CHANDLER A. PHILLIPS, M.D., P.E.

Departments of Engineering and Physiology
Wright State University
Dayton, Ohio

and

JERROLD S. PETROFSKY, Ph.D.

Departments of Engineering and Physiology
Wright State University
Dayton, Ohio

CHARLES C THOMAS • PUBLISHER
Springfield • Illinois • U.S.A.

Published and Distributed Throughout the World by

CHARLES C THOMAS • PUBLISHER
2600 South First Street
Springfield, Illinois 62717, U.S.A.

ISBN 0-398-04721-9

Library of Congress Catalog Card Number: 82-5673

With THOMAS BOOKS *careful attention is given to all details of manufacturing and design. It is the Publisher's desire to present books that are satisfactory as to their physical qualities and artistic possibilities and appropriate for their particular use.* THOMAS BOOKS *will be true to those laws of quality that assure a good name and good will.*

Printed in the United States of America

I-R5-1

Library of Congress Cataloging in Publication Data
Main entry under title:

Mechanics of skeletal and cardiac muscle.

Bibliography: p.
Includes index.
1. Striated muscle. 2. Heart–Muscle. 3. Biomechanics.
I. Phillips, Chandler A. II. Petrofsky, Jerrold Scott.
[DNLM: 1. Muscles–Physiology. 2. Heart–Physiology.
3. Biomechanics. WE 500 M485]
QP321.M339 612'.74 82-5673
ISBN 0-398-04721-9 AACR2

In memory of my father,
in recognition of my mother,
and thanks to Janie
C.A.P.

To my family,
Sherry and Missy
J.S.P.

CONTRIBUTORS

Dhanjoo N. Ghista, Ph.D.
Departments of Medicine and
Mechanical Engineering
McMaster University
Hamilton, Ontario, Canada

G. Jayaraman, Ph.D.
Department of Mechanical Engineering
and Engineering Mechanics
Michigan Technological University
Houghton, Michigan 49931

Jerrold S. Petrofsky, Ph.D.
Departments of Engineering and Physiology
Wright State University
Dayton, Ohio 45435

Chandler A. Phillips, M.D., P.E.
Departments of Engineering and Physiology
Wright State University
Dayton, Ohio 45435

Frank C.P. Yin, M.D., Ph.D.
Departments of Medicine and Physiology
Johns Hopkins University
Baltimore, Maryland 21205

George I. Zahalak, Ph.D.
Department of Mechanical Engineering
Washington University
St. Louis, Missouri 63130

PREFACE

THIS book is somewhat different from other muscle biomechanics books in that it covers both skeletal muscle mechanics and cardiac muscle mechanics. It is a state of the art review of the literature (through 1981) with an emphasis on concepts developed during the preceding decade. It also presents the current thinking of the various contributors, based upon their research experience which previously appeared only as original journal articles, but now unfolds in the more narrative style (and larger perspective) that a book permits.

The book is aimed at a broad, but scientifically sophisticated, audience. Individuals who would potentially profit from one or more of the chapters will include muscle physiologists, physical medicine specialists, cardiologists, and biomedical engineers. A fairly well developed level of mathematical understanding is required to fully appreciate all of the material presented, but (as editors and also as contributors) we have attempted to make a majority of the concepts reasonably accessible to those interested readers who may not have the necessary mathematical expertise. However, any reader must approach the various chapters with a mathematical orientation; for, to paraphrase the famous remark of Lord Kelvin, If you cannot measure it; if you cannot describe it with numbers; you have scarcely advanced the level of science.

As the title of the book implies, the "mechanics" of skeletal and cardiac muscle refers to the essentially mechanistic orientation by which the various contributors have approached their interpretations of skeletal and cardiac muscle function. The individual chapters often elucidate the physiological and anatomical basis for muscle function but proceed to describe muscle function in terms of mathematical, biophysical, mechanical, and electrical engineering principles.

As editors, we have organized this book as a progressive narrative. Each contributor, having worked in his field for the past several years, synthesizes his results, conclusions, and expectations. The original journal publications are extensively referenced for the interested reader. In the course of this unfolding narrative, it is hoped that new insights will be developed by the reader. It is certainly expected that some controversy will be generated. Indeed, there is no uniformity of agreement even among the various contributors to this book.

The individual chapters are organized for their cumulative effect. Skeletal and cardiac muscle anatomy are compared and contrasted, but are functionally unified by considering muscle as an "energy convertor" (Chapter 1). Muscle is then shown to be much more complex than a simple "motor," but its characteristic "force-velocity" relationship is still very amenable to mathematical analysis (Chapter 2). As well as being "mechanical," muscle is "electrical" and as such, this electrical activity can be analyzed with respect to what it tells us about the muscle's mechanical state (Chapter 3). Next, the mechanical aspects and electrical aspects of muscle must be integrated in order to understand the dynamics of skeletal muscle *in vivo* (Chapter 4). Finally, our understanding of and ability to control skeletal muscle electro-mechanics achieves its ultimate application with the functional electrical stimulation of paralyzed muscle (Chapter 5).

The individual chapters then change to concentrate on cardiac muscle. Initially, there is an overview that progresses through four levels of engineering analysis, literally from "molecules" to "movement" (Chapter 6). One of these levels, myocardial wall stress, is then examined in more detail and placed in both its historical and contemporary perspective (Chapter 7). Another of these levels, ventricular pumping performance, is further examined with an emphasis on myocardial (and valvular) material mechanics (Chapter 8). Finally, the narrative concludes with what "noninvasive material mechanics" can contribute to our understanding of two major categories of cardiac pathology: hypertensive cardiac disease and coronary artery disease (Chapter 9).

If this book is successful, it will expose the reader to new horizons. It will unfold a panorama in which the mathematical, physical, and engineering sciences are assisting the life sciences in understanding, explaining, and even controlling muscle function and dysfunction. The analytical techniques of thermodynamics, static and dynamic mechanics, biophysics, biomathematics, material mechanics, electronics and electrical engineering, control theory, and computer technology are all arrayed in a combined operation. However, there is no enemy to be subdued, there is only knowledge to be acquired. As any contemplative reader will soon understand, the knowledge so acquired is useful. It is knowledge that can be applied to the understanding of and the diagnosis and treatment of skeletal and cardiac muscle disorders.

The material presented in the following pages is our interpretation as editors of the current state of the art. As scientists who can express only cautious optimism, we request that our readers regard this publication as simply a first edition. Our eyes will continue to gaze toward the future; our feet (of necessity) are planted in the present; and in our hearts we acknowledge our debt to the past.

C.A.P.
J.S.P.

CONTENTS

MECHANICS OF SKELETAL AND CARDIAC MUSCLE

Chapter 1

MUSCLE

Structural and Functional Mechanics

C.A. PHILLIPS and J.S. PETROFSKY

THE beginning of understanding of any complex subject, whether a language, an insect, or an electromechanical system (such as muscle) must begin with a basic delineation of structure and function. Structure is the skeleton, a framework upon which the entire system is built. In the case of a language, it is the syntax and vocabulary. In the case of a mechanical system, whether a steam engine, an expansion bridge, or a muscle, it is the nuts and bolts of the subject, i.e. the parts and their arrangement. This book should acquaint you with the *what, where, when, why,* and *how* of muscle mechanics in a broad range of applications. The purpose of this introductory chapter is to acquaint you with the *what* upon which understanding can build.

This chapter begins with the structural mechanics of muscle, specifically comparing and contrasting the structure of skeletal and cardiac muscle (the third type of muscle present in living systems, i.e. smooth muscle, is not treated in this study). The chapter concludes with the functional mechanics of muscle by developing an operational definition of what muscle is (regardless of whether skeletal, cardiac, or smooth).

Three types of muscle are known to exist in man and animals:

Special appreciation is extended to Braunwald et al. (1976) for their very helpful review of muscle structure and to Doctor John E. Minardi through whom the authors first gained a quantitative understanding of muscle as an energy convertor.

skeletal muscle, cardiac muscle, and smooth muscle. Skeletal muscle connected to the bones and joints allows locomotion (or movement in general) and the performance of work (whether internal or external). Cardiac muscle is responsible in part for the pumping of blood within the circulatory system, and smooth muscle (predominantly visceral and vascular) affects motility and flow respectively.

STRUCTURAL MECHANICS OF MUSCLE

This section compares and contrasts the structural organization and arrangement of skeletal and cardiac muscle. The approach is not intended to be exhaustive, but rather is intended to emphasize some significant similarities as well as differences between skeletal and cardiac muscle.

Skeletal muscle cells (or fibers) are notably larger than ventricular myocardial fibers. A typical skeletal muscle cell would be 100 to 150 Å in diameter and may have a length of up to several centimeters (Fig. 1-1). However, a ventricular myocardial cell is typically 10 to 20 Å in diameter and 40 to 100 Å in length (Fig. 1-2).

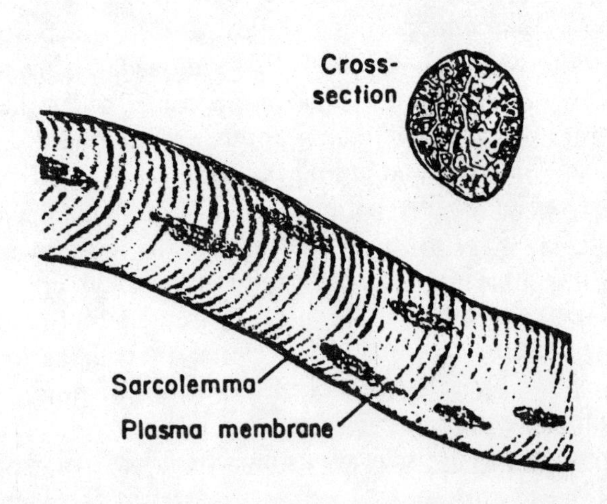

Figure 1-1. A skeletal muscle fiber viewed from the side and in cross-section. From A.C. Guyton, *Textbook of Physiology*, 5th Ed., 1976. Courtesy of W.B. Saunders Company, Philadelphia.

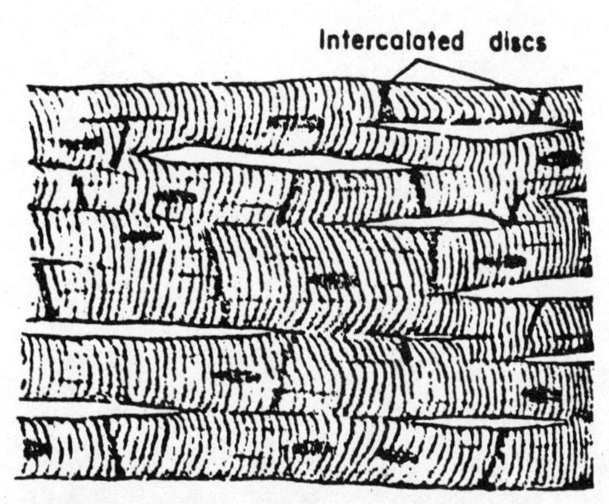

Figure 1-2. A population of cardiac muscle fibers indicating their arrangement. From A.C. Guyton, *Textbook of Physiology*, 6th Ed., 1981. Courtesy of W.B. Saunders Company, Philadelphia.

Both skeletal and cardiac muscle fibers are composed of numerous myofibrils running throughout the length of the fiber. These myofibrils are very discrete in white (or "fast") skeletal muscle; however, the myofibrils are nonuniform in diameter and partially fuse in both red (or "slow") skeletal muscle and cardiac muscle (Stenger and Spiro, 1961; Fawcett and McNutt, 1969). In both skeletal and cardiac muscle, the myofibrils are composed of sarcomeres in serial register, and the sarcomeres, in turn, are composed of the contractile proteins (or myofilaments). These sarcomeres compose 90 percent of the mass of a skeletal muscle fiber, but only about 50 percent of the mass of a cardiac muscle fiber.

With respect to cellular organelles, skeletal muscle fibers are multinucleate, while cardiac muscle fibers are uninucleate (or occasionally binucleate). Mitochondria are very plentiful in cardiac muscle, accounting for 25 to 30 percent of the cell mass (Page et al., 1972) but less than 10 percent of the mass of skeletal muscle cells.

Skeletal muscle fibers are organized in long, parallel arrays usually extending from tendinous to tendinous insertion (Fig. 1-1). Cardiac muscle fibers, on the other hand, characteristically will branch, interdigitate and form longitudinal connections via intercalated discs (Sommer and Johnson, 1968; McNutt and Weinstein, 1973) (Fig. 1-2).

Skeletal muscle cells have tubular invaginations of the sarcolemma, defined as transverse (or T) tubules that are about 400 Å in di-

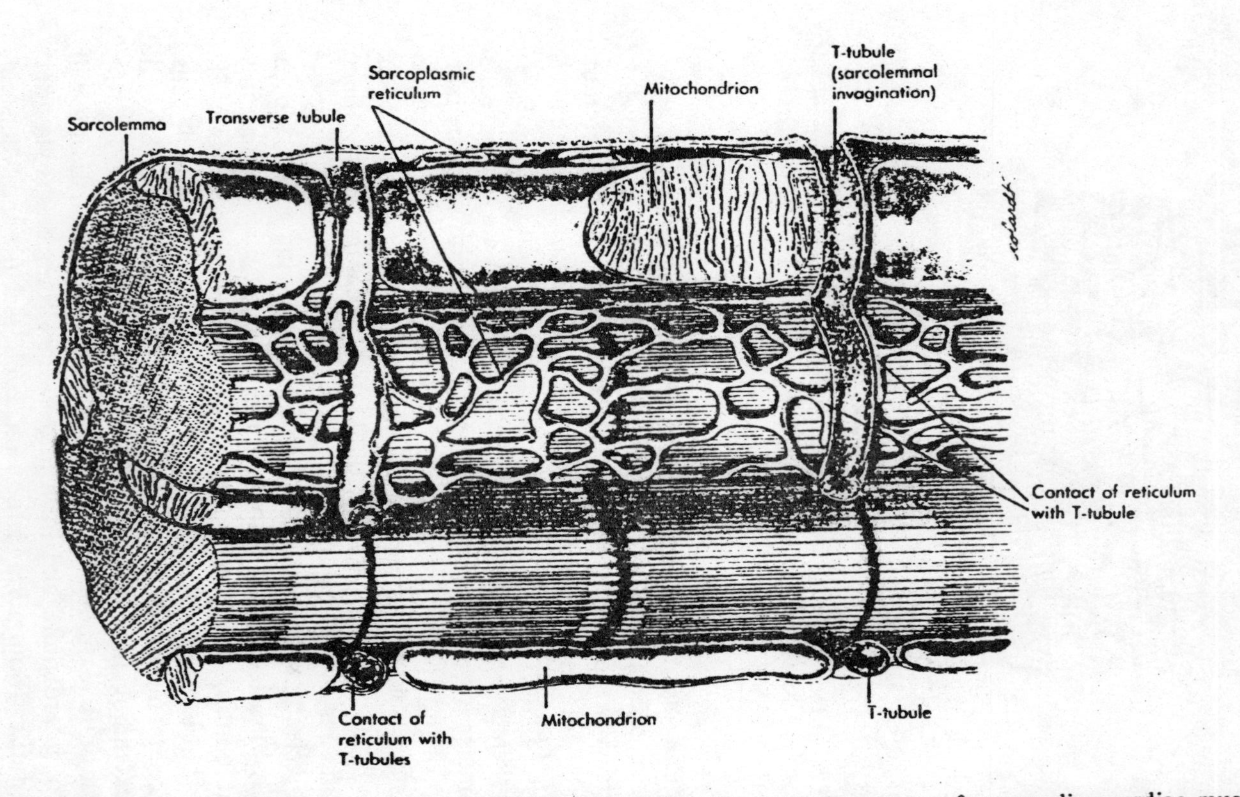

Figure 1-3. Three-dimensional view of the sarcoplasmic reticulum and T-tubule system of mammalian cardiac muscle. From W. Bloom and D.W. Fawcett, *A Textbook of Histology*, 1975. (Adapted from Fawcett and McNutt, 1969.) Courtesy of W.B. Saunders Company, Philadelphia.

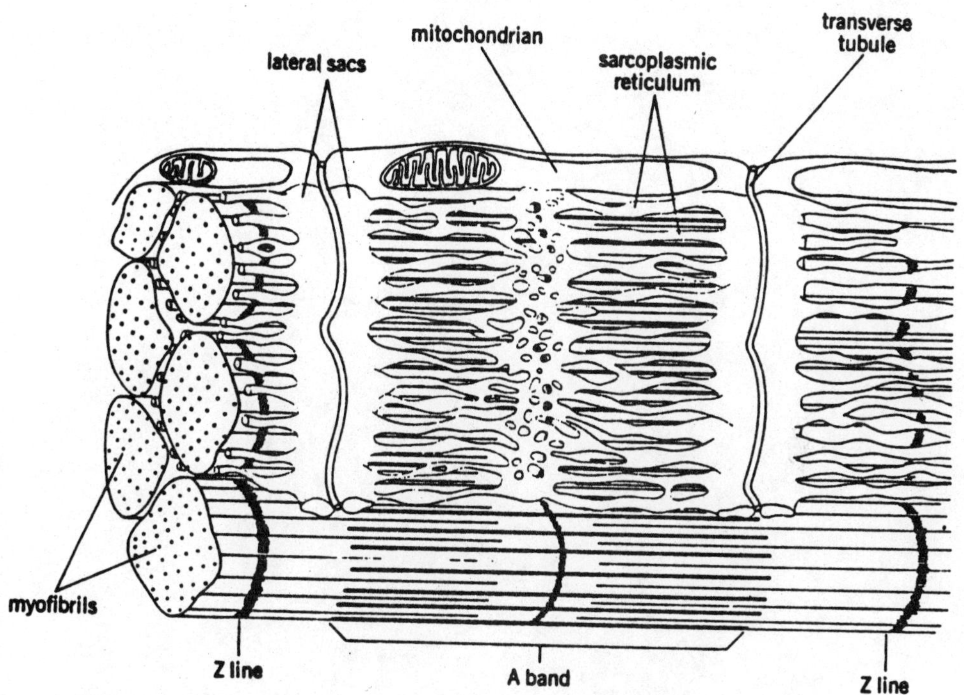

Figure 1-4. Three-dimensional view of the sarcoplasmic reticulum and T-tubule system of skeletal muscle. From A.J. Vander, J.H. Sherman, and D.S. Luciano. *Human Physiology–Mechanisms of Body Function*, 1975. (Adapted from Bloom and Fawcett, 1975.) Courtesy of McGraw-Hill Book Company, New York.

ameter and not as clearly in direct continuity with the extracellular fluid as those of myocardial cells (Smith, 1966). Ventricular myocardial cells have significantly larger T-tubules, which are 1,000 Å to 2,000 Å in diameter (Nelson and Benson, 1963; Simpson, 1965) and occur at the Z-lines of each sarcomere (Fig. 1-3). In contrast, the T-tubules of skeletal muscle fibers extend to the A-I junction of each sarcomere (Fig. 1-4).

In both skeletal (Smith, 1966; Franzini-Armstrong, 1970) and cardiac muscle (Sommer and Johnson, 1968; Fawcett and McNutt, 1969), the sarcoplasmic reticulum (SR) is a complex, anastomosing network of intracellular channels about 200 to 400 Å in diameter. However, the SR is more profuse in skeletal than in cardiac muscle.

In skeletal muscle, two terminal cisternae of the SR form a well-organized junction with the T-tubule of the sarcolemma termed a "triad." In cardiac muscle, only one terminal cisterna appears to be

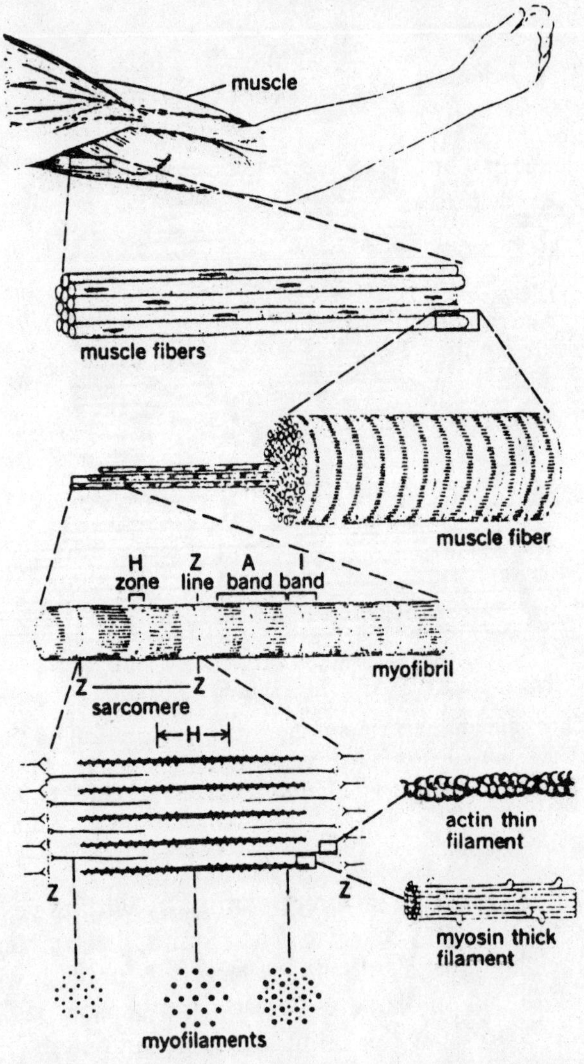

Figure 1-5. Levels of organization within a muscle fiber. The lower third of the figure shows the myofilament organization within a sarcomere and the pattern of the myofilaments in cross-section. From A.J. Vander, J.H. Sherman, and D.S. Luciano. *Human Physiology–Mechanisms of Body Function*, 1975. (Adapted from Bloom and Fawcett, 1975.) Courtesy of McGraw-Hill Book Company, New York.

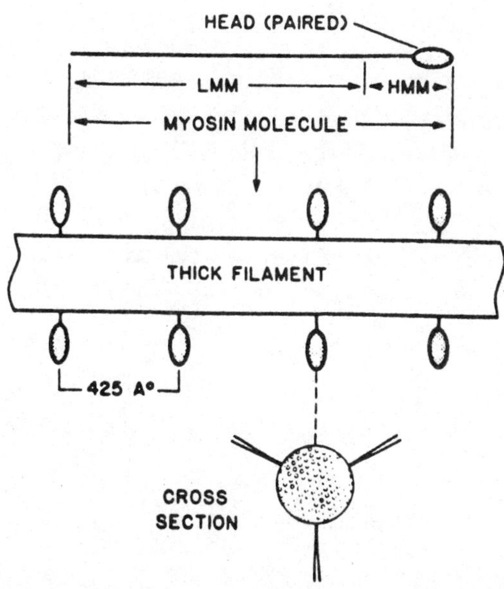

Figure 1-6. The thick filament is composed of bundles of individual myosin molecules. Each myosin molecule, in turn, consists of a rodlike light meromyosin (LMM) component and a heavy meromyosin (HMM) component. The HMM has paired heads at its end which form the radially disposed crossbridges. Adenosine triphosphatase activity is localized within the heads of the HMM molecules, which also form cross-linkages with the helical thin filaments during muscular contraction. Adapted from D. Jensen, *The Principles of Physiology*, 1980. Courtesy of Appleton-Century-Crofts, Englewood Cliffs, New Jersey.

flattened against a T-tubule forming a less well-organized "diad."

Two sets of myofilaments exist and are the same for both skeletal and cardiac muscle. The "thick" (myosin) filaments are about 100 Å in diameter and 1.5 to 1.6μ in length (Page and Huxley, 1963). The "thin" filaments are about 50 Å in diameter and 1μ in length (Page and Huxley, 1963). In cross section, the two sets of filaments interdigitate in a hexagonal array (Fig. 1-5).

Fast (white) skeletal muscle has the fastest intrinsic myosin ATP-ase activity. Slow (red) skeletal muscle has a lower intrinsic myosin ATP-ase activity and is similar in rate to cardiac myosin ATP-ase activity (Mueller et al., 1964).

Polypeptides (or "light" chains) constitute about 10 percent of the weight of the myosin molecule and have been isolated from the S1 subfragment (Taylor, 1972) (Fig. 1-6). There are three species of light chains on the myosin heads of fast (white) muscle, but only two types of light chains on slow (red) muscle and cardiac muscle (Lowey and Risby, 1971; Weeds and Frank, 1973).

The actin of fast and slow skeletal muscle and cardiac muscle appears to be similar in amino acid composition, physical properties, size, and shape. Tropomysin of skeletal and cardiac muscle is about 20 to 30 Å in diameter and 400 Å in length and is similar for both types of muscle (Fig. 1-7).

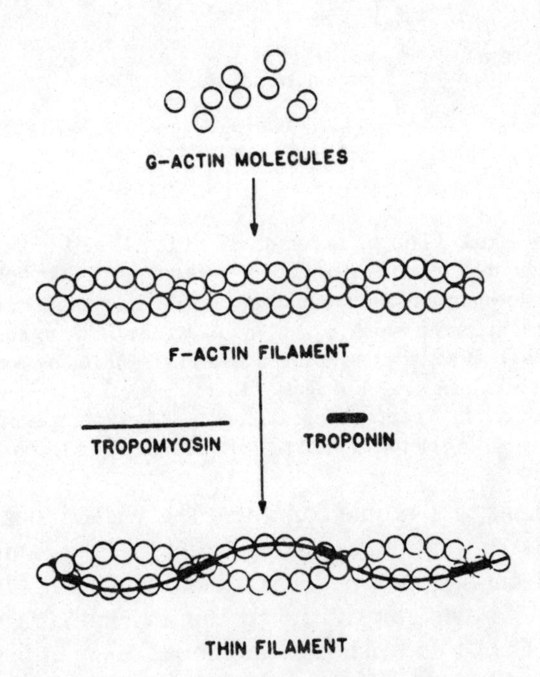

Figure 1-7. Morphologic relationships among the proteins that comprise a thin filament. Adapted from D. Jensen, *The Principles of Physiology*, 1980. Courtesy of Appleton-Century-Crofts, Englewood Cliffs, New Jersey.

Troponin is now known to consist of a three-component complex (also Fig. 1-7): a calcium sensitizing factor (TN-C), an inhibitory factor (TN-I), and an attachment factor (TN-T). Similar three-component troponin complexes have been identified for both skeletal and cardiac muscle (Dabrowska et al., 1973; Gergely, 1974; McCubbin et al., 1967). There is evidence that the troponin system of fast skeletal muscle has the highest activity, and a somewhat reduced activity and calcium sensitivity in both slow skeletal and cardiac muscle (Ebashi and Tonomura, 1973; Tsukui and Ebashi, 1973; Ebashi et al., 1974).

Finally, there are significant differences between the action potentials of a skeletal muscle cell and ventricular myocardial cell. Skeletal muscle has a "spike" action potential in which depolarization is followed rapidly by repolarization, the overall duration not exceeding 5 to 10 milliseconds (Fig. 1-8). The ventricular myocardial cell has a relatively long refractory period (phase 2) between depolarization. This refractory period is frequently greater than 100 milliseconds (Fig. 1-9) and is associated with a calcium current (Reuter, 1975) and relatively low potassium permeability (Trautwein, 1973; Beeler and Reuter, 1970; Weidmann, 1974).

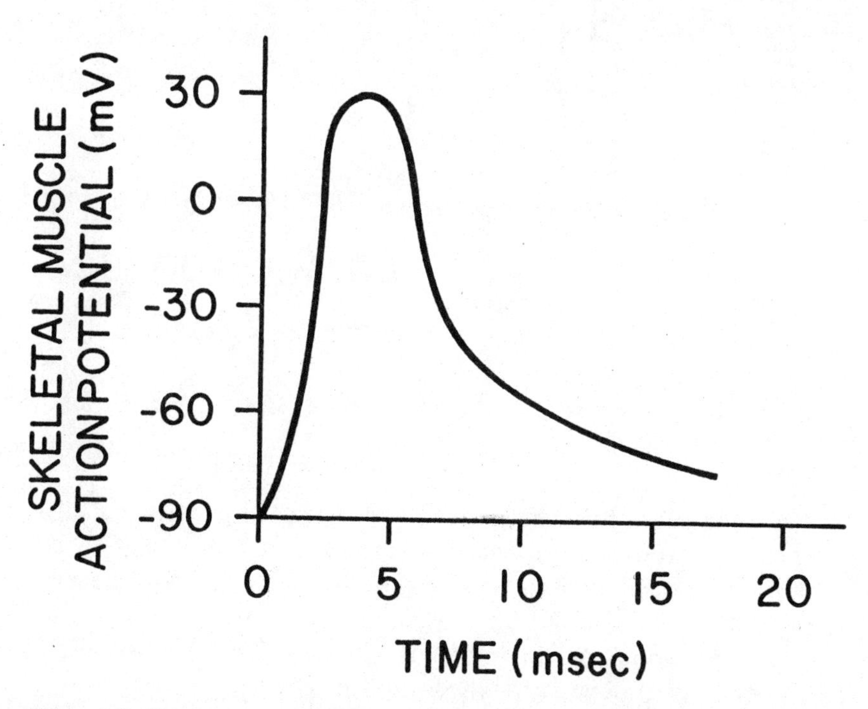

Figure 1-8. The action potential of a skeletal muscle fiber as a function of time.

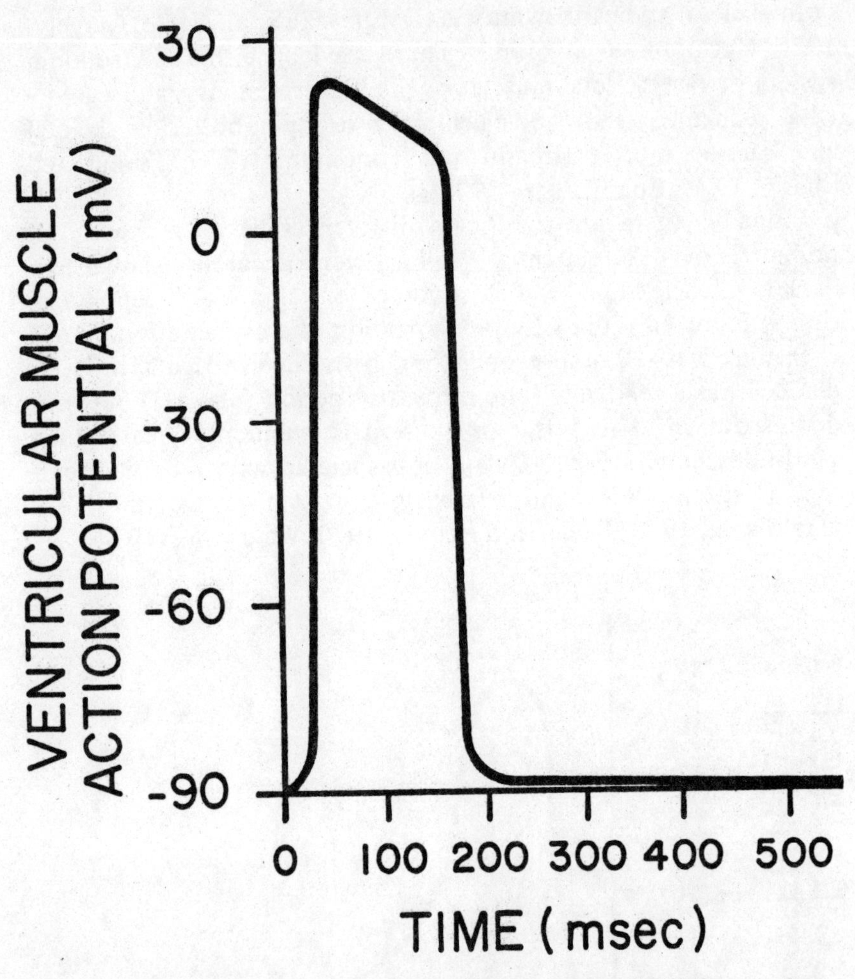

Figure 1-9. The action potential of a ventricular myocardial fiber as a function of time.

In conclusion, this review of a few of the similarities and differences between skeletal and cardiac muscle is not meant to be exhaustive. Excellent texts on skeletal muscle structure and function (Ebashi et al., 1980) and cardiac muscle structure and function in the normal and failing heart (Leyton, 1974) are available and highly recommended for the interested reader.

FUNCTIONAL MECHANICS OF MUSCLE

In muscle, there is a direct connection between the work output and the chemical driving source. This is an operational definition of muscle as a direct energy convertor. An overly simple analogy is shown in Figure 1-10. A battery is shown connected via a switch to a d.c. motor, and a load supported by a cord is attached to the motor shaft. When the switch is closed, current flows through the wire proportional to the battery voltage and inversely proportional to the d.c. resistance of the motor field coils. The load is pulled upward at a certain velocity dependent upon the torque of the motor shaft and the weight of the load.

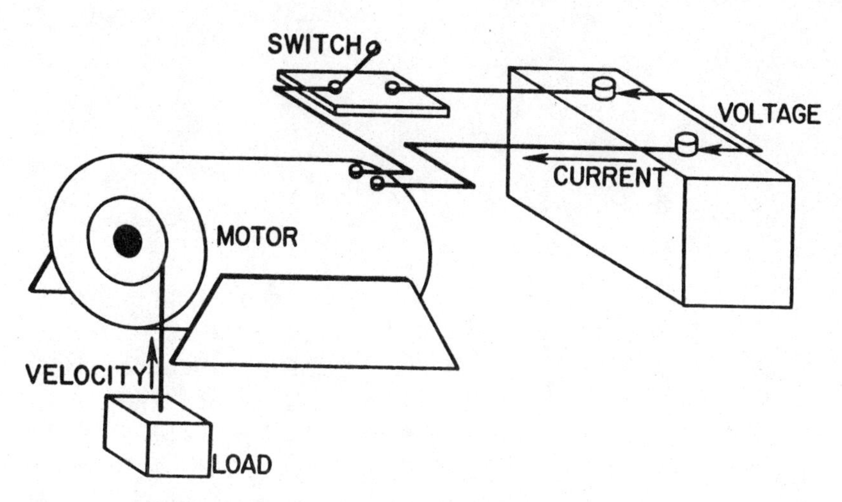

Figure 1-10. Simple battery and motor system that represents a direct energy convertor.

It is readily apparent that without knowing any internal details as to the construction of the battery or the motor and measuring only externally observable behavior, the battery voltage (E), current (I), rate of lifting the load (V), and weight of load (P), a relationship between electrical variables and mechanical variables can be defined:

$$V = L_{11}(-P) + L_{12}(E) \quad (1)$$

$$I = L_{21}(-P) + L_{22}(E). \quad (2)$$

Thus, velocity that is physically related to load is also related through the coupling of the motor to the applied voltage; current that is physically related to the voltage is also related, again through the coupling of the motor, to the applied load. The L's are transport coefficients, and, as can be seen, are in the nature of partial derivatives expressing the relationship between the thermodynamic fluxes (V and I) and thermodynamic forces (E and P).

$$L_{11} = -\left(\frac{\delta V}{\delta P}\right) \qquad E = \text{constant} \tag{3}$$

$$L_{12} = -\left(\frac{\delta V}{\delta E}\right) \qquad P = \text{constant} \tag{4}$$

$$L_{12} = -\left(\frac{\delta I}{\delta P}\right) \qquad E = \text{constant} \tag{5}$$

$$L_{22} = -\left(\frac{\delta I}{\delta E}\right) \qquad P = \text{constant} \tag{6}$$

Since the battery is a chemical cell, the mechanical events relate directly to the chemical events. Instead of E, which represents the electrical potential, the chemical affinity (A) can be used, which represents the chemical potential. Instead of I, the rate of change in electrical charge with time, chemical flux (v) can be defined, which is either the rate (in moles per second) that a chemical reactant disappears in the battery or that a chemical product is formed. The value of v will represent that rate at which the chemical events occur in the system under consideration.

Now, the linear equations become

$$V = L_{11}(-P) + L_{12}(A) \tag{7}$$

$$v = L_{21}(-P) + L_{22}(A). \tag{8}$$

Note that now, velocity is also related to the chemical affinity and chemical flux is also related to the applied load. Therefore, the overall device (battery plus motor) represents an example of a device that directly converts chemical free energy to mechanical work.

As previously explained, the transport coefficients could be rewritten as partial derivatives in terms of load (P), velocity (V), chemical flux (v), and chemical affinity (A), and these coefficients represent the basic characteristics of the convertor.

This familiar example of a simple direct energy convertor is a gross oversimplification of the contractile properties and behavior of muscle. In muscle, the switch (for example) consists of an α-motor neuron, myoneural junction and sarcolemma. The chemical battery consists of ATP, PC, Ca^{++}, troponin, tropomyosin, a Ca^{++} ATP-ase pump. The motor consists of actin, myosin, myosin ATP-ase, Mg^{++}, tendons, bones, and joints. In essence, muscle can be thought of as a very complex *electrically activated chemo-mechanical energy convertor.*

Nonetheless, it is amenable to a thermodynamic analysis when appropriate simplifying approximations are made. The transport coefficients for skeletal muscle have been characterized by Bornhorst and Minardi (1970a, 1970b) and the transport coefficients for cardiac muscle have been determined by Phillips et al. (1979a, 1979b). It is to be emphasized that such analyses are predicated primarily on the approximation that the actomyosin cross-bridge of muscle is a linear energy convertor, and as such is just a general description of the muscle system. Knowledge concerning the particular molecular details is not involved.

Molecular details are contained in operationally defined quantities called transport coefficients and in some of the driving forces that appear in the resulting thermodynamic equations. The real value of the thermodynamic equations is that they relate the different types of experimental measurements (mechanical, chemical, and heat) and are very helpful in interpreting these measurements. A thermodynamic approach and a molecular approach are complementary; therefore, such an approach is augmented by, though not dependent on, a molecular view of muscle. For example, a molecular view might be used to provide a theoretical basis for an estimate of the variation of the transport coefficients for a whole muscle. It is not dependent on this view, however, since the variation of the transport coefficients could be determined directly from experimental observations. Clearly, if the molecular view does agree with the experimental variation of these transport coefficients, then a greater understanding of muscle is obtained than if the variations of these coefficients is based solely on macroscopic observations.

REFERENCES

Beeler, G.W., Jr., and Reuter, H. (1970) Membrane calcium current in ventricular myocardial fibers. *J. Physiol.* (Lond.), 207: 191-209.

Bloom, W. and Fawcett, D.W. (1975) *A Textbook of Histology*, 10th Edition, W.B. Saunders Co., Philadelphia.

Bornhorst, W.J. and Minardi, J.E. (1970a) A phenomenological theory of muscle contraction. I. Rate equations at a given length based on irreversible thermodynamics. *Biophys. J.*, 10:137-154.

Bornhorst, W.J. and Minardi, J.E. (1970b) A phenomenological theory of muscle contraction. II. Generalized length variations. *Biophys. J.*, 10:155-171.

Braunwald, E., Ross, J., Jr., and Sonnenblick, E.H. (1976) *Mechanisms of Contraction of the Normal and Failing Heart*. Little, Brown and Co. Boston. Pp. 1-38.

Dabrowska, R., Dydynska, M., Szpacenko, A., and Drabikowski, W. (1973) Comparative studies of the composition and properties of troponin from fast, slow and cardiac muscles. *Int. J. Biochem.*, 4:189-194.

Ebashi, S., Maruyama, K., and Endo, M. (eds.), (1980) *Muscle Contraction: Its Regulatory Mechanisms*. Japan Scientific Societies Press, Tokyo; Springer-Verlag, New York.

Ebashi, S., Masaki, T., and Tsukui, R. (1974) Cardiac contractile proteins. *Adv. Cardiol.*, 12:59-69.

Ebashi, S. and Tonomura, Y. (1973) Proteins of the myofibril. In Bourne, G.H. (ed.) *The Structure and Function of Muscle*. Academic Press. New York. Pp. 286-362.

Fawcett, D.W. and McNutt, N.S. (1969) The ultrastructure of the cat myocardium: I. Ventricular papillary muscle. *J. Cell. Biol.*, 42:1-45.

Franzini-Armstrong, C. (1970) Studies of the triad: I. Structure of the junction in frog twitch fibers. *J. Cell. Biol.*, 47:488-499.

Gergley, J. (1974) Some aspects of the role of sarcoplasmic reticulum and the tropomyosin-troponin system in the control of muscle contraction by calcium ions. *Circ. Res.*, 34 (Suppl. 3): 74-81.

Guyton, A.C. (1981) *Textbook of Medical Physiology*, 6th Edition, W.B. Saunders Co., Philadelphia.

Guyton, A.C. (1976) *Textbook of Medical Physiology*. 5th Edition, W.B. Saunders Co., Philadelphia.

Jensen, D. (1980) *The Principles of Physiology*, 2nd Edition, Appleton-Century-Crofts, New York.

Leyton, R.A. (1974) Cardiac ultrastructure and function in the normal and failing heart. In Mirsky, I., Ghista, D.N., and Sandler, H. (eds.) *Cardiac Mechanics: Physiological, Clinical and Mathematical Considerations*. John Wiley and Sons. New York. Pp. 11-65.

Lowey, S. and Risby, D. (1971) Light chains from fast and slow muscle myosins. *Nature*, 234:81-85.

McCubbin, W.D., Kuoba, R.F., and Kay, G.M. (1967) Physicochemical studies on bovine cardiac tropomyosin. *Biochemistry*, 6:2417-2425.

McNutt, N.S. and Weinstein, R.S. (1973) Membrane ultrastructure at mammalian intracellular junctions. *Prog. Biophys. Mol. Biol.*, 20:45-101.

Mueller, H., Franzen, J., Rice, R.B., and Olson, R.E. (1964) Characterization of cardiac myosin from the dog. *J. Biol. Chem.*, 239:1447-1456.

Nelson, D.A. and Benson, E.S. (1963) On the structural continuities of the transverse tubular system of rabbit and human myocardial cells. *J. Cell. Biol.*, 16:297-313.

Page, E., Polimerri, P.I., Zak, R., Early, J., and Johnson, M. (1972) Myofibrillar mass in rat and rabbit heart muscle. *Circ. Res.*, 30:430-439.

Page, S.G. and Huxley, H.E. (1963) Filament lengths in striated muscle. *J. Cell. Biol.*, 19:369-390.

Phillips, C.A., Grood, E.S. Scott, W.J., and Petrofsky, J.S. (1979a) Cardiac chemical power: 1. Derivation of the chemical power equation and determination of equation constants. *Med. Biol. Engrg. Comp.*, 17:503-509.

Phillips, C.A., Grood, E.S., Scott, W.J., and Petrofsky, J.S. (1979b) Cardiac chemical power: 2. Application of chemical power, work and efficiency equations to characterize left ventricular energetics in man. *Med. Biol. Engrg. Comp.*, 17:510-517.

Reuter, H. (1975) Inward calcium current and activation of contraction in mammalian myocardial fibers. In Dhalla, N.S. (ed.) *Recent Advances in Cardiac Structure and Metabolism*. University Park Press. Baltimore. Pp. 13-18.

Simpson, F.O. (1965) The transverse tubular system in mammalian myocardial cells. *Am. J. Anat.*, 117:1-18.

Smith, D.R. (1966) Organization and function of sarcoplasmic reticulum and T-system of muscle cells. *Prog. Biophys. Mol. Biol.*, 16:107-142.

Sommer, J.R. and Johnson, E.A. (1968) Cardiac muscle: A comparative study of Purkinje fibers and ventricular fibers. *J. Cell. Biol.*, 36:497-526.

Stenger, R.J. and Spiro, D. (1961) The ultrastructure of mammalian cardiac muscle. *J. Biophys. Biochem. Cytol.*, 9:325-351.

Taylor, E.W. (1972) Chemistry of muscle contraction. *Ann. Rev. Biochem.*, 41:577-616.

Trautwein, W. (1973) Membrane currents in cardiac muscle fibers. *Physiol. Rev.*, 53:793-835.

Tsukui, R. and Ebashi, S. (1973) Cardiac troponin. *J. Biochem.*, 73:1119-1121.

Vander, A.J., Sherman, J.H., and Luciano, D.S. (1975) *Human Physiology: The Mechanisms of Body Function*. 2nd Edition, McGraw-Hill Book Co., New York.

Weeds, A.G. and Frank, G. (1973) Structural Studies on the Light Chains of Myosin. *Cold Sprg. Harb. Symp. Quant. Biol.*, 37: 9-14.

Weidmann, S. (1974) Heart: Electrophysiology. *Ann. Rev. Physiol.*, 36:155-169.

Chapter 2

THE FORCE-VELOCITY RELATIONSHIP IN SKELETAL MUSCLE

J.S. PETROFSKY and C.A. PHILLIPS

WHEN an action potential propagates across the surface of a muscle fiber, a complex series of events takes place that ultimately results in the mechanical contraction of the fiber. If a muscle such as the frog medial gastrocnemius muscle is isolated from the body as shown in Figure 2-1 and the motor nerve is left intact, a contraction can be elicited by applying a stimulus through a pair of electrodes as shown (Fig. 2-1). If one end of the muscle is fixed by a steel pin to the table and the other end of the muscle is connected to some type of velocity transducer, then the velocity of shortening of the muscle can be measured. A typical recording from such an experiment is shown in Figure 2-2.

As can be seen in this figure, following the application of the stimulus to the motor nerve, there is a time delay of between 5 and 7 milliseconds before the muscle begins to contract. This latency period is caused by a number of factors. First of all, there is an electromechanical time delay following the application of a stimulus before action potentials will develop on the motor nerve. Next, there is a delay of between 1 and 2 milliseconds, which is required for action potentials to propagate across the neuromuscular junction. This is because the electrical wave of depolarization (action potential) that arises at the presynaptic terminal of the neuromuscular junction must be transduced into the release of a neurotransmitter (acetylcholine) which then diffuses across the neuromuscular junction. After binding to acetylcholine receptors, a new wave of the depolarization begins at the neuromuscular junction and propagates across the muscle. A further delay is caused by the propagation of action potentials deep in the muscle through the transverse tubules and the eventual coupling of this wave of depolarization into the release of calcium and diffusion of calcium into the myofibrils. However, once calcium is released from the terminal cisternae of the longitudinal reticulum of skeletal muscle, the initial reaction of the muscle is relaxation and not contraction. As can be seen in Figure 2-2, the relaxation is small. This initial relaxation is called latency relaxation. The mechanism of latency relaxation is unknown.

Following latency relaxation the muscle contracts rapidly. As

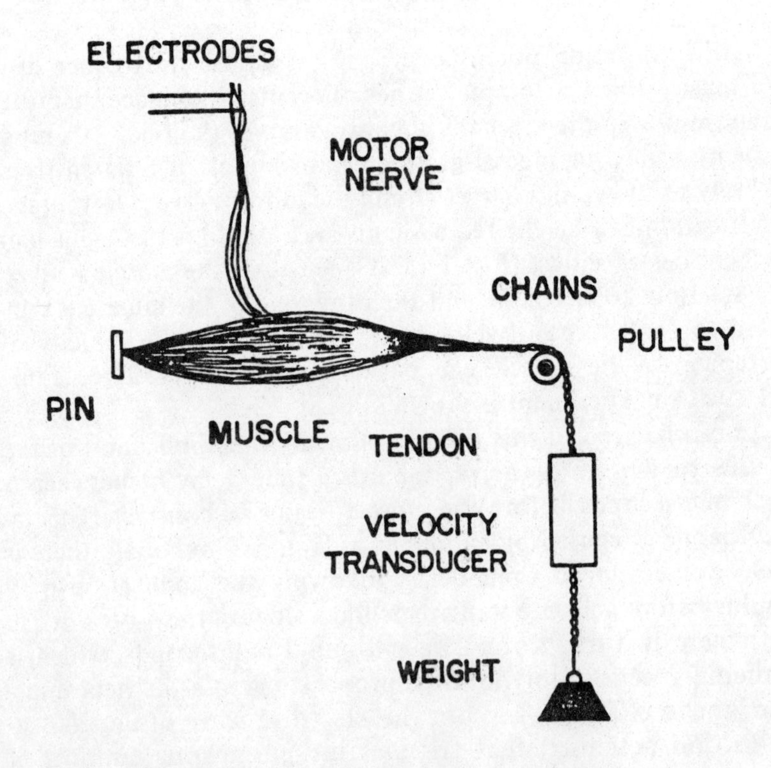

Figure 2-1. Experimental set-up for velocity measurements.

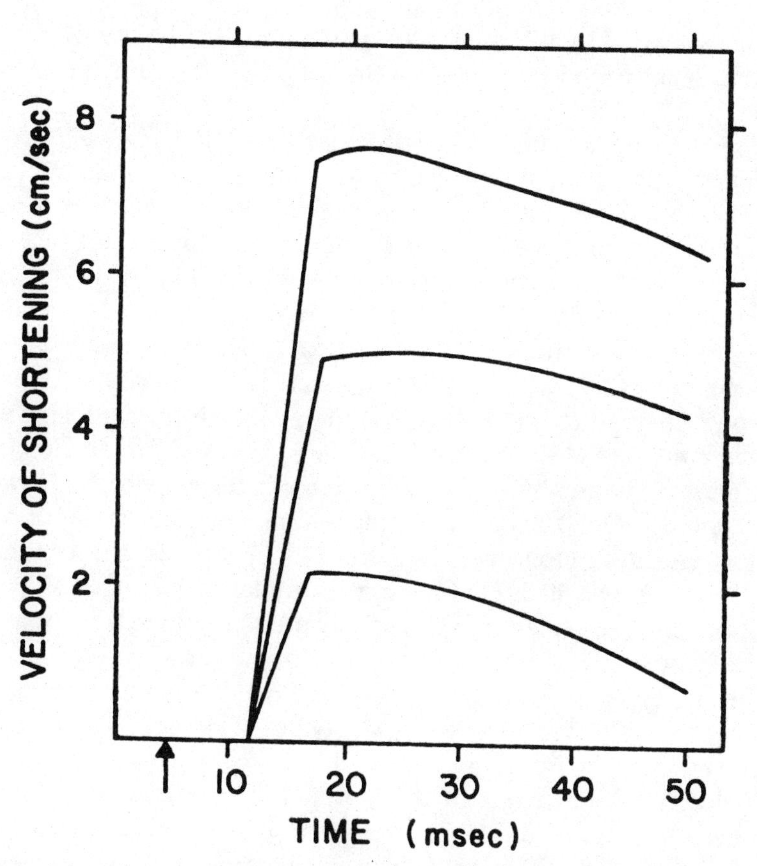

Figure 2-2. Velocity of the medial gastrocnemius during contractions induced at three different levels of recruitment. The vertical arrow on the abscissa shows the time of stimulation. From J.S. Petrofsky and C.A. Phillips, Constant-velocity Contractions in Skeletal Muscle by Sequential Stimulation of Muscle Efferents. *Medical & Biological Engineering & Computing, 17*:583-592, 1979.

seen in Figure 2-2, the muscle achieves its maximum velocity of contraction in between 10 and 15 milliseconds after stimulation. As the muscle begins to contract, the velocity of contraction first peaks and then begins to fall slightly until the contraction is terminated. If skeletal muscle is prestretched to the same length prior to stimulation and then allowed to contract against various loads, the relationship between force and velocity of contraction can be assessed. In this type of experiment, the velocity of contraction is measured as the peak velocity of contraction as shown in Figure 2-2.

The velocity of contraction of skeletal muscle is inversely proportional to the load applied. In a wrestling match, for example, the lightweight wrestlers (those wrestlers with a very light load applied to their muscles) move very quickly. In contrast, the heavy weight wrestlers (those wrestlers with a heavy load applied to their muscles) move very slowly. The exact relationship between force and velocity of contraction was first described in a classic series of experiments by A.V. Hill (1938). A typical force-velocity curve is shown in Figure 2-3. With no load applied to the muscle, the muscle contracts at its highest velocity. As the load on the muscle is increased, the muscle begins to shorten at a lower velocity. However, once the load exceeds the maximum isometric strength of the muscle, the muscle will not shorten. This point, the x-intercept, is usually called P_0 (maximum isometric strength of the muscle). The y-intercept of the graph is usually termed V_{mx} (maximum velocity of shortening of the unloaded muscle). V_{mx} is an important parameter to be measured in both skeletal and cardiac muscle since it is considered to reflect the ATP-ase activity of the actin and myocin in the muscle. If the muscle is truly unloaded, the velocity of shortening of the muscle (V_{mx}) should be proportional to the rate of cross-bridge formation and breaking (rate of hyrolysis of ATP). Therefore, in both experimental and clinical studies of both skeletal and cardiac muscle, V_{mx} is used as an index of both pathological disease states that effect actomyocin ATP-ase activity and the effect of drugs that alter the contractile state of skeletal and cardiac muscle (Fitts and Holloszy, 1977).

However, V_{mx} is a hard quantity to measure because to truly measure V_{mx}, the muscle must be fully unloaded. In experiments on skeletal muscle the inertia of the velocity measuring transducer usually is sufficient to keep true V_{mx} from being measured. Therefore, in classic skeletal muscle experiments where the force-velocity relationship is measured, the transducer is connected to the muscle through a complex series of lever arms to minimize the inertia of the transducer system. Even with these maneuvers, V_{mx} is usually estimated by extrapolating the data of the unloaded

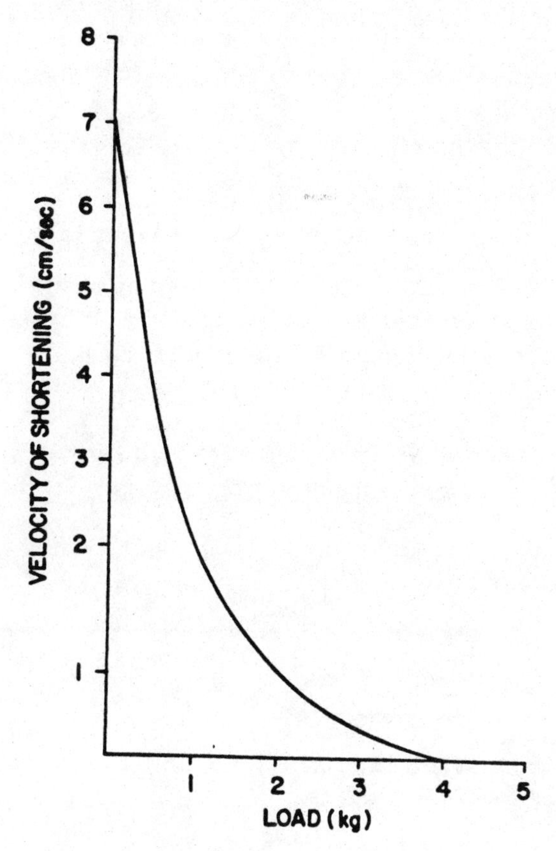

Figure 2-3. A typical force-velocity diagram.

muscle with the transducer attached and correcting for the mass of the transducer system.

A.V. Hill first measured and described the force-velocity relationship in frog skeletal muscle. He found that he could describe the force-velocity relationship for skeletal muscle by the following equation:

$$V = \frac{(P_o - P)\,b}{P + a}$$

Where

V = velocity of shortening
a = the Hill a coefficient
b = the Hill b coefficient
P = the load applied to the muscle.
P_o = maximum isometric strength of the muscle.

From this equation, the velocity of shortening of the loaded muscle (V) can be calculated for any skeletal muscle by knowing the maximum isometric strength of the muscle (P_o) and the Hill b and a coefficients.

THE EFFECT OF MUSCLE LENGTH ON THE FORCE-VELOCITY RELATIONSHIP

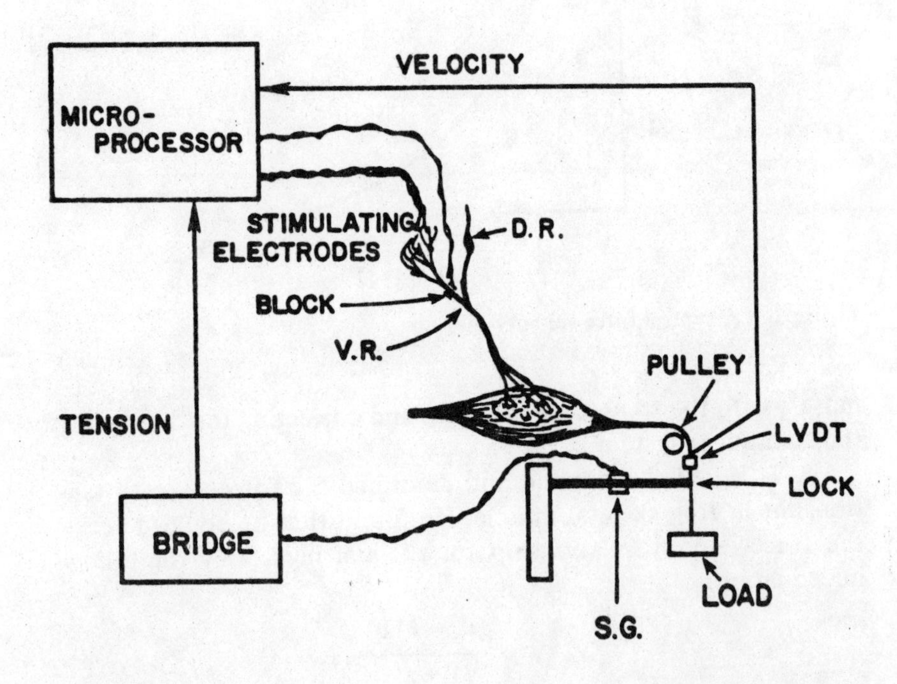

Figure 2-4. Experimental set-up used to measure velocity of contraction at different muscle lengths. From J.S. Petrofsky and C.A. Phillips, Constant-velocity Contractions in Skeletal Muscle by Sequential Stimulation of Muscle Efferents. *Medical & Biological Engineering & Computing, 17*:583-592, 1979.

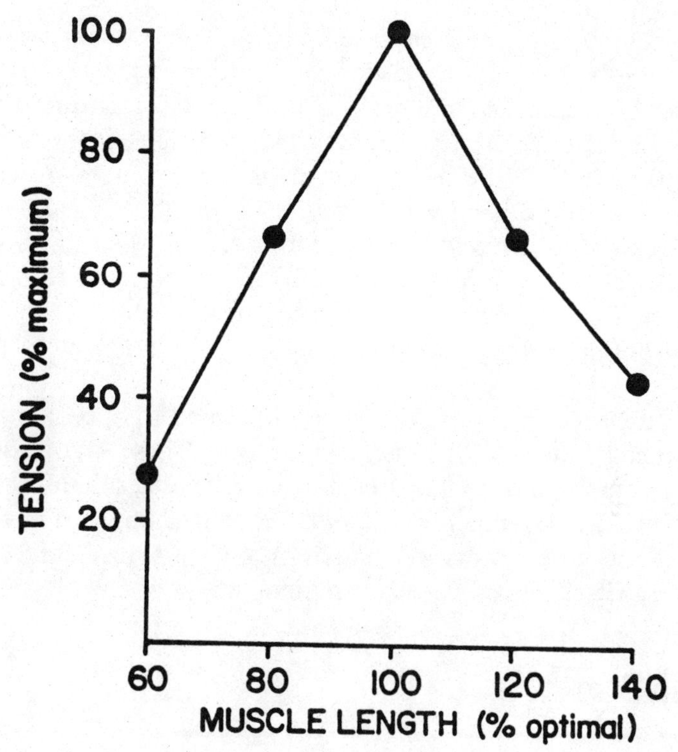

Figure 2-5. The relationship between muscle length and maximum isometric tension.

Muscle length has a dramatic influence on the force velocity relationship. The maximum isometric strength that can be developed by a muscle depends on its length. For example, when a muscle is removed from the body, the muscle contracts to a shorter length because of the elastic components in the sarcoplasmic membranes. This length is typically called the cut length. If one end of the muscle is fixed and the other is connected to an isometric force transducer (Fig. 2-4) and the muscle is stimulated through its motor nerve or directly to the muscle, the muscle will contract and the force that is developed can be measured. If a force is applied to the end of the muscle prior to stimulation, the passive elastic element will be stretched prior to contraction. If various preloads are applied prior to stimulation, the muscle will initiate contraction at different lengths; the relationship between muscle length and tension development during isometric contractions can be measured and graphically illustrated as shown in Figure 2-5. For both skeletal and cardiac muscle there exists one muscle length at which the muscle develops its maximum isometric strength when stimulated electrically. This length is called the optimal length of the muscle. When the muscle is allowed to contract at lengths above or below this length the contraction tension is much less. The mechanism of this phenomenon lies in the geometric relationship between actin and myocin within the muscle sarcomere.

A muscle cell or muscle fiber is composed of approximately 1,000 parallel strands of muscle myofibrils. These myofibrils in turn are composed of a characteristic and repeating subunits called sarcomeres. The sarcomere is the smallest contracting unit in skeletal muscle and represents the source of any developed force. The sarcomere itself is composed predominately of two types of

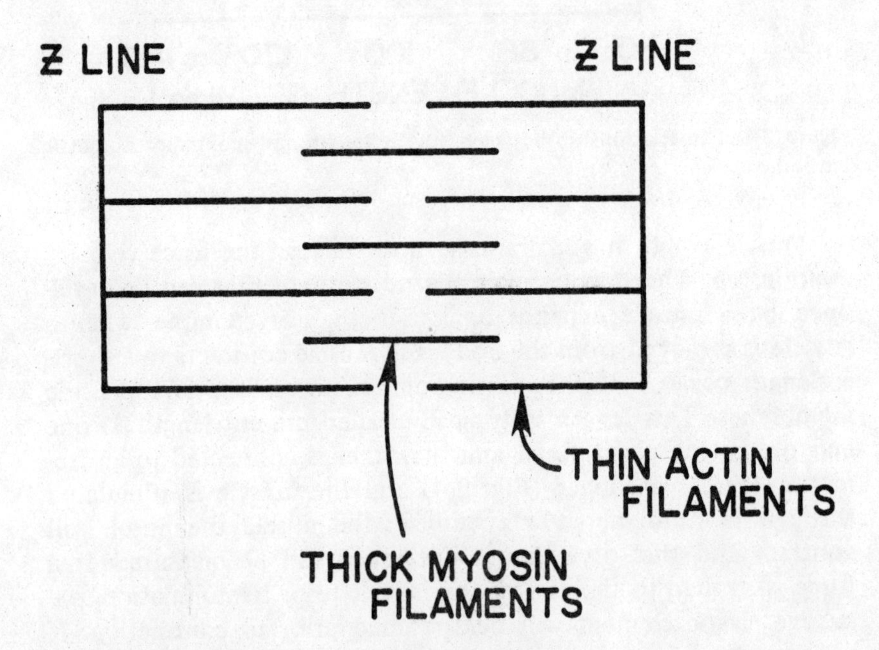

Figure 2-6. The structure of the muscle sarcomere.

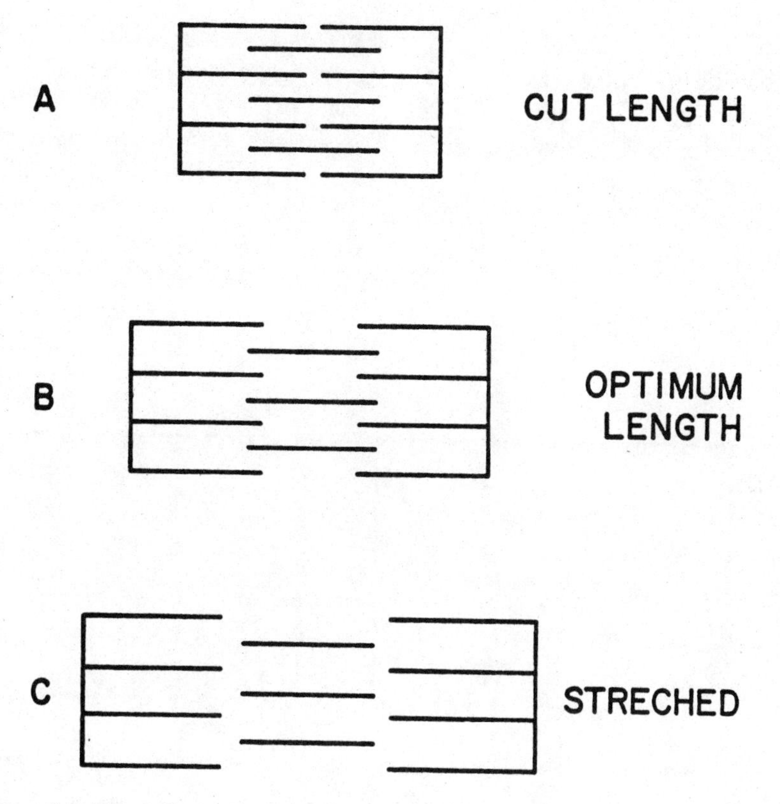

Figure 2-7. The effect of passive stretch on the muscle sarcomere (*see* text).

protein (Fig. 2-6). A thin strandlike protein (actin) originates at the end of the sarcomere (Z line) and extends inward toward the middle. In parallel, but originating in the middle of the sarcomere, is a second and more dense protein called myocin. When the muscle is at its cut length, due to the elastic nature of the sarcoplasmic membranes, the Z lines are brought close together causing the actin filaments to overlap as shown in Figure 2-7. The force developed in skeletal muscle is directly proportional to the number of actin and myocin cross-bridges that form during electrical stimulation. If a stimulus is applied to the muscle at this length, there is considerable steric interference between the thin actin filaments resulting in low force development by the muscle. If the muscle is stretched to a length just beyond the cut length, some of the interference between the actin filaments is reduced, allowing the force development to be greater following electrical stimulation. If more stretch is applied to the muscle prior to electrical stimulation, more of the interference between the actin filaments is reduced and force development is even higher. This is graphically illustrated in panel B of Figure 2-7. However, if sufficient passive force is applied to the muscle to pull the Z lines far enough apart to prevent some cross-bridges from forming between the actin and myocin as shown in panel C of this figure, less cross bridges will form following electrical stimulation and force development will

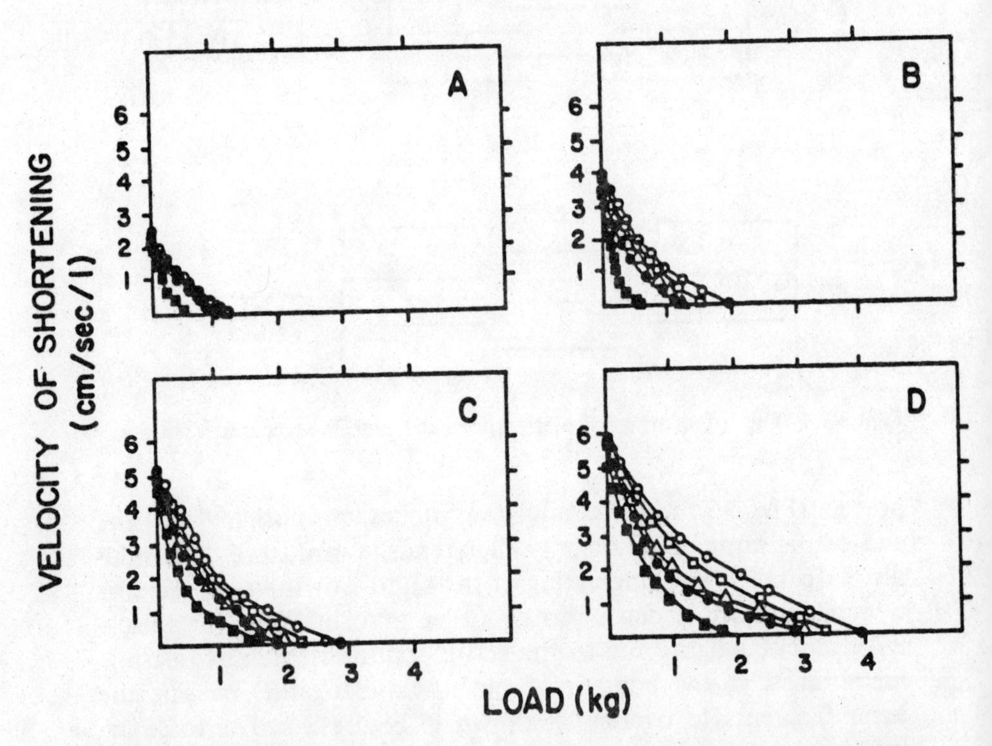

Figure 2-8. The force-velocity relationship of the muscles when prestretched prior to the isotonic contraction to their optimum length (o) and to 0.75 (■), 0.9 (△), 1.1 (□), and 1.25 (●) times that length during contractions where the activation averaged 0.35 (panel a), 0.54 (panel b), 0.718 (panel c) and 1.00 (panel d). Each point in this figure represents the mean of four experiments. During these experiments the muscle temperature was kept at *38 degrees centigrade*.

be reduced once again.

Obviously, the ATP-ase of actin and myocin should not be affected by the resting length of the muscle. Therefore it is not surprising that V_{mx} is the same at any muscle length. However, since P_o is a function of muscle length, the force velocity relationship is shifted to the left at any muscle length above or below optimal length as shown in Figure 2-8. It therefore becomes very important to make sure that muscle always starts contracting at the same length when measuring the force-velocity relationship for skeletal muscle.

When a muscle is stretched, the load imposed to stretch the muscle is called a preload, which must be overcome before the muscle will begin to contract. The load applied to the muscle to measure the force-velocity relationship must be applied after the muscle begins to contract. The reason for this is quite simple. If a heavy load was applied to the muscle before it contracted, it would stretch the elastic elements of the muscle and change the resting length. Therefore, measuring the true force-velocity relationship for skeletal muscle is a complex task. In A.V. Hill's initial experiments, he used an elegant system whereby the muscle was stretched to optimal length almost immediately prior to the onset of contraction. He termed these experiments "quick stretch experiments."

THE INFLUENCE OF TEMPERATURE ON THE FORCE-VELOCITY RELATIONSHIP

The reaction rate of enzymes is highly dependent on their temperature. This holds true for all enzymes including actin and myocin in skeletal muscle. Therefore, it is not surprising that V_{mx} for both skeletal and cardiac muscle is highly dependent upon the temperature of the contracting musculature. For example, a reduction in muscle temperature from 38 to 28 degrees centigrade will cause approximately a 50 percent reduction in V_{mx} in the medial gastrocnemius muscle of the cat (Petrofsky, Weber, and Phillips, 1980).

The effect of muscle temperature on isometric strength is much less pronounced. In man, for example, over the temperature range of 28 to 38 degrees centigrade, the maximum isometric strength that can be exerted by muscle is fairly constant (Clarke, Hellon, and Lind, 1958). However, when the muscle temperature rises above 42 degrees centigrade or falls below 28 degrees centigrade, muscle strength deteriorates rapidly. Therefore, the effect of temperature on the force-velocity relationship is nearly the opposite that for muscle length described above. Here, when the force-velocity relationship is measured at various muscle temperatures within the physiological range (22 to 38°C), the x-axis intercept of the force-velocity relationship stays fairly constant while the y-intercept changes dramatically as shown in Figure 2-9.

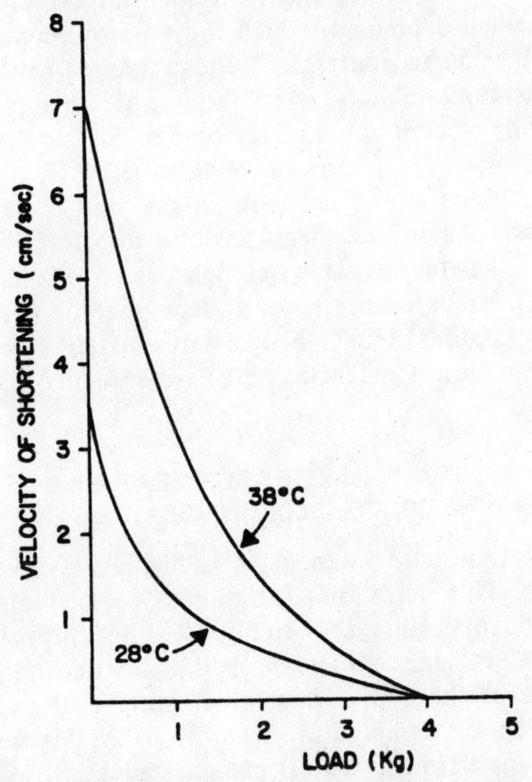

Figure 2-9. The force-velocity relationship at muscle temperatures of 28 to 38 degrees centigrade in the cat medial gastrocnemius muscle.

THE EFFECT OF RECRUITMENT MAGNITUDE AND DIRECTION ON THE FORCE-VELOCITY RELATIONSHIP

In the original experiments conducted by A.V. Hill (1938) and later experiments by Wilkie (1950), the force-velocity relationship in skeletal muscle was always examined with all motor units stimulated and with the frequency of stimulation exceeding the normal physiological frequency range. However, during voluntary activity, it is only on rare occasions that all motor units are recruited. Further, the frequency of motor unit discharge rarely exceeds

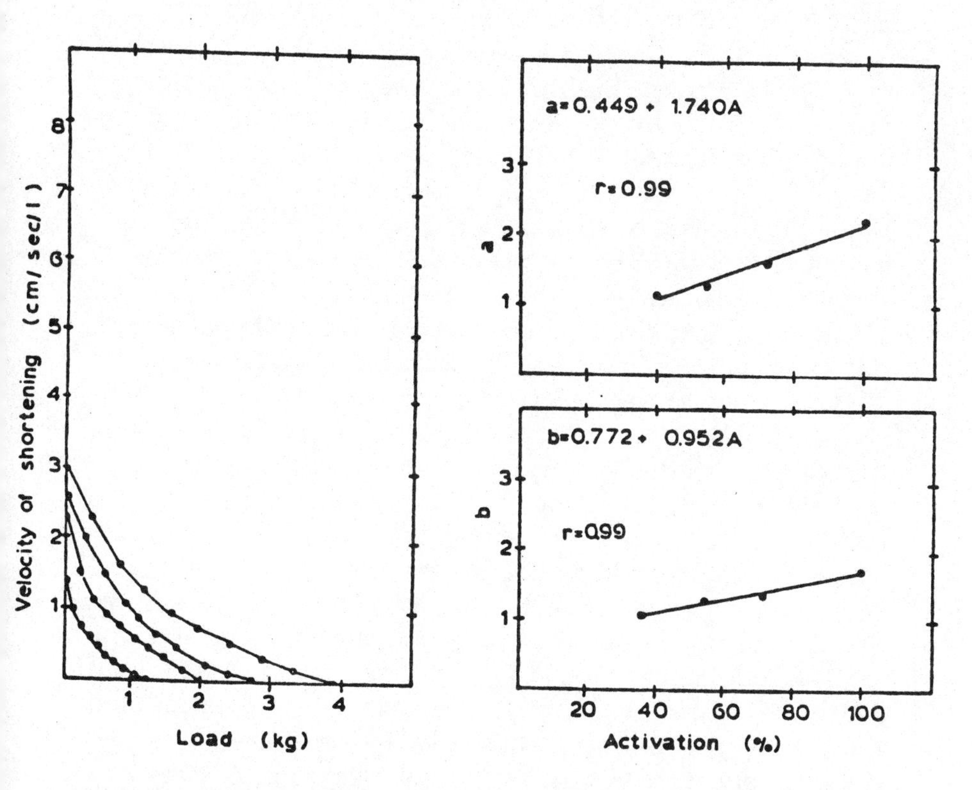

Figure 2-10. The force-velocity relationship in the partially and fully activated *soleus* muscle with recruitment proceeding from the *slowest* to *fastest* motor units, respectively. From J.S. Petrofsky and C.A. Phillips, Constant-velocity Contractions in Skeletal Muscle by Sequential Stimulation of Muscle Efferents. *Medical & Biological Engineering & Computing, 17*:583-592, 1979.

60 Hertz (Milner-Brown and Stein, 1975). Therefore, in some of our own recent work and that of others (Zahalak et al., 1976) we have reexamined the force-velocity relationship to see the effect recruitment order has on both V_{mx} and the force-velocity relationships at various submaximal loads. For a muscle that is composed entirely of the same type of motor unit, the degree of recruitment should only affect the velocity of shortening at submaximal loads and P_0. Therefore, if enough motor units are stimulated to recruit 25, 50, 75, and 100 percent of the motor units (in a muscle such as the soleus of the cat), four force-velocity relationships result under these conditions as shown in Figure 2-10. V_{mx} was only found to be reduced slightly at the lowest level of recruitment. This is probably because when only a small proportion of the motor units in the soleus muscle are contracting, the other motor units in the muscle act as a parallel damping force or load the muscle must overcome to contract. Therefore, true V_{mx} is not being recorded under these experimental conditions since a small load is applied to the muscle.

However, the majority of all muscles in man do not fit into this simple model since most muscles in man are a heterogeneous mixture of muscle fibers (contain both fast and slow skeletal muscle fibers). Historically, in terms of staining properties and velocity of contraction, the muscle fibers were divided into two categories: slow twitch muscle and fast twitch muscle. Slow twitch motor units are associated with a low V_{mx} and a high aerobic capacity (Close, 1972). In contrast, fast twitch motor units are associated with a high V_{mx} but low aerobic capacity: these motor units derive their energy predominately from anaerobic pathways. In recent years, it has been found that these two simple classifications greatly oversimplify the great variability found in the biochemical and physiological properties of individual muscle fibers. Actually, the speed of contraction of motor units and their aerobic capacity is a continuum within a muscle between the two extremes. However, some muscles are composed predominately of muscle fibers that can be classified as fast twitch while others are composed predominately of muscle fibers that can be classified as slow twitch. The typical force-velocity relationship of slow twitch muscle (soleus) and largely fast twitch muscle (medial gastrocnemius muscle) of the cat is shown in Figure 2-11. In addition to V_{mx} being lower in slow twitch muscles, the isometric strength (P_0) is lower as well resulting in the entire force-velocity relationship for a slow twitch muscle being shifted far to the left of that of a fast twitch muscle.

During voluntary activity in human and animal experiments, it has been found that motor units are not recruited in a random order. To vary the tension generated or velocity of movement during muscle activity, motor units are generally recruited according to size from the smallest to largest, respectively. Since the smallest motor units are associated with slow twitch muscle and the largest are associated with fast twitch muscle, motor unit recruitment proceeds from the slow twitch to fast twitch units, respectively, to increase the tension or velocity of shortening of muscle (Bigland and Lippold, 1954; Milner-Brown and Stein, 1975; Olson

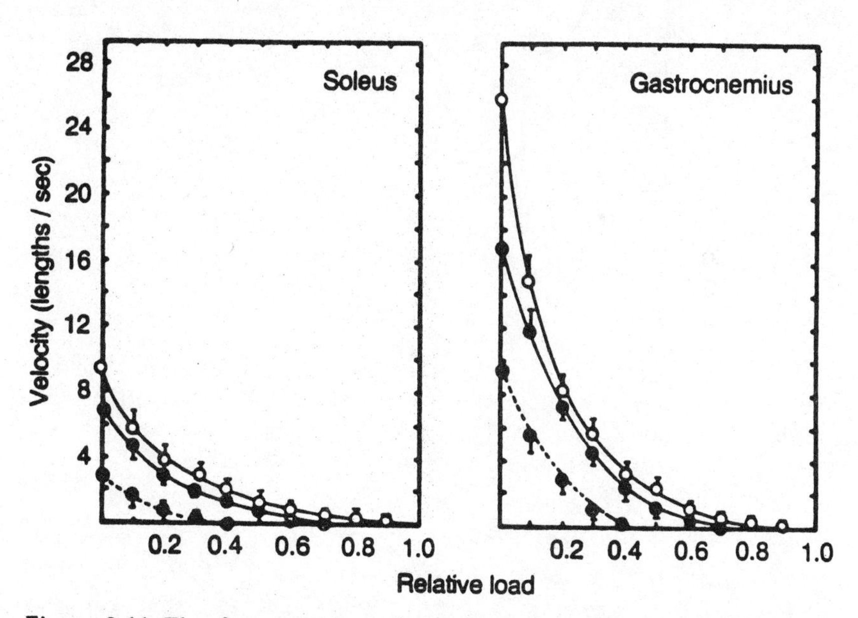

Figure 2-11. The force-velocity relationship in the soleus and medial gastrocnemius muscles of the cat at 38 degrees centigrade. From J.S. Petrofsky and C.A. Phillips, The Influence of Recruitment Order and Fibre Composition on the Force-velocity Relationship and Fatigability of Skeletal Muscle in the Cat. *Medical & Biological Engineering & Computing, 18*:381-390, 1980.

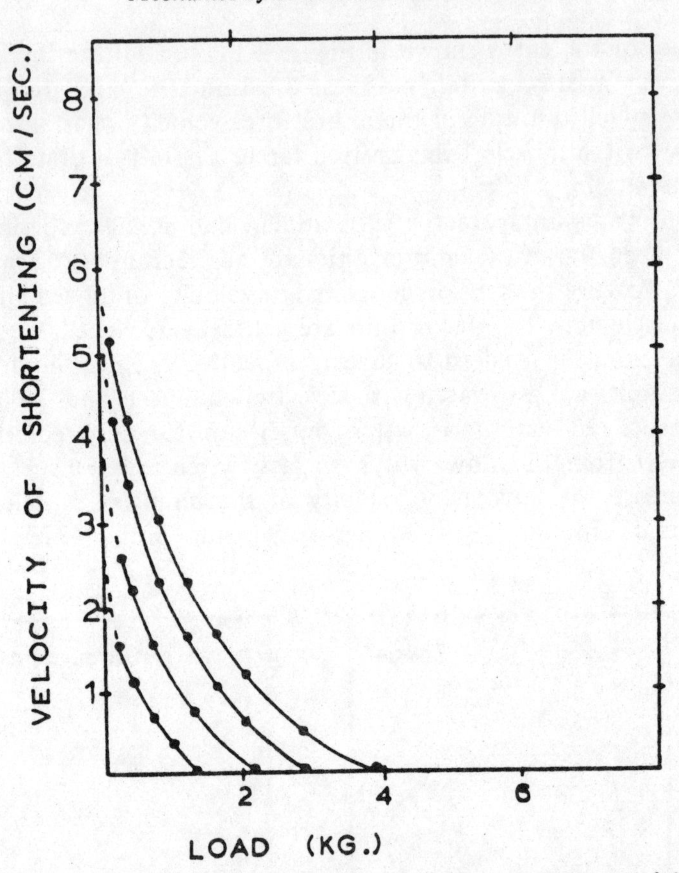

Figure 2-12. The force-velocity relationship in the partially and fully activated *medial gastrocnemius* muscle with recruitment proceeding from the *slowest* to *fastest* motor units, respectively. From J.S. Petrofsky and C.A. Phillips, The Influence of Recruitment Order and Fibre Composition on the Force-velocity Relationship and Fatigability of Skeletal Muscle in the Cat. *Medical & Biological Engineering & Computing, 18*:381-390, 1980.

et al., 1968). Since both P_0 and V_{mx} are a function of the speed of contraction of the motor unit, this makes the real relationship between force and velocity and recruitment far different from the idealized examples shown in the previous figures. If, for example, a muscle such as the medial gastrocnemius is used and recruitment is varied from the slowest to fastest motor units (that order normally observed in man) and the same experiment cited

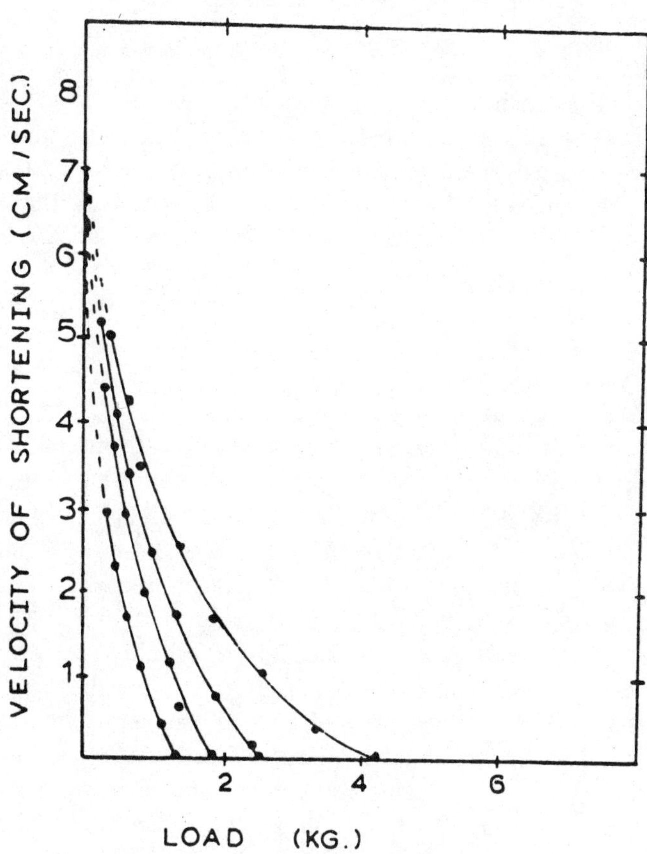

Figure 2-13. The force-velocity relationship in the partially and fully activated *medial gastrocnemius* muscle with recruitment from the *fastest* to *slowest* motor units. From J.S. Petrofsky and C.A. Phillips, The Influence of Recruitment Order and Fibre Composition on the Force-velocity Relationship and Fatigability of Skeletal Muscle in the Cat. *Medical & Biological Engineering & Computing, 18*:381-390, 1980.

above is repeated, we find a series of force velocity curves for recruitment of 25, 50, 75, and 100 percent of the motor units as shown in Figure 2-12. When 25 percent of the motor units are initially recruited, since only slow twitch motor units have been predominately recruited, both V_{mx} and P_0 are lower than that observed when all motor units are recruited together. However, as recruitment is increased, both P_0 and V_{mx} increase to the maximum observed for the muscle.

In some situations where activity has been preceded by prior muscle fatigue or proprioreceptive afferents from the skin, it has been shown that the normal order of recruitment is reversed. In other words, recruitment proceeds from the fastest to the slowest motor units respectively (Grimby and Hannerz, 1976; Hannerz and Grimby, 1979). If this were the case, the force-velocity relationship would be altered in relation to the recruitment order as shown in Figure 2-13. Here again, the medial gastrocnemius muscle was used but in this case recruitment was set from the fastest to slowest motor units. Therefore, although only 25 percent of the motor units are recruited, the maximum isometric tension (P_0) developed by the muscle is only 25 percent of the maximum P_0. The velocity of shortening at zero load (V_{mx}) with only 25 percent of the motor units recruited, is higher than the maximal velocity (V_{mx}) observed when all the motor units in the muscle are stimulated together. In the former case, it is only fast twitch motor units that have been recruited. When slow twitch motor units are recruited with fast twitch motor units, the V_{mx} of the fast twitch motor units is slowed due to the damping effect of active slow twitch motor units. It can be anticipated that for muscles with different compositions of fast and slow twitch motor units the force-velocity curves would be altered appropriately.

THE REVISED FORCE-VELOCITY RELATIONSHIP

The Hill equation describing the force-velocity relationship has stood the test of time and still serves as the most accurate mathematical description of a complex physiological process. Over the years, a number of modelers have tried to redefine the Hill equation using a variety of analytical techniques, but none of the investigators have derived an equation that more accurately depicts the force-velocity relationship. However, the original equation derived by A.V. Hill was determined from experiments on frog skeletal muscle where all motor units had been recruited. The real relationship between force and velocity in skeletal muscles in mammals must take into consideration recruitment order, fiber composition, muscle temperature, and muscle length at a minimum to fully understand and predict the relationship between force and velocity in any mammalian muscle.

In our work in recent years (Petrofsky and Phillips, 1979; 1980; 1981; Phillips and Petrofsky, 1980), we have tried to redefine the basic Hill equation to take into consideration the above mentioned variables. This work was done using the soleus (a slow twitch muscle) and medial gastrocnemius (fast twitch muscle) of the lower leg of the cat.

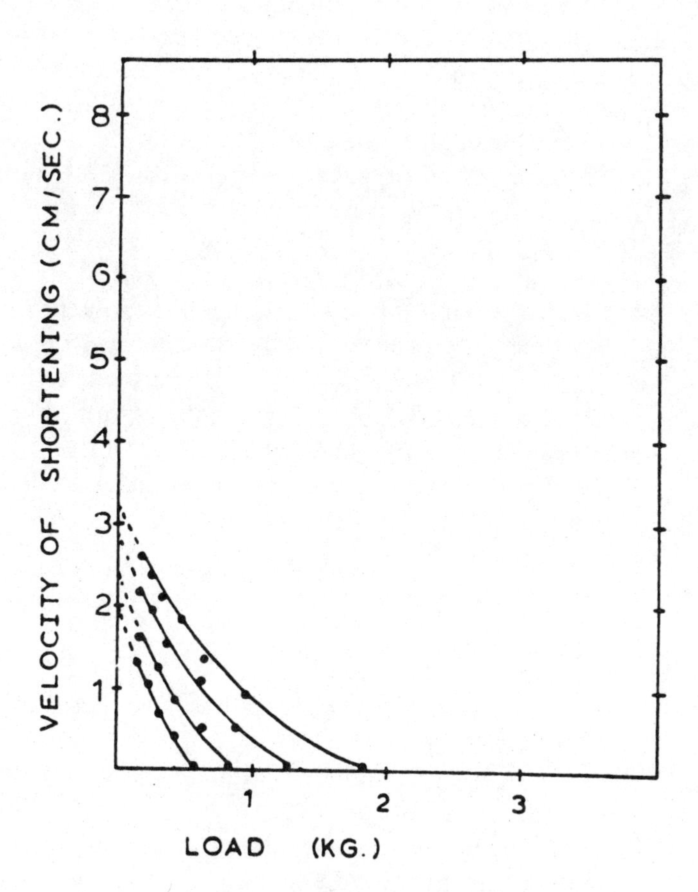

Figure 2-14. The force-velocity relationship in the partially and fully activated *soleus* muscle with recruitment from the *fastest* to *slowest* motor units. From J.S. Petrofsky and C.A. Phillips, The Influence of Recruitment Order and Fibre Composition on the Force-velocity Relationship and Fatigability of Skeletal Muscle in the Cat. *Medical & Biological Engineering & Computing, 18*:381-390, 1980.

Interrelationships Between Recruitment Magnitude and Direction, Load, and Velocity of Contraction in Fast and Slow Muscle

The velocity of contraction of the soleus and the medial gastrocnemius muscles at various levels of recruitment and various loads is shown in Figures 2-10 and 2-12 (recruitment from the slowest to the fastest motor units) and in Figures 2-13 and 2-14 (recruitment from the fastest to the slowest motor units) respectively. For the soleus muscle, a muscle considered to be homogeneously composed of all slow twitch motor units (Ariano et al., 1973), both the estimated maximum velocity of shortening ($V_{mx\ p}$) and the maximum isometric strength ($P_{mx\ p}$) at submaximal levels of recruitment increase with a progressive increase in the recruitment of the motor units in the muscle.

From these data, Hill (1938) coefficients a and b can be calculated, and are listed in Table 2-I for both directions of recruitment. The relationship between these coefficients and the magnitude of activation of the muscle (A) is then approximated with a polynomial fit of the data, which minimizes the "summed squared error," e.g. Phillips and Petrofsky (1980). Here, and throughout the remainder of this chapter, the activation (or magnitude of recruitment) is defined mathematically as

$$A = \frac{P_{mx\ p}}{P_{mx\ 100}} \tag{1}$$

where

A = the fraction of the muscle that is activated at a given stimulation voltage

$P_{mx\ p}$ = the maximum isometric strength of the muscle with some of the motor units recruited

$P_{mx\ 100}$ = the maximum isometric strength of the muscle with all of the motor units recruited.

The equations relating the a and b coefficients of the Hill equation to the activation for recruitment from the slowest to the fastest motor units in the soleus are

$$a = -0.308 + 2.937A - 0.995A^2 \tag{2}$$

$$b = 1.088 + 3.430A - 1.696A^2 \tag{3}$$

Table 2-1

ISOMETRIC LOAD, MAXIMAL (ZERO LOAD) VELOCITY, AND HILL a AND b COEFFICIENTS WITH RESPECT TO MUSCLE TYPE, ACTIVATION LEVEL AND RECRUITMENT ORDER

Soleus-slowest to fastest

A	P_{mx} kg	V_{mxp} cm/s	a	b
0.31	0.56	2.20	0.51	2.00
0.53	0.95	2.45	0.96	2.40
0.79	1.42	2.80	1.40	2.77
1.00	1.80	3.10	1.63	2.81

Soleus-fastest to slowest

A	P_{mxp} kg	V_{mxp} cm/s	a^1	b^1
0.33	0.60	2.10	0.60	2.10
0.44	0.80	2.30	0.84	2.40
0.71	1.30	2.92	1.98	4.46
1.00	1.82	3.15	2.90	5.10

Medial gastrocnemius-slowest to fastest

A	P_{mx} kg	V_{mxp} cm/s	a	b
0.36	1.40	2.60	1.30	2.40
0.54	2.10	4.00	1.57	3.00
0.72	2.80	5.20	1.90	3.54
1.00	3.90	5.85	2.40	3.90

Medial gastrocnemius-faster to slowest

A	P_{mx} kg	V_{mxp} cm/s	a^1	b^1
0.24	0.95	7.00	0.63	4.66
0.41	1.65	6.50	0.86	3.39
0.73	2.90	6.00	1.10	2.29
1.00	4.00	5.90	1.47	2.16

where

a = the Hill a coefficient
b = the Hill b coefficient
A = the fractional activation of the muscle.

The Hill a and b coefficients are best approximated by a second order polynomial expression.

When the direction of the motor unit recruitment is reversed, the resultant equations relating a and b to A are

$$a = -0.621 + 3.550A \tag{4}$$

$$b = -1.284 + 11.243A - 4.809A^2 \tag{5}$$

The Hill a coefficient (in this situation and subsequent situations) is best approximated by a first order polynomial expression. The equations relating the a and b coefficients to A for recruitment from the slowest to fastest motor units in the medial gastrocnemius muscle are

$$a = 0.657 + 1.734A \tag{6}$$

$$b = 0.571 + 6.007A - 2.673A^2 \tag{7}$$

For recruitment from the fastest to slowest motor units,

$$a = 0.388 + 1.054A \tag{8}$$

$$b = 6.870 - 10.828A + 6.136A^2 \tag{9}$$

Substituting these equations for a and b into the Hill equation, the equation relating the velocity of contraction for the soleus with recruitment from the slowest to fastest motor units becomes

$$V = \frac{(P_{mx100} - P)\,(1.088 + 3.430A - 1.696A^2)}{P - 0.308 + 2.937A - 0.955A^2} \tag{10}$$

and for recruitment from the fastest to slowest motor units

$$V = \frac{(P_{mx100} - P)(-1.284 + 11.243A - 4.809A^2)}{P - 0.621 + 3.550A} \tag{11}$$

For the medial gastrocnemius muscle, the equation for V with recruitment direction proceeding from the slowest to fastest motor units becomes

$$V = \frac{(P_{mx100} - P)(0.571 + 6.007A - 2.673A^2)}{P + 0.657 + 1.734A} \tag{12}$$

and, for the reverse recruitment direction,

$$V = \frac{(P_{mx100} - P)(6.870 - 10.828A + 6.136A^2)}{P + 0.388 + 1.054A} \tag{13}$$

For equations 10 through 13,

V = velocity of contraction
A = degree of activation of the muscles
P_{mx100} = maximum isometric strength of the muscle with all the motor units recruited
P = load imposed on the muscle

Table 2-II gives the calculated values of a and b for the two different muscles at the two different recruitment directions with the four levels of activation.

The Effect of Temperature and Recruitment Magnitude on the Force-Velocity Relationship in the Medial Gastrocnemius Muscle

The force-velocity relationship of the medial gastrocnemius muscle at a temperature of 38 degrees centigrade is shown in Figure 2-15. Each point in this figure and in Figure 2-16 represents the mean of the results on four different muscles contracting at their maximum isometric strength and at ten submaximal loads. The average maximum isometric strength of these muscles with all of the motor units recruited is 3.95 kg, while the isometric strength of these same muscles at the three submaximal levels of stimulation averages 1.38, 2.13, and 2.84 kg. These levels of recruitment then correspond to 35, 54, and 72 percent of the maximum isometric strength of the muscles.

Table 2-II

CALCULATED HILL a AND b COEFFICIENTS
WITH RESPECT TO MUSCLE TYPE, RECRUITMENT ORDER
AND ACTIVATION LEVEL

Soleus-slowest to fastest		
A	a	b
0.31	0.51	1.99
0.53	0.97	2.43
0.79	1.39	2.74
1.00	1.63	2.82

Soleus-fastest to slowest		
A	a^1	b^1
0.33	0.55	1.90
0.44	0.94	2.73
0.71	1.90	4.27
1.00	2.93	5.15

Medial gastrocnemius-slowest to fastest		
A	a	b
0.36	1.28	2.39
0.54	1.59	3.04
0.72	1.91	3.51
1.00	2.39	3.91

Medial gastrocnemius-faster to slowest		
A	a^1	b^1
0.24	0.64	4.62
0.41	0.82	3.46
0.73	1.16	2.24
1.00	1.44	2.18

Associated with an increase in activation then, there is a progressive increase in both the isometric strength (as would be expected) and the V_{mx} (maximum velocity of shortening of the unloaded muscle). From these data, the a and b coefficients of the Hill equation can be calculated and are shown on the right hand panels of Figure 2-15. The relationship between the a coefficient and the activation is linear and therefore could be represented by

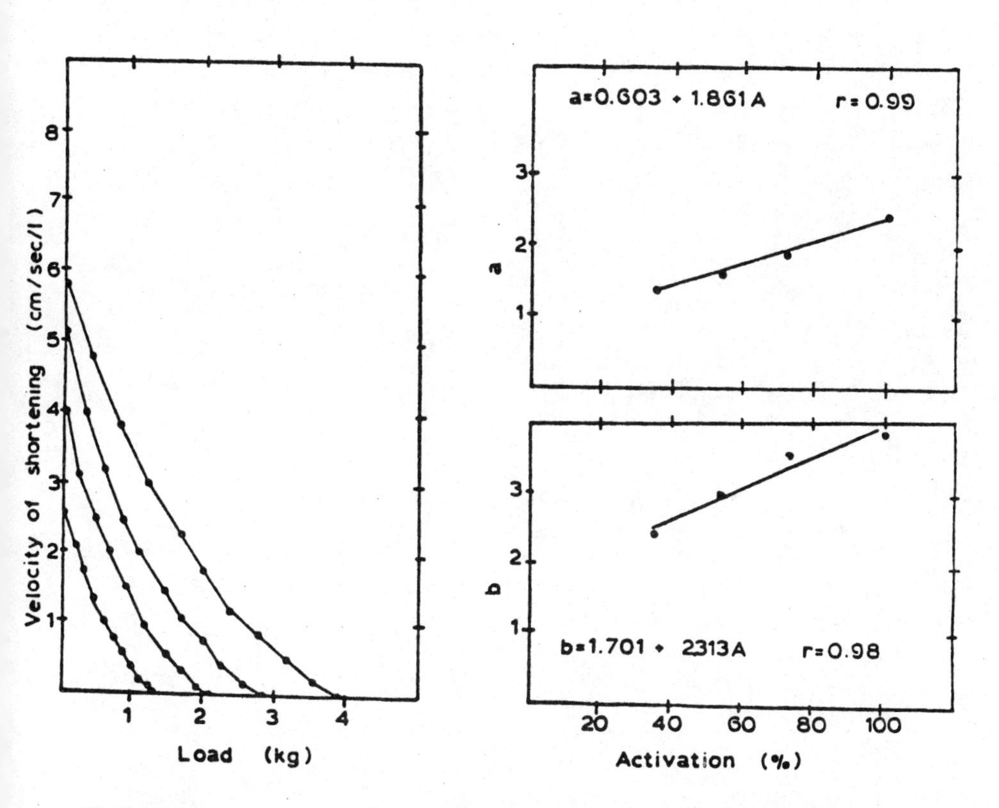

Figure 2-15. The force-velocity relationship for the medial gastrocnemius muscle of the cat at three levels of submaximal and maximal activation (see text) and the resultant Hill a and b coefficients. Each point in the figure represents the mean of four experiments. The muscle temperature was kept constant here at *38 degrees centigrade*. From J.S. Petrofsky and C.A. Phillips, The Influence of Recruitment Order and Fibre Composition on the Force-velocity Relationship and Fatigability of Skeletal Muscle in the Cat. *Medical & Biological Engineering & Computing, 18*:381-390, 1980.

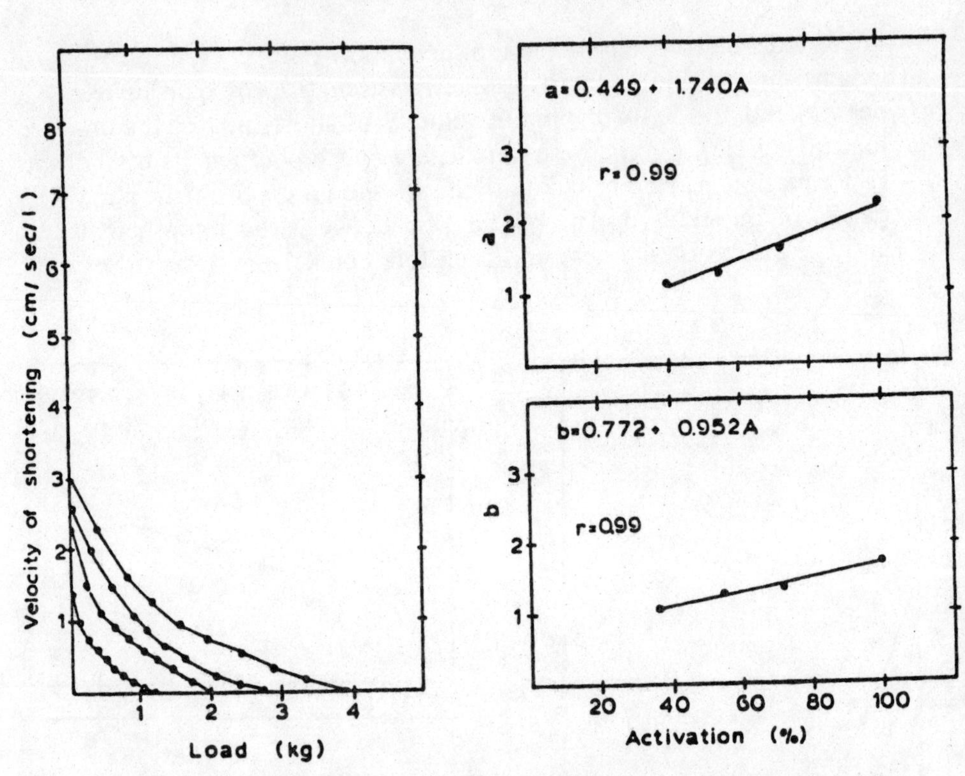

Figure 2-16. The force-velocity for the medial gastrocnemius muscle of the cat at three levels of submaximal and maximal activation (see text) and the resultant Hill a and b coefficients. Each point in the figure represents the mean of four experiments. The muscle temperature was kept constant here at *28 degrees centrigrade*. From J.S. Petrofsky and C.A. Phillips, The Influence of Recruitment Order and Fibre Composition on the Force-velocity Relationship and Fatigability of Skeletal Muscle in the Cat. *Medical & Biological Engineering & Computing, 18*:381-390, 1980.

the equation

$$a = 0.603 + 1.861A \qquad (r = 0.996) \qquad (15)$$

The relationship between the b coefficient and A is also linear and was best represented by the equation

$$b = 1.701 + 2.313A \qquad (r = 0.977) \qquad (16)$$

where, for equations 15 and 16,

a, b = the a and b coefficients of the Hill equation
A = the activation of the muscles

When the temperature of the muscles is reduced below 38 degrees centigrade, there is a progressive reduction in the velocity of shortening at any load or degree of activation. For example, the force-velocity relationship at these same four levels of activation is shown in Figure 2-16 when the temperature of the muscles is maintained at 28 degrees centrigrade throughout the experiments. At any level of activation, there is a small reduction in the isometric strength of the muscles compared to the results recorded at 38 degrees centigrade. In contrast, the maximum velocity of shortening of the unloaded muscle is reduced by almost 50 percent at any level of activation used here. This results in a small change in the a coefficient but a large change in the Hill b coefficient due to the reduction in muscle temperature. The calculated equations relating a and b to A are both nearly linear and therefore are represented by the following equations:

$$a = 0.449 + 1.740A \qquad (r = 0.999) \qquad (17)$$

$$b = 0.772 + 0.952A \qquad (r = 0.989) \qquad (18)$$

where

a, b = the a and b Hill coefficients
A = the activation of the muscles.

When these results were compared to those at muscle temperatures of 35 and 32 degrees centigrade, it is found that the reduction in the P_{mx} and V_{mx} at any level of activation is linear over the temperature range of 38 to 28 degrees centigrade. Therefore, from these data at all four temperatures, the equations representing the a and b coefficients in terms of both A and T (see Table 2-III) become

$$a = 0.085 + 1.265A + 0.017T + 0.012AT \qquad (19)$$

$$b = -1.819 - 2.939A + 0.092T + 0.139AT \qquad (20)$$

Table 2-III

COMPARISON OF MEASURED HILL COEFFICIENTS (a,b) TO CALCULATED COEFFICIENTS (a′ AND b′) AT FOUR DIFFERENT ACTIVATION LEVELS (A) AND FOUR DIFFERENT TEMPERATURES (T)

a	a'	b	b'	A	T
1.10	1.12	1.10	1.09	0.35	28
1.32	1.43	1.29	1.27	0.54	28
1.72	1.71	1.46	1.44	0.72	28
2.20	2.16	1.72	1.71	1.00	28
1.25	1.09	1.51	1.55	0.28	32
1.42	1.54	1.90	1.95	0.55	32
1.83	1.88	2.25	2.27	0.76	32
2.31	2.29	2.59	2.63	1.00	32
1.18	1.08	1.75	1.86	0.24	35
1.51	1.52	2.40	2.36	0.50	35
1.90	1.88	2.90	2.77	0.71	35
2.41	2.37	3.25	3.33	1.00	35
1.30	1.33	2.40	2.50	0.35	38
1.57	1.66	3.00	2.94	0.54	38
1.90	1.97	3.54	3.36	0.72	38
2.50	2.45	3.90	4.02	1.00	38

where

a, b = the a and b coefficients of the Hill equation
A = the activation level
T = the temperature of the muscle in degrees centigrade.

Interrelationships Between Muscle Length, Temperature, and Recruitment Magnitude and the Force-Velocity Relationship

The velocity of shortening at loads up to the maximum isometric strength of the medial gastrocnemius muscles at activation levels of 0.35 (panel a), 0.54 (panel b), 0.718 (panel c) and 1.00 (panel d) and at five different initial lengths are shown in Figure 2-8. Each point in this figure shows the mean of the responses of four cats where the muscle temperature was kept constant at 38 degrees centigrade. For all four degrees of activation, the V_{mx} is independent of the initial length of the muscles. In contrast, for the partially or fully loaded muscle, the fastest velocity of shortening when the muscle lifted a set load was found when the muscle started the contraction at its optimal length. These same experiments are repeated at a muscle temperature of 28 degrees centigrade; the results of these determinations on the same four cats is shown in Figure 2-17. The results recorded at this muscle temperature closely parallel those recorded at a muscle temperature of 38 degrees centigrade. The only difference is that the velocity of contraction is reduced by about half under any experimental condition.

From these data then, the a and b coefficients of the Hill equation can be calculated. These results are shown in Figure 2-18 (muscle temperature of 38° C) and Figure 2-19 (muscle temperature of 28° C). The length of the muscle in both of these figures has been normalized in terms of the length of the muscle at which it developed its largest isometric strength when the muscle is fully activated. Both the a and b coefficients show their largest values for this muscle length and a marked reduction for other muscle lengths. From these relationships, the equations relating the a and b coefficients of the Hill equation to the initial length and temperature of the muscle have been calculated.

Because of the nonlinearity of the relationships in Figures 2-18 and 2-19, these data and the resultant equations are piece-wise linearized at the midpoint (L = 1.0). For muscle lengths below the optimal length (L < 1.0)

$$a_{0.99} = -0.071 - 4.087A + 0.245L - 0.051T + 0.010AT + 0.062LT + 5.545AL \quad (21)$$

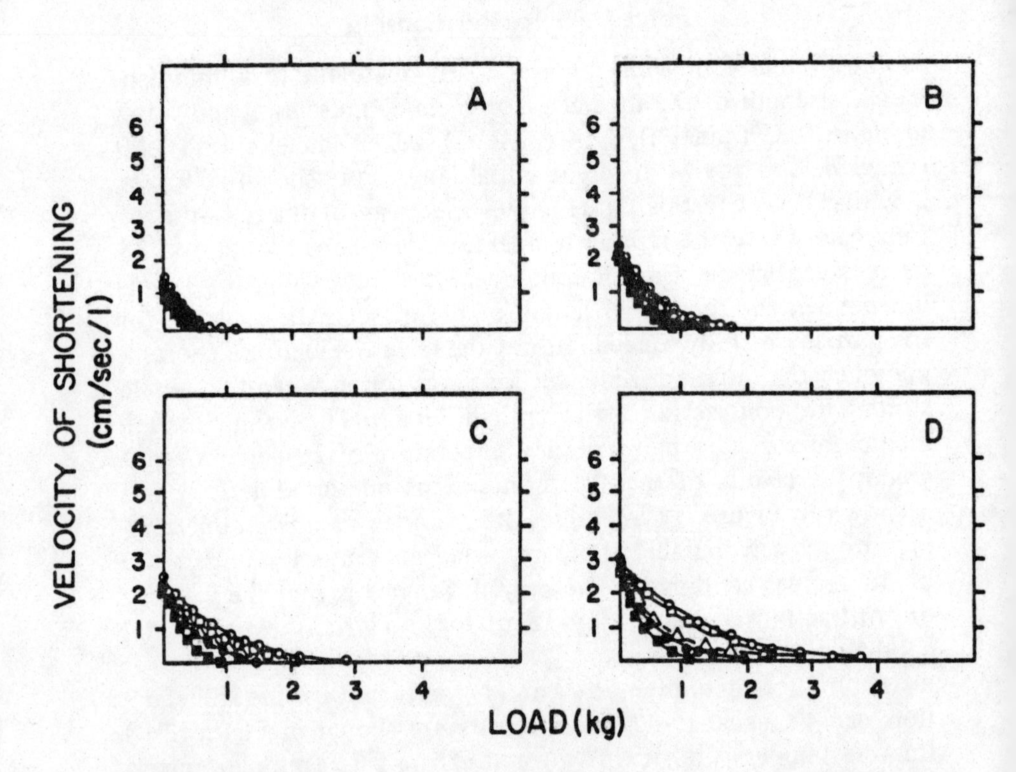

Figure 2-17. The force-velocity relationship of the muscles when prestretched prior to the isotonic contraction to their optimum length (o) and to 0.75(■), 0.9(△), 1.1(□), and 1.25(●) times that length during contractions where the activation averaged 0.35 (panel a), 0.54 (panel b), 0.718 (panel c) and 1.00 (panel d). Each point in this figure represents the mean of four experiments. During these experiments the muscle temperature was kept constant at *28 degrees centigrade*. From J.S. Petrofsky and C.A. Phillips, The Influence of Recruitment Order and Fibre Composition on the Force-velocity Relationship and Fatigability of Skeletal Muscle in the Cat. *Medical & Biological Engineering & Computing, 18*:381-390, 1980.

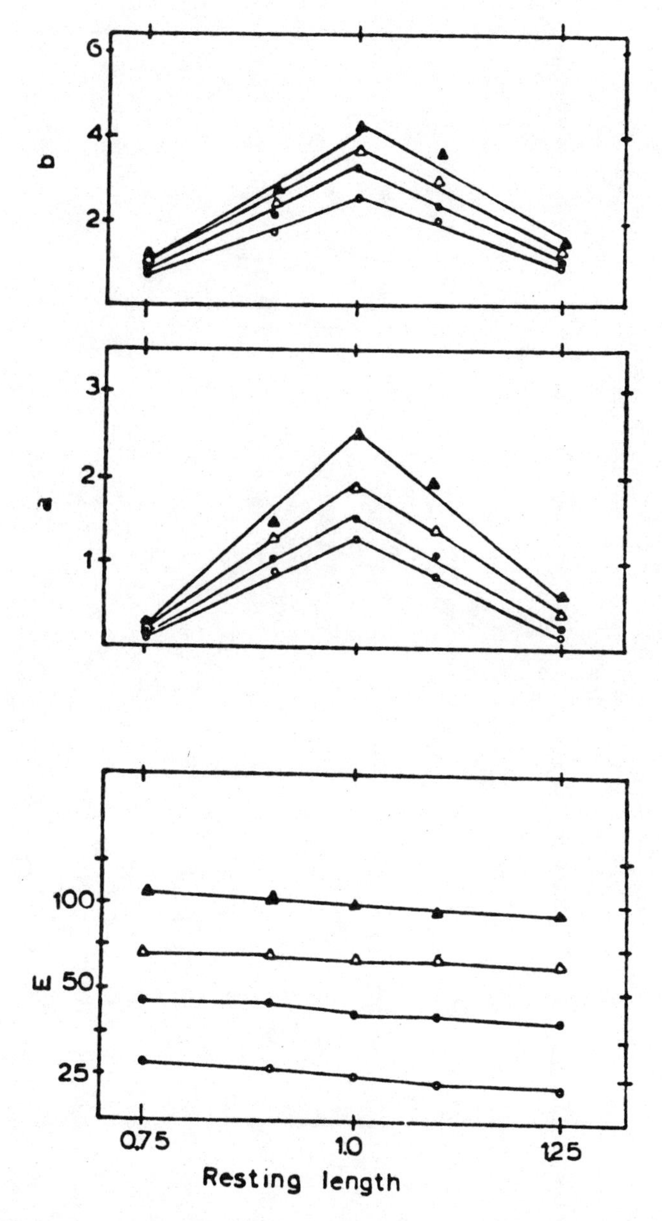

Figure 2-18. The relationship between the Hill a and b coefficients and the RMS EMG amplitude (E) and the length of the muscle (normalized in terms of the optimum length of the muscle). Each point in this figure represents the mean of four experiments. The muscle temperature was kept constant at *38 degrees centigrade*. From J.S. Petrofsky and C.A. Phillips, The Influence of Recruitment Order and Fibre Composition on the Force-velocity Relationship and Fatigability of Skeletal Muscle in the Cat. *Medical & Biological Engineering & Computing, 18*:381-390, 1980.

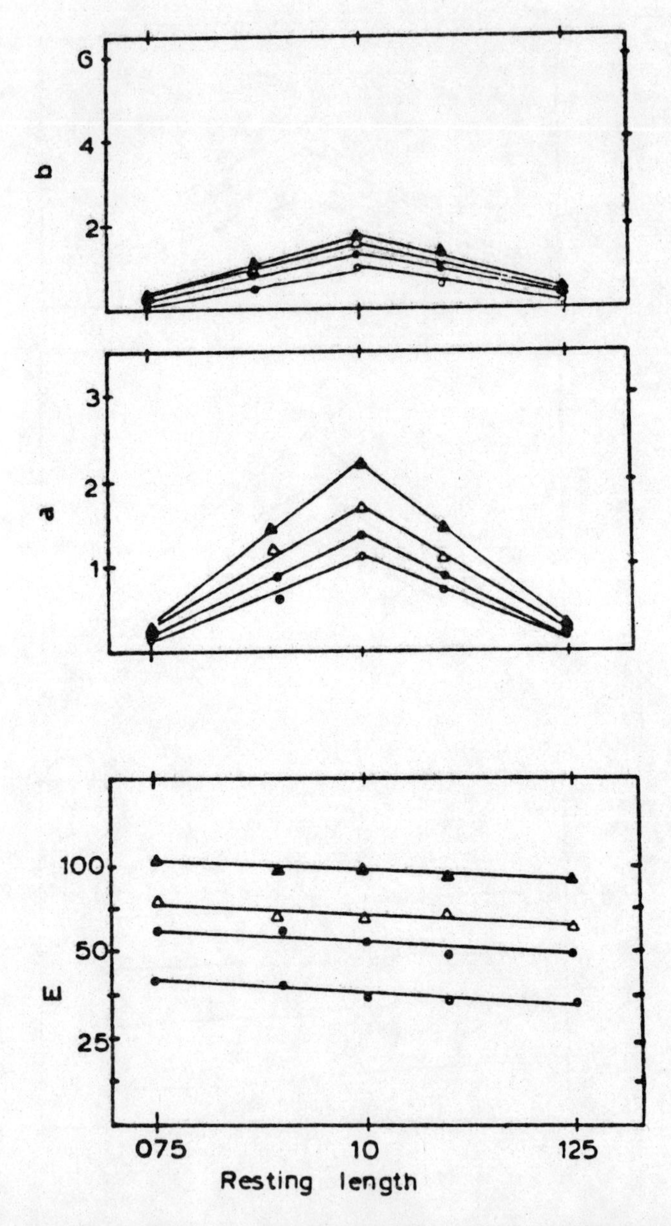

Figure 2-19. The relationship between the Hill a and b coefficients and the RMS EMG amplitude (E) and the length of the muscle (normalized in terms of the optimum length of the muscle). Each point in this figure represents the mean of four experiments. The muscle temperature was kept constant at *28 degrees centigrade*. From J.S. Petrofsky and C.A. Phillips, The Influence of Recruitment Order and Fibre Composition on the Force-velocity Relationship and Fatigability of Skeletal Muscle in the Cat. *Medical & Biological Engineering & Computing, 18*:381-390, 1980.

$$b_{0.99} = 7.378 - 4.240A - 11.077L - 0.301T + 0.043AT + 0.450LT + 4.389AL \quad (22)$$

For muscle lengths equal to or greater than the optimal length $(L \geqslant 1.0)$

$$a_{1.0} = 1.146 + 5.286A - 0.540L + 0.055T + 0.044AT - 0.057LT - 4.899AL \quad (23)$$

$$b_{1.0} = 8.955 + 1.263A + 6.494L + 0.416T + 0.121AT - 0.307LT - 3.405AL \quad (24)$$

where

a, b = the a and b coefficients of the Hill equation
A = the activation of the muscle
L = the initial length of the muscle prior to isotonic contractions
T = the temperature of the muscle in degrees centigrade

Table 2-IV gives the calculated Hill's coefficients (a′, b′) and compares them with the measured Hill's coefficients (a,b) for lower muscle lengths (L = 0.75, 0.90, 1.00).

Table 2-V gives the calculated Hill's coefficients (a′, b′) and compares them with the measured Hill's coefficients (a, b) for higher muscle lengths (L = 1.00, 1.10, 1.25).

The basic Hill (1938) equation that relates muscle load to velocity is

$$(P + a)(V) = b(P_0 - P) \quad (25)$$

which can be rearranged to solve for velocity in terms of load and the a and b coefficients

$$V = \frac{(P_0 - P)b}{P + a} \quad (26)$$

Substituting equations 21 and 22 into 26 yields the velocity for muscle lengths less than the optimal length

$$V_{0.99} = \frac{(P_0 - P)b_{0.99}}{P + a_{0.99}} \quad (27)$$

Table 2-IV

COMPARISON OF MEASURED HILL COEFFICIENTS (a AND b) TO CALCULATED COEFFICIENTS (a' AND b') AT FOUR ACTIVATION LEVELS (A) AND TWO TEMPERATURES (T) FOR LOWER MUSCLE LENGTHS (L < 1.0)

a	a'	b	b'	A	L	T
0.10	0.11	0.10	0.18	0.35	0.75	28
0.19	0.18	0.25	0.23	0.54	0.75	28
0.26	0.24	0.35	0.28	0.72	0.75	28
0.31	0.34	0.42	0.35	1.00	0.75	28
0.63	0.70	0.50	0.64	0.35	0.90	28
0.89	0.92	0.75	0.81	0.54	0.90	28
1.12	1.14	1.00	0.99	0.72	0.90	28
1.44	1.47	1.20	1.23	1.00	0.90	28
1.10	1.09	1.11	0.95	0.35	1.00	28
1.46	1.42	1.30	1.20	0.54	1.00	28
1.73	1.73	1.45	1.45	0.72	1.00	28
2.20	2.22	1.71	1.83	1.00	1.00	28
0.10	0.10	0.80	0.70	0.35	0.75	38
0.20	0.19	0.90	0.83	0.54	0.75	38
0.30	0.27	1.10	0.95	0.72	0.75	38
0.40	0.40	1.15	1.14	1.00	0.75	38
0.75	0.78	1.80	1.83	0.35	0.90	38
0.90	1.02	2.00	2.09	0.54	0.90	38
1.20	1.26	2.20	2.33	0.72	0.90	38
1.50	1.62	2.40	2.70	1.00	0.90	38
1.30	1.24	2.40	2.59	0.35	1.00	38
1.57	1.58	3.00	2.93	0.54	1.00	38
1.90	1.92	3.54	3.25	0.72	1.00	38
2.50	2.43	3.90	3.75	1.00	1.00	38

Table 2-V

COMPARISON OF MEASURED HILL COEFFICIENTS (a AND b) TO CALCULATED COEFFICIENTS (a′ AND b′) AT FOUR ACTIVATION LEVELS (A) AND TWO TEMPERATURES (T) FOR HIGHER MUSCLE LENGTHS (L > 1.0)

a	a'	b	b'	A	L	T
1.10	1.12	1.11	1.03	0.35	1.00	28
1.46	1.42	1.30	1.26	0.54	1.00	28
1.73	1.72	1.45	1.49	0.72	1.00	28
2.20	2.17	1.71	1.84	1.00	1.00	28
0.75	0.73	0.50	0.70	0.35	1.10	28
0.89	0.95	1.00	0.87	0.54	1.10	28
1.10	1.15	1.13	1.03	0.72	1.10	28
1.45	1.47	1.30	1.29	1.00	1.10	28
0.15	0.15	0.15	0.20	0.35	1.25	28
0.26	0.23	0.27	0.29	0.54	1.25	28
0.32	0.30	0.38	0.35	0.72	1.25	28
0.40	0.41	0.45	0.46	1.00	1.25	28
1.30	1.25	2.40	2.54	0.35	1.00	38
1.57	1.64	3.00	3.01	0.54	1.00	38
1.90	2.01	3.54	3.45	0.72	1.00	38
2.50	2.59	3.90	4.14	1.00	1.00	38
0.80	0.81	1.90	1.90	0.35	1.10	38
1.20	1.11	2.20	2.31	0.54	1.10	38
1.50	1.40	3.00	2.69	0.72	1.10	38
2.00	1.83	3.50	3.28	1.00	1.10	38
0.15	0.15	1.00	0.95	0.35	1.25	38
0.25	0.30	1.25	1.25	0.54	1.25	38
0.40	0.45	1.50	1.54	0.72	1.25	38
0.65	0.69	1.80	1.99	1.00	1.25	38

Substituting equations 23 and 24 into equation 26 yields the velocity for muscle lengths equal to or greater than the optimal length

$$V_{1.0} = \frac{(P_0 - P)b_{1.0}}{P + a_{1.0}} \tag{28}$$

APPROACH TO A FINAL SYNTHESIS

The astute reader will now recognize that the velocity of muscle shortening (V) has been defined with respect to recruitment in the forward direction (V) and in the reverse direction (V^1) as a function of isometric load (P_0) and isotonic afterload (P) via the Hill equation (1938). Furthermore, V and V^1 have been related to muscle activation (A), temperature (T), and length (L) via the Hill a and b coefficients, so that two system equations have resulted of the form:

$$V = f(P_0, P, A, L, T) \tag{29}$$

$$V' = f(P_0', P, A, L, T) \tag{30}$$

The approach to a final synthesis of these equations requires that we recognize the following:

1. P_0 is also a dependent variable, i.e. $P_0 = f(A)$, $P_0 = f(L)$, and $P_0 = f(T)$
2. V and V^1 are also a function of an additional independent variable (X), i.e. the fraction of slow motor units, $X = 1.00$ for soleus and $X = 0.25$ for medial gastrocnemius.

In a study that accounted for these two items, Phillips and Petrofsky (1980) studied lateral gastrocnemius ($X = 0.18$), medial gastrocnemius ($X = 0.25$), and soleus ($X = 1.00$) of the cat as an independent variable as well as A (activation level or recruitment magnitude) between 0.20 and 1.00. These investigators found that

$$P_0 = 4.655A - 2.860XA \tag{31}$$

$$P_0' = 4.731A - 2.907XA \tag{32}$$

When they solved for the Hill a and b coefficients as a function of X and A, the results for the three muscles (i.e., X = 0.18, 0.25 and 1.00) at four activation levels (i.e., A = 0.25, 0.50, 0.75 and 1.00) by the "least summed square" error method were

$$a = 1.707 + 10.032X + 2.955A - 0.345XA - 8.551X^2 - 0.782A^2 \quad (33)$$

$$b = 1.024 + 1.972X + 4.383A - 0.385XA - 2.100X^2 - 2.029A^2 \quad (34)$$

$$a' = 1.667 - 6.466X + 0.069A + 2.633XA + 4.433X^2 + 0.600A^2 \quad (35)$$

$$b' = 8.310 - 14.766X - 7.911A + 7.971XA + 8.393X^2 + 3.363A^2 \quad (36)$$

The details of the derivation of equations 31 to 36 are given in Appendices 1 and 2 of Phillips and Petrofsky (1980).

Therefore, the Hill force-velocity relationship must ultimately be defined in terms of recruitment direction, recruitment magnitude (activation), fiber composition, initial length, and muscle temperature as well as the load on the muscle. Mathematically, this means that the following comprehensive equations must be developed for skeletal muscle:

$$V = f(P_0, A, X, L, T, P) \quad (37)$$

$$V' = f(P_0', A, X, L, T, P) \quad (38)$$

where

$$P_0 = f(A, X, L, T) \quad (39)$$

$$P_0' = f(A, X, L, T) \quad (40)$$

The work presented in this chapter represents an initial approach. The final equation to be derived will require that at least two conditions be satisfied:

1. An adequate "data bank" of experimental variables, each varying individually over the physiological range, while the others are held constant so that characteristic "sets" of vari-

ables are defined.

2. An appropriate "piece-wise" linear analysis of the experimental sets of variables, so that the methods of linear regression can be applied to equations of the mixed polynomial form.

CONCLUDING REMARKS

In summation, we have reexamined the classical Hill force-velocity relationship in terms of recruitment direction and magnitude, fiber composition, initial length, and muscle temperature. We have proposed the direction of future research (which these initial studies have indicated). However, still more variables must be considered before our understanding is complete. These include the passive elasticity of muscle (whatever its ultimate anatomical basis) and its effect upon the force-velocity relationship (Phillips and Petrofsky, 1981) as well as the effects of muscle fatigue (Fitts and Holloszy, 1977; Petrofsky, Weber, and Phillips, 1980). This understanding may well prove decisive when functional electrical stimulation is applied to control muscle movement in the paralyzed individual (see Chapter 5).

REFERENCES

Ariano, M., Armstrong, R.B., and Edgerton, V.R. (1973) Hindlimb muscle fiber populations of 5 animals. *J. Hist. and Cytomchem.*, 21:51-55.

Bigland, B. and Lippold, O.J.C. (1954) The relation between force velocity and integrated electrical activity in human muscles. *J. Physiol.*, 123:214-224.

Clarke, R.D.J., Hellon R.F.R., and Lind A.R. (1958) The duration of sustained contraction of the human forearm at different muscle temperatures. *J. Physiol.*, 143:454-463.

Close, R. (1972) Dynamic properties of skeletal muscles. *Physiol. Rev.*, 52:129-187.

Fitts, R.H. and Holloszy, J.O. (1977) Contractile properties of rat soleus muscle: effects of training and fatigue. *Am. J. Physiol.*, 233(3):C86-C91.

Grimby, L. and Hannerz, J. (1976) Disturbance of recruitment order in low and high frequency motor units on blockade of proprioreceptive activity. *Acta Physiol. Scand.*, 96:207-216.

Hannerz, J. and Grimby, L. (1979) The afferent influence on the voluntary firing range of individual motor units in man. *Muscle and Nerve*, 2:414-422.

Hill, A.V. (1938) The heat of shortening and dynamic constants of muscle contraction. *Proc. Roy. Soc. Ser. B.*, 126:136-152.

Milner-Brown, H.S. and Stein, R.B. (1975) The relationship between the surface electromyogram and muscular force. *J. Physiol.*, 246:549-561.

Olson, C.B., Carpenter, D.O., and Henneman, E. (1968) Orderly recruitment of muscle action potentials. *Arch. Neurol.*, 19: 591-597.

Petrofsky, J.S. and Phillips, C.A. (1979) Constant velocity contractions in skeletal muscle by sequential stimulation of muscle efferents. *Med. Biol. Eng. Comp.*, 17:583-592.

Petrofsky, J.S., Weber, C., and Phillips, C.A. (1980) Mechanical and electrical correlates of isometric muscle fatigue in skeletal muscle in the cat. *Europ. J. Physiol.*, 387:33-38.

Petrofsky, J.S. and Phillips, C.A. (1980) The influence of recruitment order and fibre composition on the force-velocity relationship and fatigability of skeletal muscle in the cat. *Med. Biol. Eng. Computing*, 18:381-390.

Petrofsky, J.S. and Phillips, C.A. (1981) The influence of temperature, initial length, and electrical activity on the force-velocity relationship of the medial gastrocnemius muscle of the cat. *J. Biomech.*, 14:297-306.

Phillips, C.A. and Petrofsky, J.S. (1980) Velocity of contraction of skeletal muscle as a function of activation and fiber composition: a mathematical model. *J. Biomech.*, 13:549-558.

Phillips, C.A. and Petrofsky, J.S. (1981) The passive elastic force-velocity relationship of cat skeletal muscle: influence upon maximal contractile element velocity. *J. Biomech.*, 14:399-403.

Wilkie, D.R. (1950) The relation between force and velocity in human muscle. *J. Physiol.*, 110:249-280.

Zahalak, G., Duffy, J., Stewart, P.A., Litchman, H.M., Hawley, R.H., and Paslay, P.R. (1976) Partially activated human skeletal muscle: an experimental investigation of force, velocity, and EMG. *J. Appl. Mech.*, 1:81-87.

Chapter 3

THE ELECTROMYOGRAM DURING STATIC AND DYNAMIC EXERCISE

J.S. PETROFSKY and C.A. PHILLIPS

AMPLITUDE OF THE ELECTROMYOGRAM DURING STATIC EXERCISE

THE amplitude of the electromyogram (EMG) during isometric contractions can clearly be subdivided into two categories. For submaximal isometric contractions, i.e. for contractions sustained at tensions less than the individual's maximum strength, there is generally a linear relationship between the amplitude of the EMG and tension for brief isometric contractions. For example, Figure 3-1 shows the amplitude of the EMG during 3 second isometric contractions of the handgrip muscles at tensions between 10 and 100 percent of the maximum isometric strength (MVC) of eight male subjects. Each point in the figure shows the mean response of the subjects ± the S.D. (standard deviation). The EMG was recorded from electrodes placed on the surface of the skin on the inner surface of the forearm over the flexor muscles that control the handgrip. As can be seen from this figure, there was a linear relationship between tension and the RMS amplitude of the surface EMG (Bigland and Lippold, 1954; Edwards and Lippold, 1956; Lippold, 1952; Milner-Brown and Stein, 1975; Petrofsky and Lind, 1980a; 1980b; Lind and Petrofsky, 1979). If the contractions are sustained to fatigue (Fig. 3-2), there is an almost linear rise in the amplitude of the surface EMG with time (Cobb and Forbes, 1923; Edwards and Lippold, 1956; Eason, 1960;

DeVries, 1968; Lloyd, 1971).

For contractions at tensions less than 70 percent of the individual's maximum strength, at the end of the contractions, the initial relationship between tension and EMG amplitude is preserved, i.e. although the amplitude of the EMG is substantially higher at the end of these contractions, the amplitude of the EMG is still linearly related to the tension exerted by the muscles (Lind and Petrofsky, 1979; Petrofsky, 1980a; Petrofsky and Lind, 1980a;

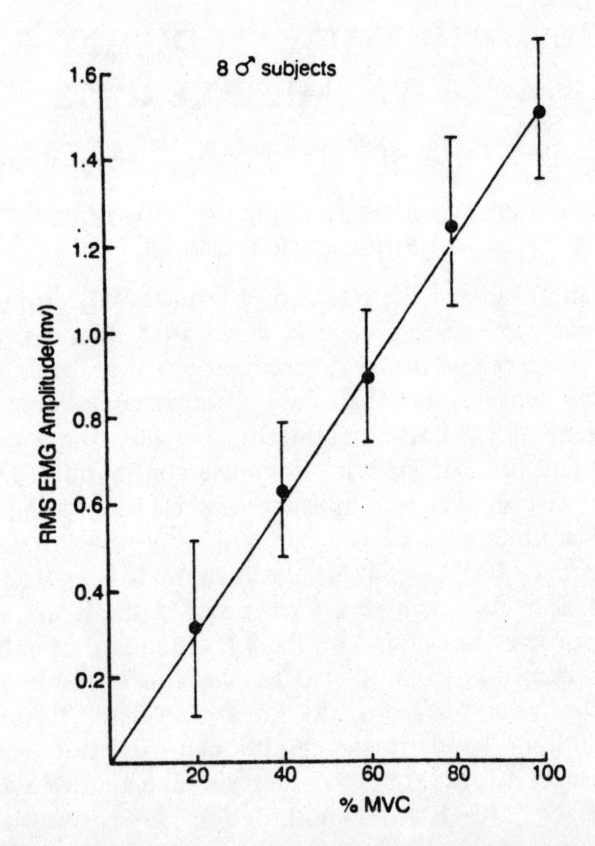

Figure 3-1. The relationship between the tension of a brief (3 seconds) handgrip contraction up to the maximal voluntary contraction (MVC) and the resultant amplitude of the surface EMG, which is normalized to the maximal value found in response to the MVC. Each point represents the mean ± S.D. for two experiments on each of eight subjects.

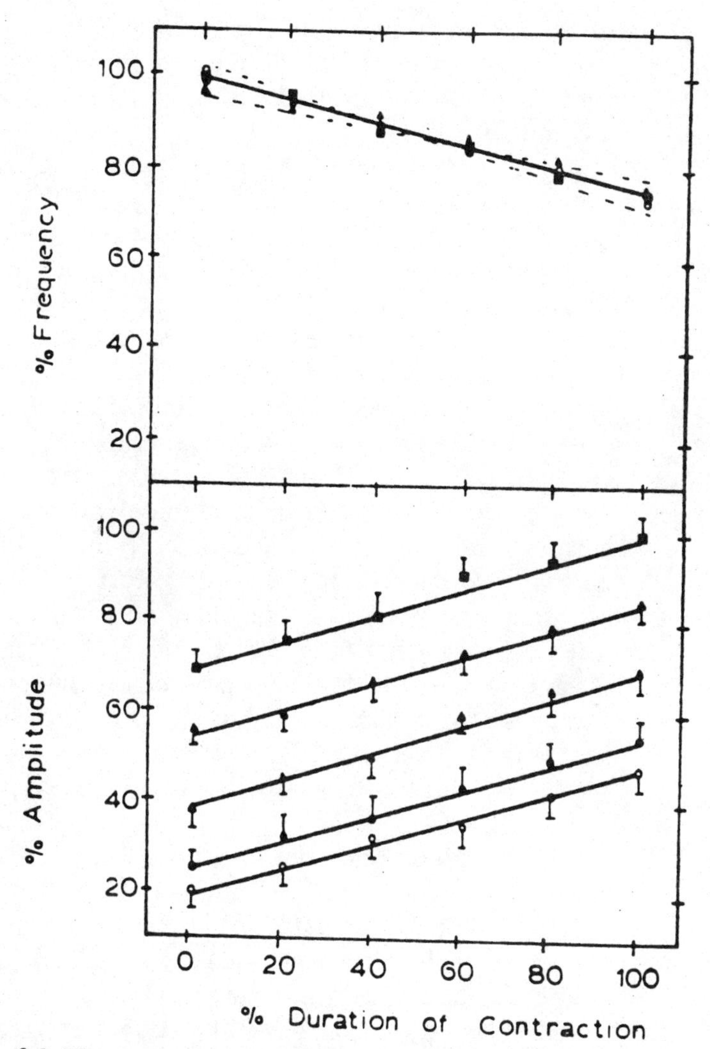

Figure 3-2. The center frequency of the power spectrum and RMS amplitude of the EMG during fatiguing isometric contractions of all six subjects at tensions of 20 (o), 25 (●), 40 (△), 55 (▲), and 70 (■) percent MVC. Each point illustrates the mean of two determinations on each of the six subjects with the respective standard deviations. Amplitudes and center frequencies have been normalized in terms of the RMS amplitude and the center frequency of the EMG for a brief MVC exerted prior to the endurance contraction. The regression line relating frequency to the duration of the contraction (solid line) is shown plus or minus the standard error of the regression (dotted lines). Reprinted with permission from *Computers in Biology & Medicine, 10*:83-

1980b). However, for contractions at 70 percent of the subject's maximum strength or above, the amplitude of the EMG at the point of fatigue is equal to that exerted during a brief maximal effort in the fresh muscle.

The difference in EMG amplitude associated with contraction at, for example, 25 percent of the subject's strength and 70 percent of the subject's strength is striking. For example, in this figure, we see that the amplitude of the EMG at the point of fatigue following an isometric contraction at 25 percent of the subject's maximum strength is only equal to 56 percent of the EMG amplitude during a brief maximal effort in the fresh muscle. This is in sharp contrast to contractions above 70 percent of the maximal strength where the EMG amplitude was maximal.

Since the EMG amplitude reflects the degree of recruitment and frequency of discharge of motor units in muscle, it has been assumed that if the amplitude of the EMG at the end of exercise is not equal to that recorded during a maximal effort in the fresh muscle there is transmission failure in the neuromuscular mechanism. It has been proposed that fatigue of transmission of action potentials from the nerve to the muscle occurs during low tension isometric contractions at the point where the motor nerve bifurcates into the many branches that innervate the muscle fibers. This region of the nerve is particularly susceptible to hypoxia (Paul, 1961).

It has also been proposed that failure occurs either in the neuromuscular junction due possibly to the depletion of acetylcholine (Brown and Burns, 1949) or in the failure of the sarcolemma to propagate action potentials (Stephens and Taylor, 1972). It is felt that this electrical failure in turn is partly responsible for the fatigue in muscle. However, what many authors have overlooked is that the amplitude of the surface EMG is a linear combination of both the degree of recruitment in muscle and the frequency of discharge of motor units. The latter factor makes interpretation of the electromyogram a complex issue. For example, if the frequency of discharge of the motor units was lower at the end of fatiguing contractions at a tension of 25 percent of the individual's maximum strength than at the end of a contraction at 70 percent

---

95, J.S. Petrofsky, Computer Analysis of the Surface EMG during Isometric Exercise, Copyright 1980, Pergamon Press, Ltd.

of the individual's maximum strength, the EMG amplitude would be lower at the point of fatigue for the contractions of 25 percent of the maximum strength. This is what appears to occur during fatiguing isometric contractions.

The alpha motorneuron cell bodies lie in the ventral horn of the spinal cord. As many as 10,000 snyapses form on the alpha motorneuron cell bodies. These synapses are both inhibitory and excitatory. At any one point in time the activity of an alpha motorneuron reflects an algebraic summation of the inhibitory and excitatory synapsis. It has been shown that proprioceptive activity causes feedback inhibition on the alpha motorneuron cell bodies. In the alpha motorneurons innervating the muscles that flex the big toe, Grimby and Hannerz (1976) have shown that proprioceptive activity can lower the firing frequency of alpha motorneurons. In our own work, we have found a similar situation to occur. In the adductor pollicis muscle, the maximum frequency of discharge of alpha motorneurons ranges between 50 and 70 Hertz during a brief maximal effort. However, at the end of a fatiguing isometric contraction at 25 percent of the subject's maximum strength, the maximum discharge frequency of the alpha motorneurons is only 30 to 50 Hertz (Fig. 3-3). In contrast, at the end of a fatiguing contraction at 55 percent of the subject's maximum strength, the discharge frequency ranges between 50 and 70 Hertz. These changes in motor unit discharge frequency are of a significantly large order of magnitude to account for the changes in EMG amplitude reported by ourselves and others between contractions at low and high isometric tensions. The mechanism of this response is probably due to some type of proprioceptive tension feedback pathway going to the alpha motorneuron cell bodies that limit motor unit discharge frequency.

Although the motor unit discharge frequency is certainly lower at the end of fatiguing contractions at low as compared to high isometric tensions, this probably has no effect on muscle fatigue. Although unfatigued muscle requires discharge frequencies as high as 40 to 50 Hertz to tetanize fully, when the muscle fatigues it has been shown in a variety of human and animal experiments

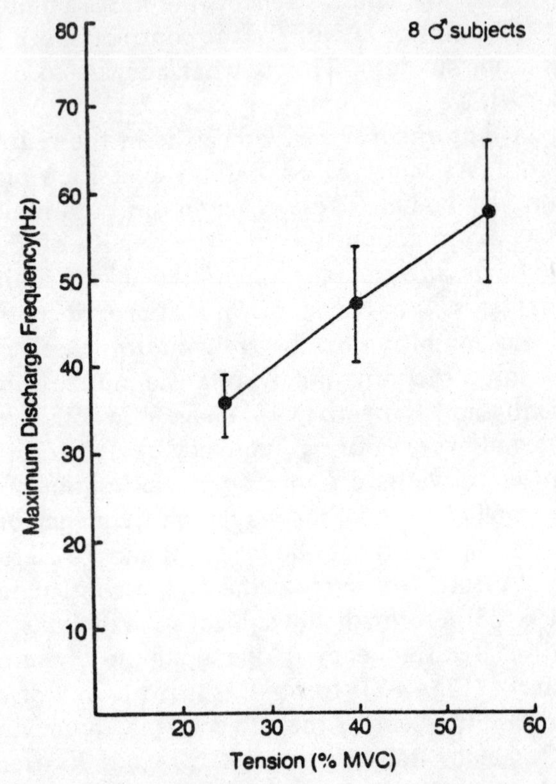

Figure 3-3. The discharge frequency of motor units at the end of voluntary contractions of the hand-grip muscles.

that there is a reduction in actomyocin ATP-ase (Petrofsky and Fitch, 1980), a reduction in ATP turnover (Edwards et al., 1977), a slowing in the conduction velocity of action potentials across the sarcolemma (Mortimer et al., 1970; Petrofsky et al., 1980; Petrofsky and Fitch, 1980), and a reduction in the tissue level of ATP and phosphocreatinine (Karlsson et al., 1975; Edwards et al., 1972; Petrofsky et al., 1980). The overall result of these changes in the biochemistry of the muscle is that the frequency tension relationship is shifted to the left when the muscle fatigues as shown in Figure 3-4 (Petrofsky, 1981).

What this means in practical terms is that the muscle will tetan-

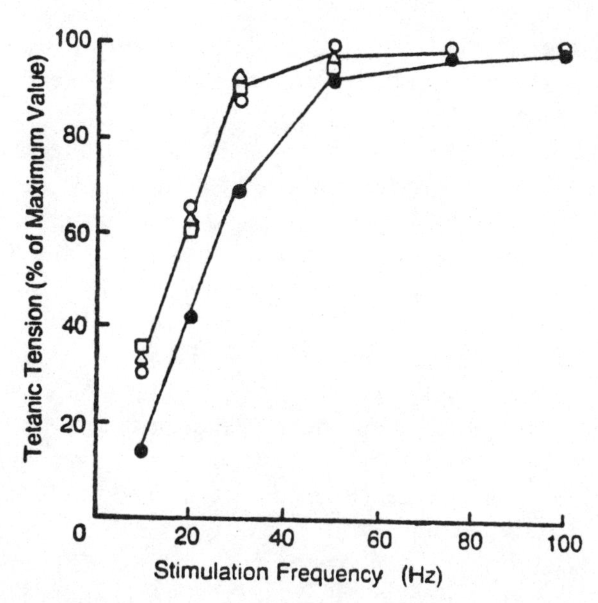

Figure 3-4. The frequency tension relationship in fresh and fatigued muscles.

ize at lower motor unit discharge frequencies when it fatigues. It is, therefore, probable that the lower discharge frequencies found in human muscle at the end of low tension isometric contractions may have no effect on the tension generated by the muscle. Although there is a small reduction in the amplitude of motor unit action potentials when muscle fatigues (Petrofsky et al., 1980), Fink & Luttgau (1976) have shown that there can be small changes in both the amplitude of duration of motor unit action potentials with no effect on the tension generated by muscle fibers. It would, therefore, appear that muscle fatigue during sustained submaximal isometric contractions is not due to electrical failure, although there is a reduction in the maximum amplitude of the EMG at the end of low tension isometric contractions.

Although, in general, there is a linear relationship between the amplitude of the EMG and tension during brief isometric contractions, some studies have reported a nonlinear relationship between these variables (Zuniga and Simmons, 1970). However, nonlinear relationships have only been reported consistently for a few muscles such as the bicep muscles. In other muscles, these nonlinear relationships appear to be artifacts of the recording technique. Lynn et al. (1978) have shown that placing electrodes too widely apart can cause a nonlinear EMG tension relationship. In a few muscles, however, such as the biceps, the pattern of motor unit recruitment is different than in other muscles (Bigland-Ritchie et al., 1978). In this muscle, motor units are recruited continuously over the range of 0 to 100 percent of the muscle's strength, causing a nonlinear EMG-tension relationship. This may be true for other muscles as well.

Sustained maximal isometric contractions differ from those types of contractions described above, both in the response of the surface EMG and in the experimental procedures used in the studies. In studies where subjects exerted sustained submaximal isometric contractions, the experimental protocol is simple. The subjects are asked to exert their maximum isometric strength. Then, a target tension is chosen as a percentage of that maximum strength measured on each subject. This tension is sustained until, through fatigue, the target can no longer be held. At this point, the contraction is terminated. The entire length of time the tension can be held is called the endurance time. In contrast, to achieve sustained maximal isometric contractions, the subject exerts his maximal isometric strength and continues to hold that strength for a set length of time. For example, as shown in Figure 3-5, when a subject maintains a maximal effort for five minutes, his maximum isometric strength falls exponentially. Associated with this exponential fall in isometric strength is an almost parallel fall in the electrical activity of the muscle for the first two to three minutes. For longer periods of time, the amplitude plateaus while the tension continues to slowly fall. Stephens and Taylor (1972) said that this showed that two different processes were occurring; first of all, during the initial phase of a sustained maximal effort (the first two to three minutes) fatigue was caused by a failure to propagate action potentials across the sarcolemma. As the contractions were sustained, electrical failure stabilized and further fatigue was caused by either failure of electromechanical coupling within the muscle or failure in the contractile components

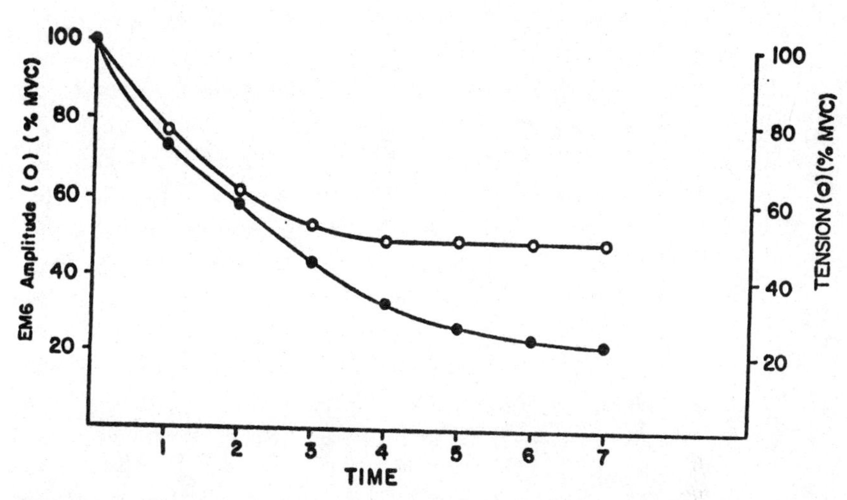

Figure 3-5. The tension developed by a muscle and EMG amplitude during a sustained maximal effort.

of the muscle themselves.

However, Bigland-Ritchie and Lippold (1980) repeated the Stephens and Taylor experiments, and although they found similar results concerning the amplitude EMG, they found some contradictory evidence as well. Bigland-Ritchie and Lippold, in repeating these experiments made one modification to the experimental protocol. In addition to having the subjects exert a sustained maximal effort, they inserted two electrodes into the active muscle. Through these electrodes they stimulated the muscle with sufficient amplitude and at a sufficient frequency to fully tetanize the muscle for brief periods during the sustained maximal effort. Therefore, if there was electrical failure in the neuromuscular junction, more tension should be developed by electrical stimulation than by maximal voluntary contractions as the muscle fatigues. These were not their findings.

They found instead that the tension was the same as that which occurred during electrical stimulation. This indicated to them that, like the sustained maximal contractions described above, electrical failure was not occurring during sustained maximal efforts. They thought that perhaps the fall in the amplitude of the surface EMG might be due to a change in motor unit discharge frequency and not due to a reduction in motor unit recruitment as proposed by Stephens and Taylor (1972). Therefore, they inserted needle electrodes into some of the muscle fibers within the muscle and recorded the motor unit discharge properties during fatiguing maximal isometric contractions. They found that at the onset of a sustained maximal effort, the motor unit discharge frequencies ranged between 60 and 70 Hertz. However, as the contractions were sustained during the first two to three minutes of a sustained maximal effort, motor unit discharge frequencies were reduced exponentially and in parallel with tension to frequencies of between 30 to 40 Hertz. Again, for the same argument given above for sustained submaximal isometric contractions, this reduction in motor unit discharge frequency may have no correlation to isometric fatigue.

FREQUENCY COMPONENTS OF THE ELECTROMYOGRAM DURING STATIC EXERCISE

The changes in the frequency components of the surface EMG during submaximal and maximal isometric contractions are striking. Piper (1912) recorded a broadening in the large waves (carrier waves) of the EMG during high tension muscle contractions. This early observation has led to sixty years of intensive effort to use this phenomenon as an index on muscle fatigue. One problem that has arisen ever since these early experiments is that the EMG is a complicated wave form and has been very difficult to analyze. In these early experiments, described by Piper and later by Cobb and Forbes (1923) and others, the EMG was examined from photographic records second by second. As you can imagine, this is a very difficult process.

In the early 1950s, several investigators tried to use automatic methods of analyzing the EMG. These included zero crossing detectors, peak detectors, and frequency bank amplifiers (Scherrer and Bourguigon, 1959; Larsson, 1971; Stalberg, 1966; Fusfeld, 1971; Lindström et al., 1970; Kogi and Hakamada, 1962; Stulen and DeLuca, 1978). Zero crossing and peak turn detectors rely on the fact that as the carrier waves become broader (longer in duration) there would be more time between waves. Therefore, the number of zero crossings of the EMG would be reduced as the muscle fatigues. The same argument holds for peak or turns detectors. The same type of analysis was attempted when investigators began to use frequency selective audio amplifiers to selectively filter the low frequency components of the EMG. However, frequency selective amplification also added new knowledge to the analysis of the EMG. When the EMG is analyzed through an audio filter bank analyzer, it has been found that muscle fatigue is associated with both an increase in the low frequency components of the EMG and a reduction in the high frequency components of the EMG.

In the 1960s, several groups of investigators tried to use a ratio of the changes in the low and high frequency components of the EMG as a more sensitive fatigue index (Kogi and Hakamada, 1962; Kaiser and Peterson, 1965). Typically, a ratio was calculated from the average or RMS amplitude of the EMG in the frequency range of 40 to 50 Hertz and 300 to 500 Hertz. This type of ratio did give a more sensitive fatigue index than simply examining either the low or high frequency components of the EMG alone. However, this type of fatigue index still exhibited a great degree of variability and was unreliable in repeated tests.

Simply using two of the frequency components of a complex wave form such as the EMG largely ignores much of the data that can be derived. A more reliable fatigue index has been found using either a more complex filter bank system (i.e. four or more filters) (Petrofsky, 1980b; Stulen and DeLuca, 1978) or by using Fourier wave form analysis (Petrofsky et al., 1976; 1977; 1979; 1980; Petrofsky, 1979; 1980b; Lloyd, 1971; Lindstrom et al., 1977; Viitasalo and Komi, 1977). Fourier analysis of the EMG allows the derivation of an infinite number of frequency selective filters through which the EMG can be processed. A typical Fourier power spectrum representing the EMG during a brief contraction at 40 percent of the maximum isometric strength is shown in Figure 3-6.

As described by Piper (1912), the majority of the power of the EMG occurs in the frequency range of 40 to 60 Hertz. Lindström and his colleagues (1970) noticed a depression in the power of the EMG at a frequency of about 150 to 200 Hertz. This depression is caused by the fact that two surface electrodes placed

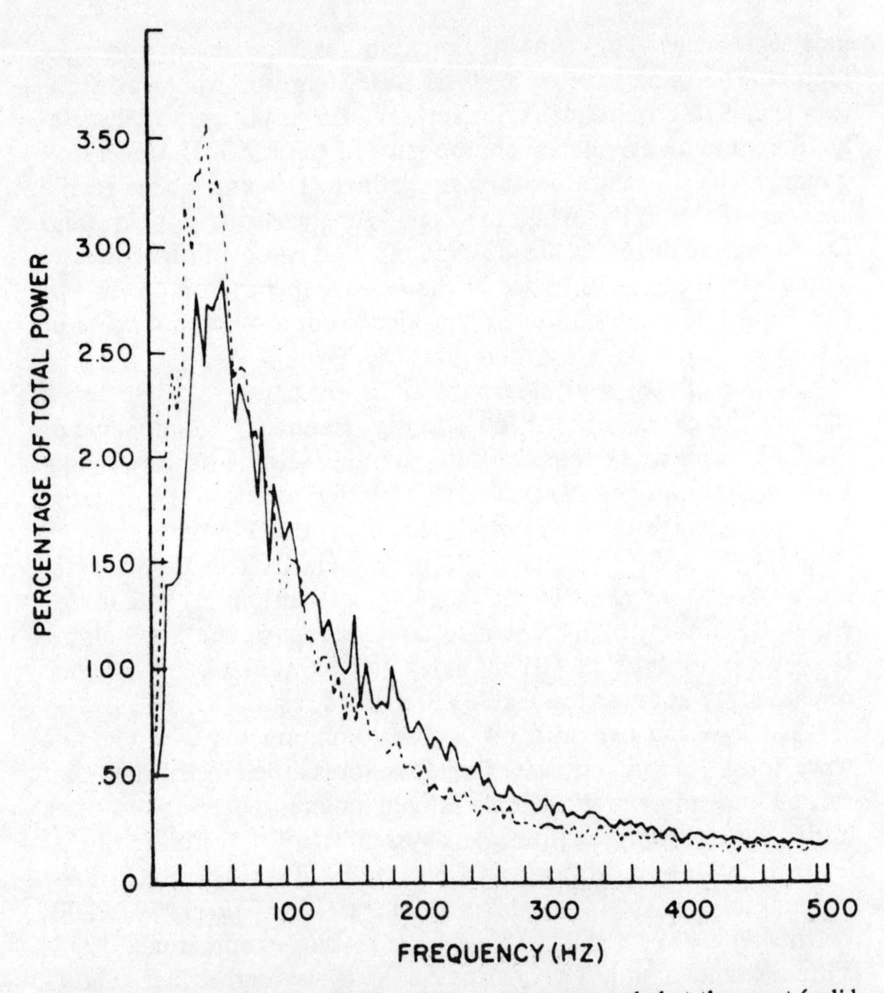

Figure 3-6. Example of an EMG power spectrum recorded at the onset (solid line) and end (broken line) of a fatiguing isometric contraction. Power is expressed as a percentage of the total spectral power recorded at all frequencies. From J.S. Petrofsky and A.R. Lind, Frequency of the EMG During Isometric Exercise. *European Journal of Applied Physiology, 43*:173-182, 1980.

over a conducting media act as a transverse filter and null at the frequency equivalent to the inverse of the conduction velocity of action potentials between the electrodes (Lynn et al., 1978). Lindström et al. (1970) called this type of analysis DIP analysis. The frequency of the DIP then is proportional to the conduction

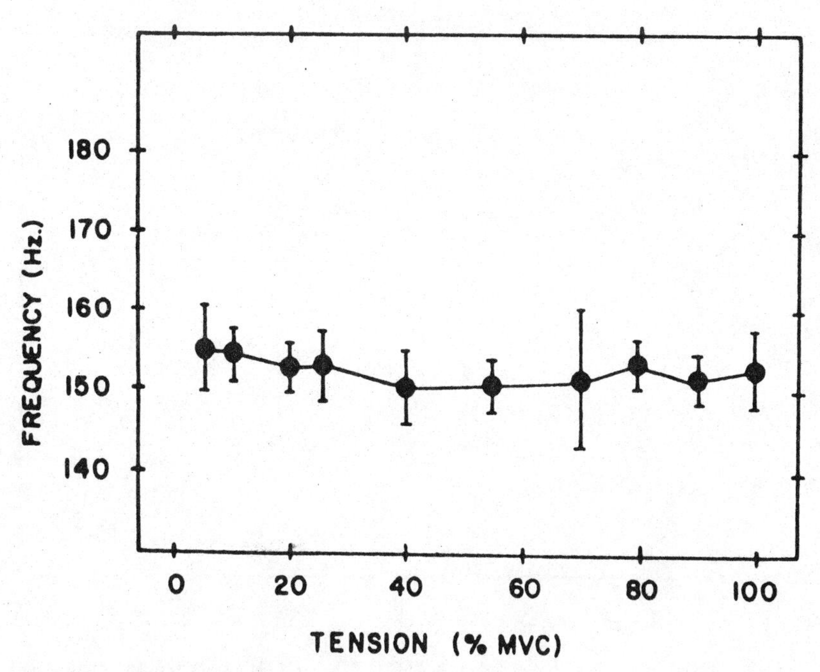

Figure 3-7. Mean center frequencies of the EMG power spectra for EMG sampled during brief isometric contractions at tensions ranging from 5 to 100 percent MVC. Each point is the mean of 16 experiments, and the bars represent ± S.D. From J.S. Petrofsky and A.R. Lind, Frequency Analysis of the EMG During Isometric Exercise. *European Journal of Applied Physiology, 43*:173-182, 1980.

velocity of action potentials across the sarcolemma of the underlying muscle. Lindström et al. (1977) tried measuring the DIP frequency and using this as a fatigue index. However, because of electrode movement and electrode placement the DIP sometimes did not occur clearly during fatiguing exercise. Further, for individuals with a thick subcutaneous fat layer under the electrodes, the DIP is hard to see. This technique therefore did not provide a very reliable fatigue index.

In Fourier analysis, one common engineering tool is to calculate the geometric center of the Fourier power spectra. This point is called the center of gravity of the power spectra. The coordinates of the center of gravity are the average frequency of the EMG and the average amplitude of the EMG. The average or center

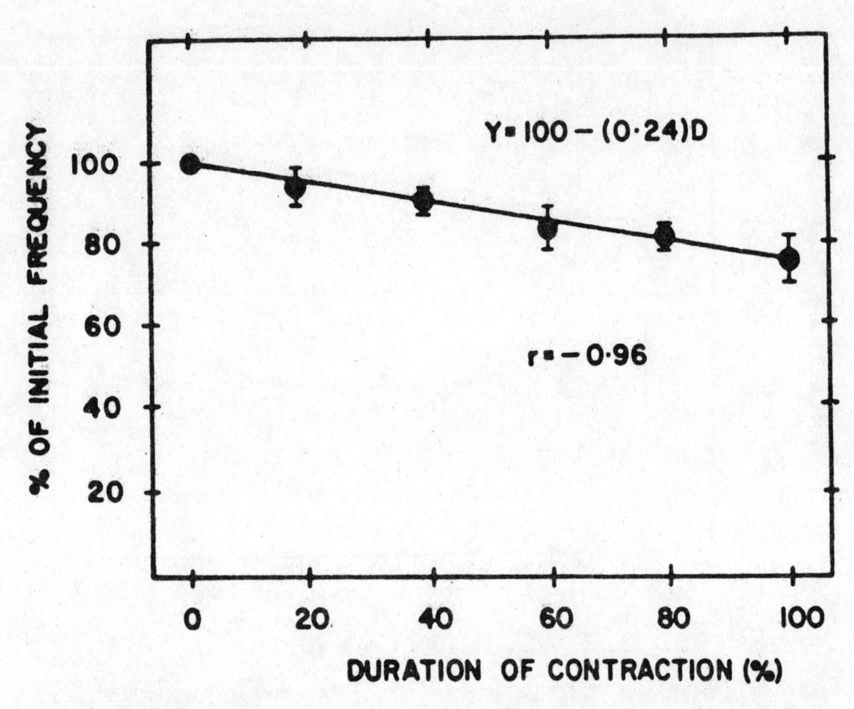

Figure 3-8. Mean center frequencies of the EMG power spectra throughout fatiguing isometric contractions at tensions of 25, 40, 55, 70, 80, and 90 percent MVC normalized in terms of the center frequency recorded at the onset of the contraction. Each point is the mean of 96 experiments ± S.D. From J.S. Petrofsky and A.R. Lind, Frequency Analysis of the EMG During Isometric Exercise. *European Journal of Applied Physiology, 43*:173-182, 1980.

frequency of the Fourier power spectra does appear to provide a fairly reliable fatigue index during isometric contractions (Petrofsky et al., 1976; 1977; 1979; 1980; Petrofsky, 1980a). The advantage of using the center frequency of the Fourier power spectrum is that it provides a single number that reflects the average of the frequency components of the EMG. In Figure 3-7 for example, the Fourier power spectra was calculated from a fundamental of frequency of 4 Hertz through the first 120 harmonics. This would be equivalent to a 120 filter analyzer.

In terms of the electronics for an equivalent circuit, the package would be very complex. However, the same type of analysis is calculated quite easily through a fast Fourier transform (FFT). As described above, during brief isometric contractions, there appears to be little effect of tension on the center frequency of the EMG. However, during fatiguing submaximal isometric contractions there is a linear reduction in the center frequency of the Fourier power spectra as shown in Figure 3-8. This figure represents the average center frequency of the Fourier power spectra ± the S.D. for eight subjects recorded throughout the duration of fatiguing isometric contractions at 25, 40, 55, 70, and 90 percent of the maximum isometric strength of the subjects.

The EMG was sampled here over 1.5 second periods. Since the contractions last for different lengths of time at different tensions, the time has been normalized during the contraction, and the EMG has been sampled at the onset and at 20, 40, 60, 80, and 100 percent of the duration of the isometric contractions. For all tensions examined here, there appears to be a linear fall in the center frequency of the Fourier power spectra of the EMG throughout the duration of the fatiguing isometric contraction. At the point of fatigue, the center frequency was also approximately at the same point independent of the tension exerted during the contractions.

This same phenomenon occurs during a sustained maximal effort. During the first few seconds of a sustained maximal effort when the tension is maintained by the subject, the center frequency falls linearly. However, as tension starts to fall, the center frequency plateaus at the same frequency as that found at the end of a sustained submaximal isometric contraction. On the average, the center frequency of the Fourier power spectra decreases in the order of about 25 percent throughout the duration of fatiguing submaximal isometric contractions.

The reduction in the center frequency of the EMG power spectra like the reduction in the DIP frequency has been attributed to a slowing in the conduction velocity of action potentials on the sarcolemma associated with muscle fatigue. Certainly, in animal experiments, due to either isometric fatigue (Petrofsky et al., 1980) or anoxia (Mortimer et al., 1970), there is a reduction in the conduction velocity of action potentials across the sarcolemma of sufficient magnitude to account for the observed changes in the center frequency of the EMG power spectra. However, the reason that these changes occur remains obscure.

Isometric endurance following fatiguing isometric contractions requires as long as twenty-four hours to recover fully (Lind, 1959; Edwards et al., 1977). In contrast, strength requires about ten minutes to recover fully following a fatiguing isometric contraction. However, the conduction velocity of action potentials across the sarcolemma as well as the center frequency of the surface EMG recovers within ten seconds to that of the fresh muscle following a fatiguing static contraction. Therefore, although the center frequency of the surface EMG does provide an index of muscle fatigue during sustained maximal or submaximal isometric contractions, its recovery does not parallel that of isometric endurance, and the biochemical pathways involved in the two responses may also be different.

It is tempting to speculate that the reduction in the center frequency of the surface EMG is linked to the accumulation of some metabolite released by the contracting muscle. This metabolite cannot be lactic acid since lactic acid takes from minutes to hours to be cleared from the muscle following a fatiguing isometric contraction (Karlsson et al. 1975). Further, although the concentration of lactic acid in the muscle is about the same following contractions at 25 and 70 percent of the muscle's maximum strength, it is nearly twice as high at the end of fatiguing contractions at 40 percent of the muscle's maximum strength (Karlsson et al., 1975; Edwards et al., 1972). Since the center frequency decreases by the same order of magnitude at all three of these tensions, the metabolite must also be found at similar concentrations at the end of contraction at all three tensions.

One candidate for the metabolite involved may be potassium. Potassium is an ion stored in high concentration intracellularly within skeletal muscle and in nerves. During vigorous nerve or muscle activity, potassium can be lost in great quantities into the blood. In marathon runners for example, the normal resting arterial potassium of 4 milliequivalents per liter can rise to as high as 9.2 milliequivalents per liter at the end of their 25 mile marathon. However, unlike dynamic exercise where the blood flow is quite high during the exercise (Astrand and Rodahl, 1970), during isometric contractions there is an increased intramuscular pressure that limits the blood flow to skeletal muscle (Lind et al., 1964; Barcroft and Edholm, 1946). For this reason, the changes in arterial potassium are very small during isometric contractions. In contrast, interstitial potassium concentrations can get quite high (Hnik et al., 1969; 1973; Petrofsky et al., 1980).

Since the amplitude and shape of the motor unit action potential depends upon a potassium concentration gradient across the sarcolemma, an increase in extracellular potassium from 4 to 8 milliequivalents per liter, for example, would reduce the potassium concentration gradient across the cell by 50 percent and have a dramatic effect on the motor unit action potential. A reduction in the potassium concentration gradient across the sarcolemma would result in a prolongation of motor unit action potentials and a corresponding reduction in the center frequency of the EMG power spectra. Once the contraction is terminated, intramuscular pressure would be reduced and the excess in extracellular potassium would be rapidly flushed out. When potassium has been measured in the venous blood during and following isometric contractions, it has been shown that the potassium returns to the preexercised values within about fifteen seconds following fatiguing isometric contraction (Lind et al., 1964). Potassium, therefore, may be a likely candidate for the metabolite causing the shift in the Fourier power spectra.

Because endurance recovers so much slower than the center frequency of the EMG power spectra, it would be difficult to consider the center frequency as a true measure of muscle fatigue. However, as a correlate of muscle fatigue, the center frequency of the EMG can be quite useful. The center frequency of the EMG always decreases to the same percentage of the center frequency of the unfatigued muscle irrespective of the tension exerted by the contraction and in successive isometric contractions. Therefore, by measuring the EMG over a set period of time, i.e. fifteen seconds, it is possible to predict the length of time a contraction could be held in terms of the rate of change of the center frequency of the surface EMG. Further, in successive isometric contractions, the center frequency of the surface EMG still ends at the same final value at the point of fatigue (24 percent less than the center frequency in the unfatigued muscle). So, by looking at the rate of decrease of EMG center frequency during a fifteen or 20 second sample of EMG and by knowing the endpoint, the center frequency may still be used as a predictive tool to predict the duration of the fatiguing isometric contractions. This can be accomplished with the following equation:

$$D = [100 - (100\, F_{c2}/F_{c1})]/76$$

where

D = percent of contraction (fraction of the total length of time a contraction can be held)
F_{c2} = current EMG center frequency
F_{c1} = EMG center frequency in the fresh muscle

For example, in Figure 3-9, subjects were asked to exert pairs of fatiguing isometric contractions at a tension of 40 percent of their maximum isometric strength. By comparing the endurance of the second contraction to the first, the degree of recovery between the two contractions can be assessed. The length of time between the contractions in this figure was varied between one and twenty minutes. For example, with a contraction interval of twenty minutes, the length of the second contraction is 90 percent that of the first, therefore the recovery is said to be 90 percent complete in the twenty minute recovery period. In comparison, during these same isometric contractions, a fifteen second sample of the EMG was taken, and, from this sample, the length of time the contraction could be sustained was calculated by this equation. As can be seen from this figure, the correlation between predicted fatigue and actual fatigue was quite high. Therefore, even though center frequency does not have the same rate of recovery as endurance, the rate of change of the center frequency during a given length of time can be used as an adequate fatigue index.

Although the center frequency of the EMG power spectra during brief isometric contractions in most people is about 150 Hertz when analyzed as described above, there are several factors that cause this figure to vary. One such factor is the position of the EMG electrodes. Since the center frequency of the EMG power spectra is affected by the conduction velocity of action poten-

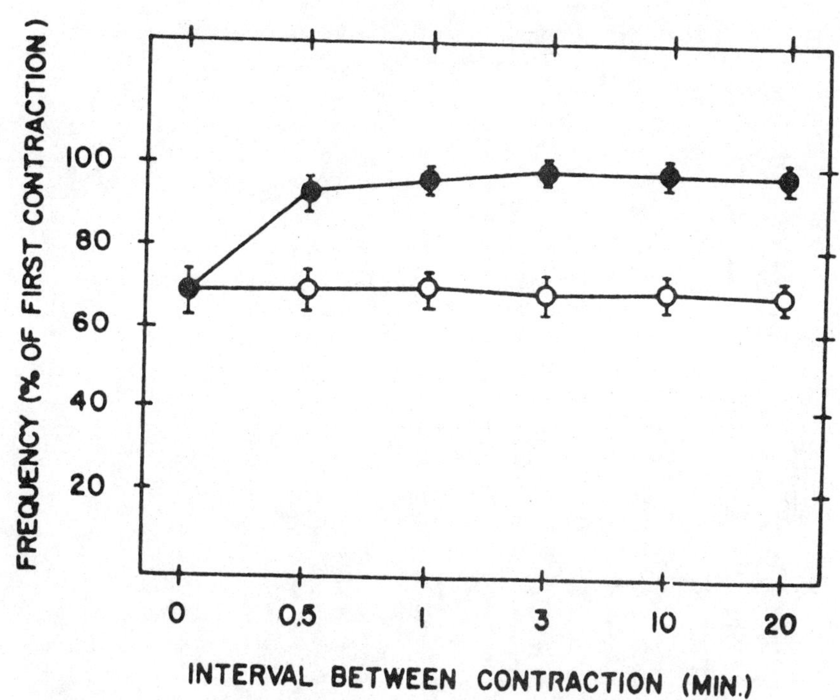

Figure 3-9. Center frequency recorded at the beginning (●) and end (○) of the 2nd of a pair of fatiguing isometric contractions compared with that of the 1st. The interval between the contractions is shown at 0, 0.5, 1, 3, 10, or 20 min. Each point represents the mean of 16 experiments ± S.D. From J.S. Petrofsky and A.R. Lind, Frequency Analysis of the EMG During Isometric Exercise. *European Journal of Applied Physiology, 43*:173-182, 1980.

tials across the sarcolemma, the orientation of the muscle fibers as well as the electrode position alters the amplitude of the frequency components of the EMG (Lynn et al., 1978). It is, therefore, very important to stabilize electrode position from day to day and from subject to subject to minimize this variation.

One factor that can dramatically influence the frequency components of the surface EMG is muscle temperature. Unlike the core tissues, muscle is a shell tissue. As such, its temperature can vary widely due to environmental factors (Hall et al., 1947; Clarke, Hellon and Lind, 1958; Petrofsky and Lind, 1975; 1980a). The conduction velocity of action potentials across the sarcolemma is directly proportional to the temperature of the muscle. A

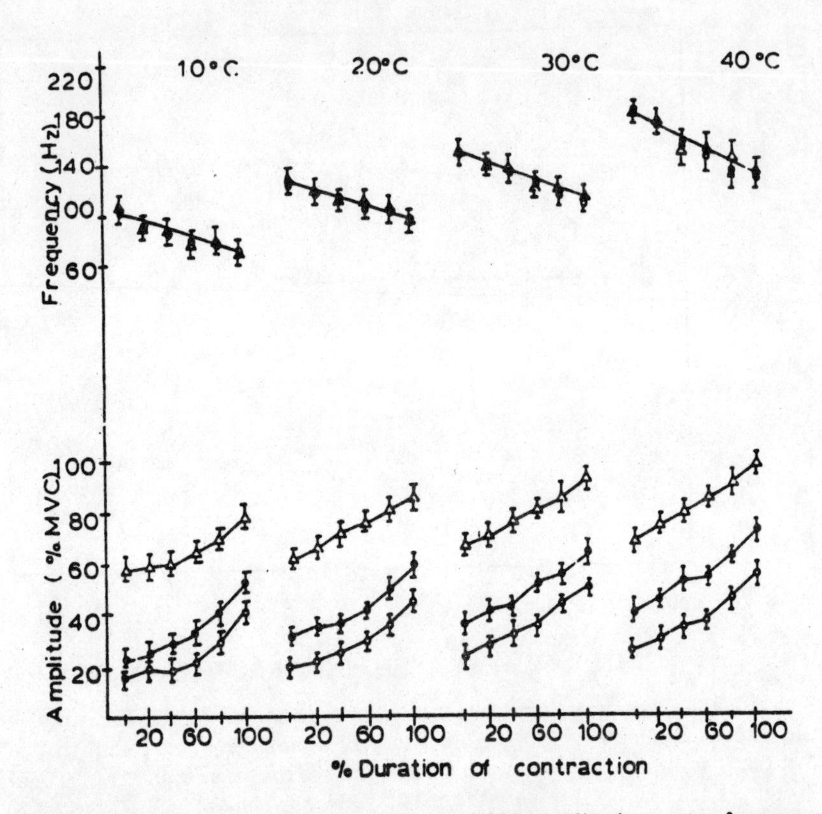

Figure 3-10. The relationship between the RMS amplitude, center frequency of the EMG power spectra and muscle temperature during a brief isometric contraction at 100 percent MVC in each of four subjects. Each point illustrates the mean of two experiments ± the S.D. on each of the subjects. From J.S. Petrofsky and A.R. Lind, Frequency Analysis of the EMG During Isometric Exercise. *European Journal of Applied Physiology, 43*:173-182, 1980.

change in muscle temperature from 28 to 38 degrees centigrade can double the conduction velocity of action potentials. A similar phenomenon occurs with the center frequency. As shown in Figure 3-10, there is a reduction in the low frequency components and a general shift into the high frequency region when the muscle temperature changes from 28 to 38 degrees centigrade during a brief isometric contraction at 40 percent MVC (Petrofsky and Lind, 1980a). The center frequency changes from 110 to 152.5 Hertz under these experimental conditions. Therefore, even

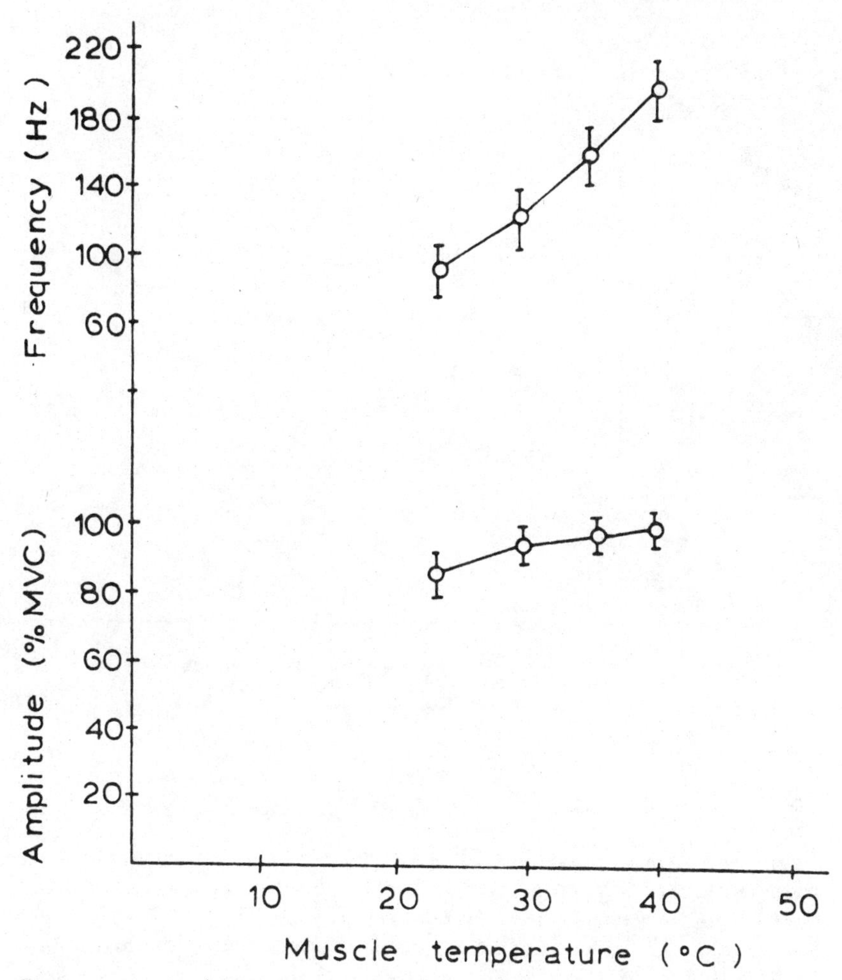

Figure 3-11. The RMS amplitude and center frequency of the EMG power spectra of the seven subjects. From J.S. Petrofsky and A.R. Lind, Frequency Analysis of the EMG During Isometric Exercise. *European Journal of Applied Physiology, 43*:173-182, 1980.

a change of a few degrees centigrade during an experiment can alter the EMG center frequency more than that which occurs due to muscle fatigue.

It is very important to stabilize muscle temperature throughout the experiment. However, since the resting muscle tempera-

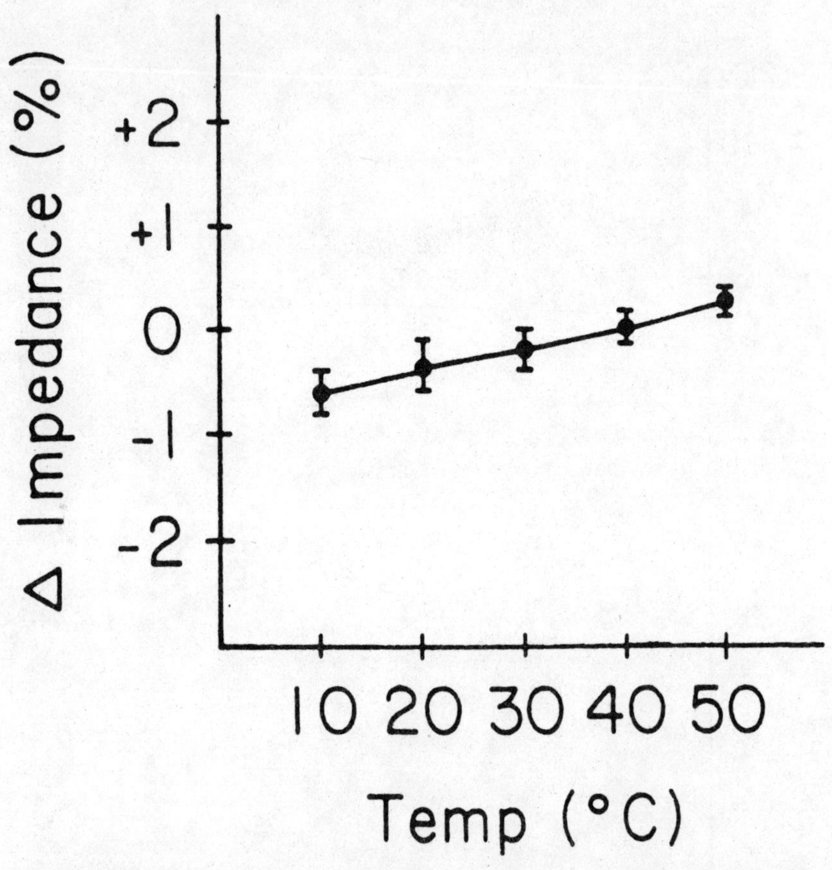

Figure 3-12. Skin to muscle impedance as a function of bath temperature in seven subjects ± the appropriate S.D. From J.S. Petrofsky and C.A. Phillips, Interactions Between Fatigue, Muscle Temperature, Blood Flow, and the Surface EMG. *IEEE NAECON Record*, 520-527, Dayton, Ohio, May 20-22, 1980. © 1980 IEEE.3/M

ture varies so widely due to environmental and inherent influences, it is equally important to know what changes occur in the center frequency during isometric contractions at widely different muscle temperatures. Fortunately, the reduction in the center frequency of the EMG remains constant at about 24 percent from the onset to the end of the contractions during a fatiguing isometric contraction irrespective of the muscle temperature. For example, in Figure 3-11, fatiguing isometric contractions were performed at muscle temperatures of 22, 28, and 37 degrees centi-

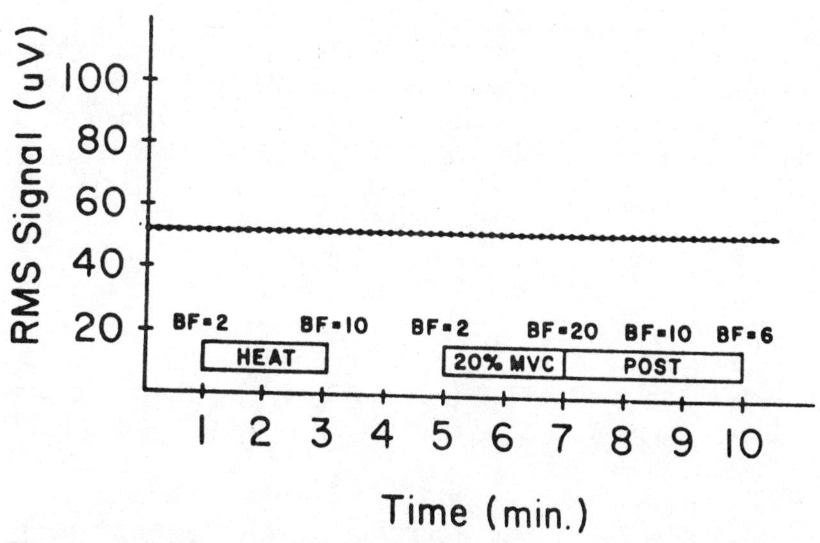

Figure 3-13. The RMS amplitude of the signal recorded on the skin generated from muscle during passive heating of the skin and a brief isometric contraction. From J.S. Petrofsky and C.A. Phillips, Interactions Between Fatigue, Muscle Temperature,, Blood Flow, and the Surface EMG. *IEEE NAECON Record*, 520-527, Dayton, Ohio, May 20-22, 1980. © 1980 IEEE. 3/M

grade. Although, the center frequency was dramatically reduced when the muscle temperature was changed from 38 to 22 degrees centigrade, the center frequency always decreased by about 25 percent during the fatiguing isometric contractions. Therefore, the center frequency of the surface EMG may provide a good fatigue index for isometric contractions.

The mechanism of these changes appear to be related to changes in the conduction velocity of action potentials on the muscle sarcolemma. Although the impedance between the muscle and skin (Fig. 3-12) is affected by temperature, if a signal is injected into the muscle by two electrodes and recorded on the skin, changing skin blood flow or muscle blood flow while altering impedance (Fig. 3-13) has no effect on the amplitude of the transmitted signal (Petrofsky and Phillips, 1980).

THE ELECTROMYOGRAM DURING DYNAMIC EXERCISE

The electromyogram during isotonic or dynamic exercise is

far more complex to record than that which can be recorded during static exercise. As mentioned above in the section under static exercise, one factor that strongly influences the surface EMG is the length of the muscle. When a muscle contracts isotonically, the thin and thick filaments slide over one another, and as a result, the muscle shortens (Astrand and Rodahl, 1970). As this occurs, the muscle can shorten to as much as 57 percent of its initial length. The sarcolemma is, therefore, forced to change shape. As the muscle contracts the sarcolemma thickens. This results in a change in the conduction velocity of action potentials across the muscle (Lynn et al., 1978). Conduction velocity of action potentials in a stretched muscle are much higher than in muscle in the contracted state.

The impact this has on the surface EMG is obvious. As the muscle is displaced away from the recording electrodes during contraction, the amplitude of the surface EMG is reduced. The most dramatic effect is on the frequency components of the surface EMG. As cited above, the frequency components of the surface EMG are derived in a large part from the conduction velocity of action potentials across the sarcolemma of the individual muscle fibers (Lindström et al., 1970, Person and Lipkind, 1970). As conduction velocity changes, so does the frequency of the surface EMG. Therefore, if two electrodes were placed, for example, over the biceps muscle and the muscle were to contract throughout its length, it has been shown that the center frequency of the surface EMG might change as much as 50 percent during the contraction of the muscle. Because of these problems with the movement of muscle, it has been very difficult to analyze the amplitude and frequency components of the EMG during dynamic exercise.

Bigland-Ritchie and Woods (1974) first looked at the amplitude of the EMG during exercise on the bicycle ergometer. They found that for low level dynamic exercise, there was a linear relationship between the intensity of the work and the amplitude of the surface EMG; the frequency components of the surface EMG were not analyzed here. These studies were expanded in our own lab to see what the relationship was between the amplitude of the EMG and tension throughout the complete range of dynamic exercise (Petrofsky, 1980a). We also wanted to examine the frequency components of the surface EMG under these same experimental circumstances. Since we wanted to look at the frequency components of the surface EMG, it was necessary to remove the motion artifacts described above. This was accomplished by only examining the surface EMG when the muscle was at the same length. To do this, the standard Monarck bicycle ergometer was modified.

The modification involved placing a light emitting diode and a photo darlington detector across the path of the pedals. The pedals only cut the beam of light over an arc of approximately 10 degrees of the 360 degree rotation of the crank shaft. The photo darlington light sensor acted as a trigger for a LINC digital computer, which was used to sample the EMG. In this manner, the EMG was sampled for only a very brief period of time. Rather than sample the EMG continously, it was sampled for 1.5 second periods as was done above for static exercise, 0.25 seconds of the EMG were sampled in this manner; six sequential samples were used for any one measurement. The amplitude and frequency components of each 0.25 second sample were calculated individually, and then the average of the six samples was taken for each measurement.

To vary the work load on a Monarck bicycle ergometer, a one kilogram weight was placed on a friction belt perpendicular to the direction of movement of the flywheel. One kilogram of resistance applied to the wheel is said to be a load of one kilopond (kp). When the load was increased, the oxygen uptake of the subject increases linearly with the load. For the four subjects used in this study, there was always a linear relationship between the load and the oxygen uptake of the muscles up to one load above which the oxygen uptake plateaued. Above this work load no further increase in oxygen uptake occurred. The oxygen uptake at which this occurs is typically called the maximum aerobic capacity of the muscles ($\dot{V}O_{2mx}$). When work exceeds this level, although the intensity can be increased, the work is of very short duration because the muscles are working anaerobically. Under these conditions, energy is produced in the muscle through the production of large quantities of lactic acid.

One of the purposes of this series of experiments was not only to examine the relationship between $\dot{V}O_2$ and the amplitude and frequency of the EMG, but to see how these compared to the amplitude and frequency components in the EMG in the same muscle groups during static exercise. To accomplish this then, the pedals of the Monarck bicycle ergometer were locked in place at the position where they cut the beam from the LED and strain gauges. Strain gauges mounted on the pedals were used to measure isometric tension. Therefore, the Monarck bicycle ergometer served both as a dynamic exercise and static exercise dynamometer.

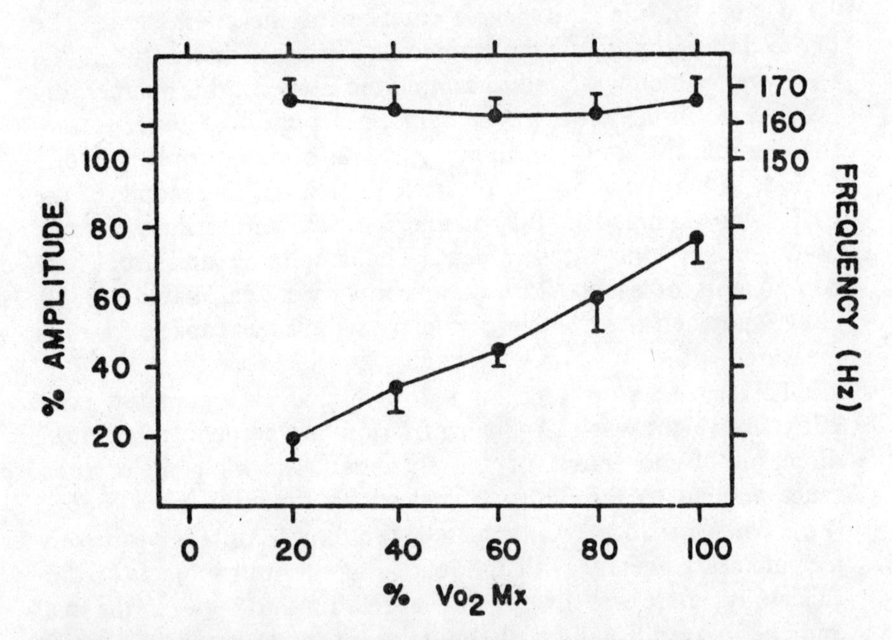

Figure 3-14. The RMS amplitude and center frequency during work up to one hour in duration at 20 to 100 percent of the $\dot{V}O_2$ max. Each point illustrates the mean of six determinations ± the standard deviations. From J.S. Petrofsky, Frequency and Amplitude Analysis of the EMG During Exercise on the Bicycle Ergometer. *European Journal of Applied Physiology, 41*:1-15, 1979.

When the amplitude of the surface EMG was related to the work load on the bicycle ergometer, there was a linear relationship between the amplitude of the surface EMG and $\dot{V}O_2$ up to 100 percent of the $\dot{V}O_{2mx}$ (Fig. 3-14). However, when the subjects performed at 100 percent $\dot{V}O_{2mx}$, the amplitude of the EMG was only about 60 percent that of the amplitude of the EMG during a maximal voluntary effort during static exercise. The implication here was that only 60 percent of the maximum strength of the muscle was needed during maximal dynamic exercise. However, during static exercise, caution must be exercised here. It is possible that the firing frequencies of motor units are much higher during static than dynamic exercise. This would account for the lower amplitude of the EMG during maximum dynamic work.

The frequency components of the surface EMG were similar to those found during static exercise for brief bouts of work. The center frequency of the Fourier power spectra was similar during the two types of exercise. During prolonged bouts of work (as long as 60 min.), the amplitude of the EMG was constant at work loads below 50 percent of the $\dot{V}O_{2mx}$ but increased continuously at higher work loads (Fig. 3-15). Of the $\dot{V}O_{2mx}$, 50 percent is considered to be the aerobic threshold (Wasserman et al. 1973). For levels of dynamic work below 50 percent $\dot{V}O_{2mx}$ in most subjects, no lactic acid is produced during the exercise (Wasserman et al., 1973). However, as the level of work increases past 50 percent $\dot{V}O_{2mx}$, energy is being utilized by the muscle faster than it can be produced by aerobic pathways (through the breakdown of oxygen). For this reason, lactic acid is produced. The production of lactic acid in muscle at levels of work between 50 and 100 percent $\dot{V}O_{2mx}$ is exponentially related to the work level and, as mentioned above, at levels above 100 percent $\dot{V}O_{2mx}$ large quantities of lactic acid are produced (Fig. 3-16).

For this reason, one would expect to find large reductions in the center frequency of the surface EMG at levels of work above 50 percent $\dot{V}O_{2mx}$; these were not our findings. What was found, was that for levels of work above 50 percent $\dot{V}O_{2mx}$, at the onset of work, there was a rise in the center frequency of the EMG followed by a slow decline. There also was a clear reduction in the center frequency of the surface EMG for work at 75 to 100 percent $\dot{V}O_{2mx}$ (Fig. 3-17). At first, this was somewhat surprising since it is hard to imagine what would cause the center frequency of the surface EMG to increase. However, during isometric exercise, the temperature of the exercising muscle usually only changed

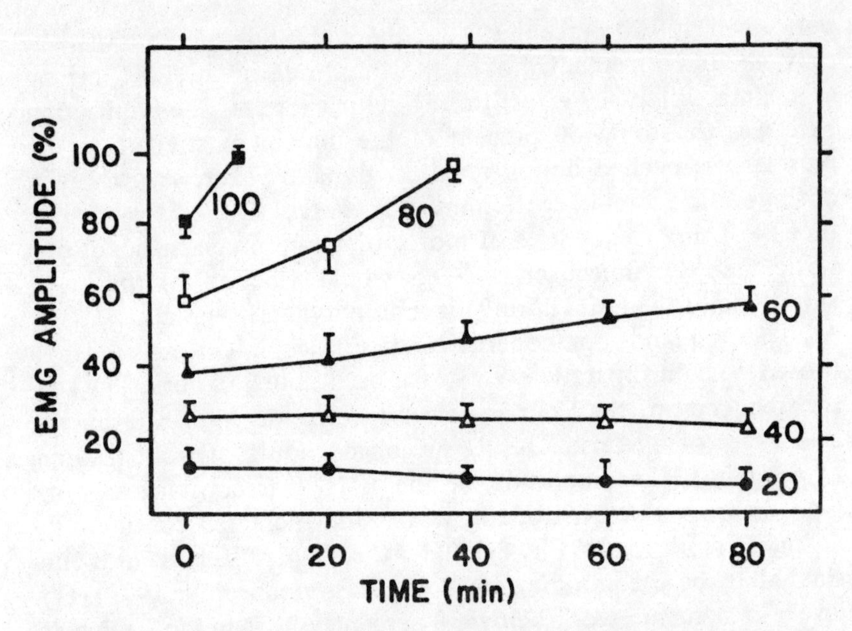

Figure 3-15. The RMS amplitude during 3 second maximal isometric contractions interposed during dynamic exercise at 20, 40, 60, 80, and 100 percent of the $\dot{V}O_2$ max for either 80 min. or carried to fatigue. Each point illustrates the mean of six determinations ± the respective standard deviations. From J.S. Petrofsky, Frequency and Amplitude Analysis of the EMG During Exercise on the Bicycle Ergometer. *European Journal of Applied Physiology, 41*: 1-15, 1979.

a few tenths of a degree centigrade. In contrast, during dynamic exercise, the muscle temperature changes dramatically (Saltin and Hermansen, 1966; Petrofsky, 1980a).

As discussed above, the resting muscle temperature of, for example, a forearm muscle such as the brachioradialus is about 32 degrees centigrade. This temperature is 6 degrees below that of the core. During dynamic exercise the blood flow to muscle will increase almost tenfold. The warm blood rushing in from the core will immediately cause an increase in muscle temperature. Therefore, at the onset of work, although the muscle temperature might be 32 degrees centigrade, a few minutes after work has begun the muscle temperature will rise to between 37 and 38 degrees. This causes an increase in the conduction velocity of

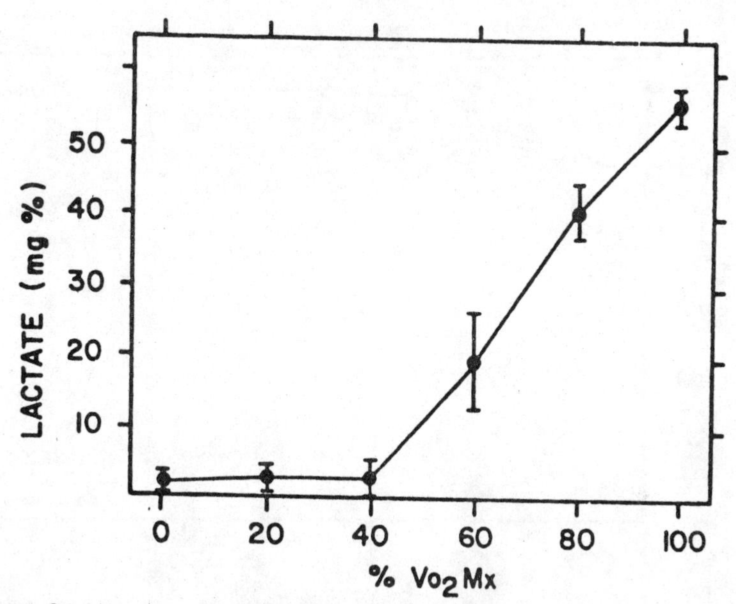

Figure 3-16. Lactates found in the arterialized blood of the subjects with the subjects at rest and after dynamic work periods at 20, 40, 60, 80, and 100 percent of the $\dot{V}O_2$ max. Each point illustrates the means of three experiments ± the respective ranges. From J.S. Petrofsky, Frequency and Amplitude Analysis of the EMG During Exercise on the Bicycle Ergometer. *European Journal of Applied Physiology, 41*:1-15, 1979.

action potentials across the muscle sarcolemma (Petrofsky, 1980a). As a result, the frequency of the surface EMG also increases (Fig. 3-18).

It is easy to see then why, during light dynamic exercise at levels up to 50 percent of the $\dot{V}O_{2mx}$, there was an increase in the center frequency of the surface EMG after the onset of work. At levels of work above 50 percent $\dot{V}O_{2mx}$, the muscle temperature rises above that of the core temperature throughout the duration of work. This increase in muscle temperature is directly proportional to the level of work sustained by the muscle. Unlike static exercise where the muscle does not change length, during dynamic exercise, as the muscle changes length, there is a great deal of internal friction found within the muscle due to the sliding of the thin and thick filaments in a very viscous intracellular medium. This results in the transduction of a large amount of energy to

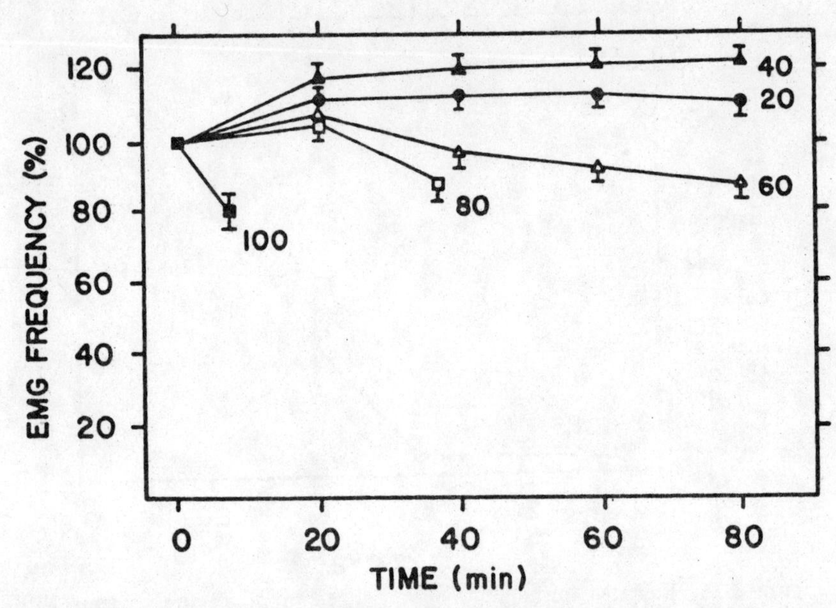

Figure 3-17. The center frequency during 3 second maximal isometric contractions interposed during dynamic exercise at 20, 40, 60, 80, and 100 percent of the $\dot{V}O_2$ max. for either 80 min. or carried to fatigue. Each point illustrates the mean of six determinations ± the respective standard deviations. From J.S. Petrofsky, Frequency and Amplitude Analysis of the EMG During Exercise on the Bicycle Ergometer. *European Journal of Applied Physiology, 41*:1-15, 1979.

heat.

During dynamic exercise at levels of work as high as 100 percent of the $\dot{V}O_{2mx}$ muscle temperature may approach 41 to 42 degrees centigrade (Saltin and Hermansen, 1966). This further increase in muscle temperature in itself should cause the center frequency of the surface EMG to increase even further. However, during these heavy bouts of work, muscle fatigue also appears to cause the center frequency of the surface EMG to decrease in a manner similar to that which occurs during static exercise.

Therefore, during dynamic exercise, there are two opposing factors: the rise in muscle temperature causing the center frequency of the surface EMG to increase and muscle fatigue causing the center frequency of the surface EMG to decrease. For this reason then, at low levels of work, the temperature becomes the

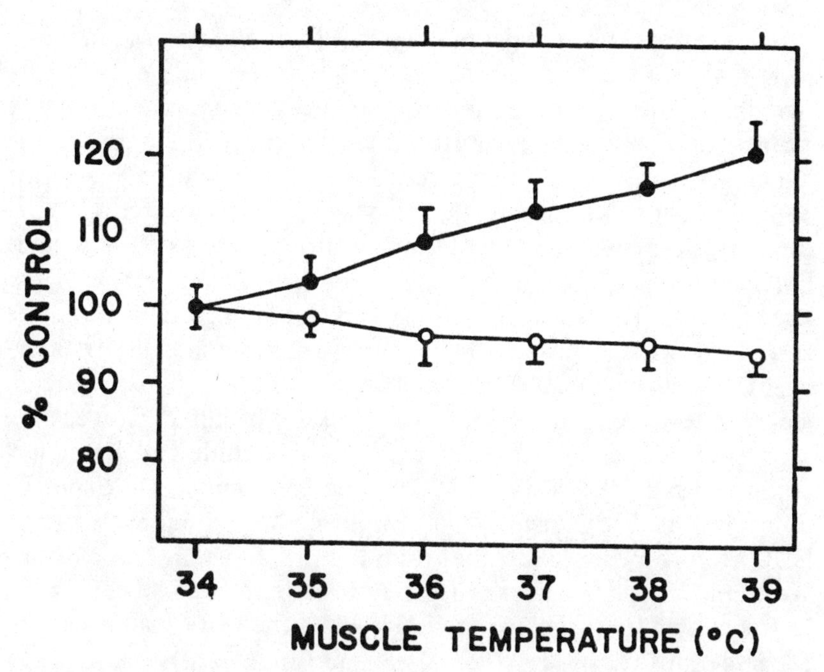

Figure 3-18. The RMS amplitude (○) and center frequency (●) during three second isometric contractions at a tension of 60 percent MVC with the muscle at temperatures of between 34 and 39 degrees centigrade. Each point shows the mean of six measurements ± the respective standard deviations. From J.S. Petrofsky, Frequency and Amplitude Analysis of the EMG During Exercise on the Bicycle Ergometer. *European Journal of Applied Physiology, 41*:1-15, 1979.

predominate factor and the center frequency of the surface EMG changes little during apparently fatiguing work. However, at high levels of work where the duration of the work is short (5 to 20 min.) the rise in temperature is small and the fatigue effect predominates, causing the center frequency of the surface EMG to drop dramatically.

The effect of temperature on the muscles can be removed by several means. The most obvious way of removing this temperature effect would be to place the appendage during the exercise in a controlled temperature water bath. However, this is quite difficult to do. Therefore, an easier way is to have the subject do light work as a "warm-up" prior to the test period. Light dynamic exercise is not fatiguing and results in an increase in blood flow that will warm the muscle to between 37 and 38 degrees centigrade. If the amplitude and the frequency components of the surface EMG are now measured and used as a reference for the dynamic work, the temperature of the muscle will only increase slightly above this level even during heavy dynamic exercise. Therefore, by normalizing the amplitude and frequency of the surface EMG in terms of a muscle that as been prewarmed, many of the temperature effects on the muscle can be removed. This was done in a study where we wanted to look at the amplitude and frequency of the surface EMG during lifting. In this study, we had subjects lift lead boxes from floor to table top height at rates as high as 60 per minute. When we studied the amplitude and frequency components of the surface EMG of the back muscles at a specific point in the lifting cycle, the amplitude and frequency of the EMG were very similar to that found during fatiguing static effort. We found a linear increase in the amplitude of the EMG and a fall in the center frequency of the EMG when the work load exceeded 50 percent of the VO_{2mx} (Petrofsky and Lind, 1978).

Therefore, although the amplitude and frequency components of the EMG during dynamic exercise are much more difficult to assess due to both thermal and mechanical problems, they can be assessed with careful experimental technique. However, unlike static exercise, very little has been done in terms of research on the amplitude and frequency components of the EMG during dynamic exercise, and much more investigation is needed.

REFERENCES

Astrand, P.O. and Rodahl, H. (1970) *Textbook of Work Physiology*. McGraw-Hill, New York 1970

Barcroft, H. and Edholm, O.G. (1946) Temperature and blood flow in the human forearm. *J. Physiol.*, 104:366-376.

Bigland, B. and Lippold, O.C.J. (1954) Motor unit activity in the voluntary contraction of human muscle. *J. Physiol.*, 125:322.

Bigland-Ritchie, B., Jones, D.A., Hosking, G.P., and Edwards, R.H.T. (1978) Central and peripheral fatigue in sustained maximum voluntary contractions of human quadricep muscle. *Clin. Sci. Mol. Med.*, 54:609-614.

Bigland-Ritchie, B. and Lippold, O.C.J. (1980) Neuromuscular block and motor firing frequencies in human muscular fatigue. *Proc. Int. Physiol. Soc.*, 14:326

Bigland-Ritchie, B. and Woods, J. (1974) Integrated EMG and oxygen uptake during dynamic contractions of human muscles. *J. Appl. Physiol.*, 46:475-479

Brown, G.L. and Burns, B.D. (1949) Fatigue and neuromuscular block in mammalian skeletal muscle. *Proc. Roy. Soc.*, 136:182

Clarke, R.S.J., Hellon, R.F., and Lind, A.R. (1958) Duration of sustained contractions of the human forearm at different muscle temperatures. *J. Physiol.*, 143:454

Cobb, S. and Forbes, A. (1923) Electromyographic studies of muscular fatigue in man. *J. Physiol.*, 65:234-251

De Vries, H.A. (1968) Method of evaluation of muscle fatigue and endurance from electromyographic curves. *Am. J. Phys. Med.*, 47:125-135

Eason, R. (1960) Electromyographic study of local and generalized muscular impairment. *J. Appl. Physiol.*, 156:479-482

Edwards, R.H.T., Harris, R.C., Hultman, E., Kaizser, L., Koh, D., and Nordesjo, L.O. (1972) Effect of temperature on muscle energy metabolism and endurance during successive isometric contractions sustained to fatigue of the quadriceps muscle in man. *J. Physiol.*, 220:335-352

Edwards, R.H.T., Hill, D.K., Jones, D.A., and Merton, P.A. (1977) Fatigue of long duration in human skeletal muscle after exercise. *J. Physiol.*, 272:769-778

Edwards, R.H.T. and Lippold, O.C.J. (1956) The relation between force and integrated electrical activity in fatigued muscle. *J. Physiol.*, 132:677-681

Fink, R. and Luttgau, H.D. (1976) An evaluation of membrane constants and potassium conductance in metabolically exhausted fibers. *J. Physiol.*, 263:215-239

Fusfeld, R.D. (1971) Analysis of electromyographic signals by measurement of wave duration. *Electromyogr. Clin. Neurophysiol.*, 30:337-344

Grimby, L. and Hannerz J. (1976) Disturbances in voluntary recruitment order of low and high frequency motor units on blockades of prospective afferent activity. *Acta Physiol.*

Scand., 96:207-216

Hall, V.E., Mendoz, E., and Fitch, B. (1947) Reduction of the strength of muscle contraction by the application of moist heat to the overlying skin. *Arch. of Phys. Med.*, 28:493-499

Hnik, P., Hudlicka, O., Kuchera, J., and Payne, R. (1969) Activation of muscle afferents by nonproprioceptive stimuli. *Am. J. Physiol.*, 217:1451-1458

Hnik, P., Kriz, N., Vyskocil, F., Smiesko, V., Mejsnar, S., Vjec, E., and Holas, M. (1973) Work-induced potassium changes in muscle venous effluent blood measured by ion-specific electrodes. *Pflugers Arch.*, 338:177-181

Kaiser, E. and Petersen, I. (1965) Muscle action potentials studied by frequency analysis and duration measurement. *Acta Neurol. Scand.*, [Suppl] 13

Karlsson, J., Funderburk, C., Essen, B., and Lind, A.R. (1975) Constituents of human muscle in isometric fatigue. *J. Appl. Physiol.*, 38:208-211

Kogi, K. and Hakamada, T. (1962) Slowing of surface electromyogram and muscle strength in muscle fatigue. *Rep. Physiol. Lab-Inst. Sci. Labour*, 60:27-41

Larsson, L.E. (1971) Frequency analysis of the electromyogram in neuromuscular disease. *Electroencephlogr. Clin. Neurophysiol.*, 30:259-261

Lind, A.R. (1959) Muscle fatigue and recovery from fatigue induced by sustained contractions. *J. Physiol.*,147:162

Lind, A.R. and Petrofsky, J.S. (1979) Amplitude of the surface electromyograms during fatiguing isometric contractions. *Muscle and Nerve*, 2:257-264

Lind, A.R., Taylor, S.H., Humphreys, P.W., Kennelly, B.M., and Donald, K.W. (1964) The circulatory effects of sustained voluntary muscle contraction. *Clin. Sci.*, 27:229-244

Lindström, L., Kadefors, R., and Petersen, I. (1977) An electromyographic index for localized muscle fatigue. *J. Appl. Physiol.*, 43:750-754

Lindström, L., Magnusson, R., and Petersen, I. (1970) Muscular fatigue and action potential conduction velocity changes studied with frequency analysis of EMG signals. *Electro-*

myography, 10:341-356

Lippold, O.C.J. (1952) The relation between integrated action potentials in a human muscle and its isometric tension. *J. Physiol.*, 117:492-499

Lloyd, A.J. (1971) Surface electromyography during sustained isometric contractions. *J. Appl. Physiol.*, 30:713-719

Lynn, P.A., Bettles, H.D., Hughes, A.D., and Johnson, S.W. (1978) Influences of electrode geometry on bipolar recordings of the surface electromyogram. *Med. Biol. Eng. Computing*, 16:651-661

Milner-Brown, H.S., and Stein, R.B. (1975) The relation between the surface electromyogram and muscular force. *J. Physiol.*, 246:549

Mortimer, T., Magnusson, R., and Petersen, I. (1970) Conduction velocity in ischemic muscle; effect on EMG power spectrum. *Am. J. Physiol.*, 131:1324-1329

Paul, D.H. (1961) The effects of anoxia on the isolated rat phrenic-nerve-diaphragm preparation. *J. Physiol.*, 155:359-371

Person, R.S. and Lipkind, M.S. (1970) Simulation of electromyograms showing interference patterns. *Electroencephalogr. Clin. Neurophysiol.*, 28:625-632

Petrofsky, J.S. (1979) Frequency and amplitude analysis of the EMG during exercise on the bicycle ergometer. *Eur. J. Appl. Physiol.*, 41:1-15

Petrofsky, J.S. (1980a) Computer analysis of the surface EMG during isometric exercise. *Computers in Biol. and Med.*, 10: 83-95

Petrofsky, J.S. (1980b) Filter bank analyzer for the automatic analysis of the EMG. *Med. Biol. Eng. Computing*, 18:585-590

Petrofsky, J.S. (1981) Motor unit recruitment patterns during submaximal isometric contractions. *J. Appl. Physiol.* (in press)

Petrofsky, J.S., Betts, W., and Lind, A.R. (1977) Quantification of the surface EMG. *Fed. Proc.*, 36:1194

Petrofsky, J.S. and Fitch, C. (1980) Contractile characteristics of skeletal muscle depleted of phosphocreatine. *Europ. J. Physiol.*, 384:123-129

Petrofsky, J.S., Guard, A., and Phillips, C.A. (1979) The effects of muscle fatigue on the isometric contractile characteristics of fast and slow twitch skeletal muscle in the cat. *Life Sciences*, 24:2287-2292

Petrofsky, J.S., LaDonne, D., Rinehart, J., and Lind, A.R. (1976) Isometric strength and endurance during the menstrual cycyle in healthy young women. *Europ. J. Appl. Physiol.*, 35:1-10

Petrofsky, J.S. and Lind, A.R. (1975) Isometric strength, endurance, and the blood pressure and heart rate responses during isometric exercise in men and women with special reference to age and body fat content. *Pflugers Arch.*, 360:49-61

Petrofsky, J.S. and Lind, A.R. (1978) Metabolic, cardiovascular, and respiratory factors in the development of fatigue in lifting tasks. *J. Appl. Physiol.*, 45:64-68

Petrofsky, J.S. and Lind, A.R. (1980a) The influence of temperature on the amplitude and frequency components of the EMG during brief and sustained isometric contractions. *Europ. J. Appl. Physiol.*, 44:198-208

Petrofsky, J.S. and Lind, A.R. (1980b) Frequency analysis of the EMG during isometric exercise. *Europ. J. Appl. Physiol.*, 43:173-182

Petrofsky, J.S. and Phillips, C.A. (1980) Interactions between fatigue, muscle temperature, blood flow, and the surface EMG. *IEEE NAECON Record,* 520-527, Dayton, Ohio, May 20-22, 1980

Petrofsky, J.S., Weber, C., and Phillips, C.A. (1980) Mechanical and electrical correlates of isometric muscle fatigue in skeletal muscle in the cat. *Eur. J. Physiol.*, 387:33-38

Piper, H. (1912) *Elektrophysiologie Menschlicher Muskein*. Springer, Berlin, Heidelberg, New York, p 126

Saltin, B. and Hermansen, L. (1966) Esophageal, rectal, and muscle temperature during exercise. *J. Appl. Physiol.*, 21:1757-1762

Scherrer, J. and Bourguigon, A. (1959) Changes in the electromyogram produced by fatigue in man. *Am. J. Phys. Med.*, 38:148-158

Stalberg, E. (1966) Propagation velocity in human muscle fibers in situ. *Acta Physiol. Scand.*, [Suppl] 287

Stephens, J.A. and Tayor, A. (1972) Fatigue of maintained voluntary contractions in man. *J. Physiol.*, 220:1-18

Stulen, F.B. and DeLuca, C.J. (1978) A non-invasive device for monitoring metabolic correlates of myoelectric signals. *ACEMB Proc.*, 20:764

Viitasalo, J. and Komi, P. (1977) Signal characteristic of EMG during fatigue. *Eur. J. Appl. Physiol.*, 37:111-121

Wasserman, K., Whipp, B.J., Koyal, S.N., and Beaver, W.L. (1973) Anaerobic threshold and respiratory gas exchange during exercise. *J. Appl. Physiol.*, 35:236-243

Zuniga, E.N. and Simmons, D.G. (1970) Nonlinear relationship between averaged electromyogram potential and muscle tension in normal subjects. *Arch. Phys. Med. Rehabil.*, 50:264-272

Chapter 4

THE DYNAMICS OF ACTIVE HUMAN SKELETAL MUSCLE *IN VIVO*

G.I. ZAHALAK

IN a monograph devoted to muscle mechanics, it is highly appropriate to point out that after fifty years of intense research the basic *engineering* problem of skeletal muscle remains unsolved, even for the case of functionally isolated muscle under controlled stimulation. This basic problem may be stated as follows: given the stimulation and the load on a muscle as functions of time, what is the resulting variation of muscle length? Alternatively one could pose the problem as follows: if the variations of stimulation and of muscle length are prescribed, how will the force generated by the muscle vary? Thus, from an engineering viewpoint, one may regard a muscle as a two input (stimulation and length) one output (force) system and ask for the quantitative input-output relation that characterizes this system and may be represented schematically as an equation

$$f(P, x, E, t) = 0 \tag{1}$$

In this equation, P represents the muscle force, x the muscle length, E the stimulation, and t the time (the system may be time-varying). It must be appreciated that, in general, equation one (1) is expected to be an integrodifferential relation as both rate, e.g.

The author's frequency-response studies described in this chapter were supported by the National Science Foundation.

force-velocity, and history, e.g. fatigue, effects are known to influence the mechanical behavior of muscle. The form of f(P, x, E, t) is not known over the entire physiological range of its arguments, although approximations are available under some restricted experimental conditions.

While the quantitative characterization of isolated muscle under controlled stimulation is a difficult and yet unsolved problem, the corresponding problem for muscle *in vivo* is much more so. The following are among the factors that complicate the study of muscle mechanics in the latter case: (1) the stimulation is difficult to measure and control; (2) in almost all *in vivo* experiments, several muscles are simultaneously involved, forming a statically indeterminate system and complicating the interpretation of experimental measurements in terms of individual muscle properties; and (3) the behavior of muscle *in vivo* is usually strongly influenced by the presence of intact neural reflex feedback loops. The last problem is particularly challenging for the investigator of *in vivo* muscle mechanics as it requires that he extract the behavior of a subsystem—the muscle—from observations of the behavior of a complicated nonlinear control system. Yet it seems unavoidable that the mechanical behavior of muscle *in vivo* must be studied, to a large extent, *in vivo*. This is because natural "stimulation" in a behaving animal consists of an increasing level of excitation of the motoneuron pool supplying the muscle and manifests itself as a simultaneous increase in motoneuron firing rate and motor unit recruitment. At present, this natural stimulation cannot be stimulated with confidence in isolated muscle preparations, and caution must be exercised in extrapolating results from controlled stimulation experiments to muscle *in vivo*.

As muscle almost never operates *in vivo* in the absence of its associated neural feedback circuits, the question arises whether it is useful to seek to separate the intrinsic mechanical behavior of muscle from that of the control systems in which it is embedded. Such a separation certainly seems desirable. As with any complex control system, detailed knowledge of the operating characteristics of its subsystems provides a deeper understanding of the overall system performance. For example, skeletal muscle interposes itself between the externally observable motion of a limb and the measurable electrical activity in the CNS during voluntary movements. The role of the CNS in the control of movement is currently a very active area of neurophysiological investigation, and one can expect that a better quantitative understanding of the input-output relations for skeletal muscle will illuminate the connections between CNS activity and the limb motion it produces. The utility of separating muscle from its associated neural circuitry is well illustrated in the analysis of stretch reflexes (Stein, 1974) where several investigators have found that a large portion of the response is not of reflex origin at all; rather it is attributable to intrinsic muscle properties and hence is termed a *pseudoreflex*. Another reason for isolating the mechanical behavior of skeletal muscle *in vivo* is that, to the extent that such knowledge may be useful in the diagnosis of neuromuscular disorders, this isolation will help to differentiate between myogenous and neurogenous dysfunctions.

The purpose of this chapter is to review the available quantitative information about the mechanical behavior, particularly under general time-varying conditions, of active normal human skeletal muscle *in vivo*. In keeping with this objective and space constraints, many important related topics will be omitted or discussed only briefly: (1) steady state (isometric and isotonic) behavior, (2) the distribution of muscle forces, (3) motor unit structure, (4) reflexes, and (5) pathological responses. Some of these subjects are treated in other chapters of this monograph. To date, the major extensions of the classic isometric and isotonic results have come via *in vivo* frequence-response studies by various investigators, and these will be emphasized. While far from complete, this work represents significant progress toward an understanding of the mechanical behavior of the muscle actuator as it normally functions in the living body.

RESPONSES OF ISOLATED AND *IN VIVO* MUSCLE TO CONTROLLED ELECTRICAL STIMULATION

As background for the subsequent discussion it is useful to review briefly some of the salient results that have emerged from mechanical measurements of functionally isolated and *in vivo* muscle under controlled electrical stimulation. Probably the most extensive studies of the former type available in the literature are those of Rack and his associates (Rack and Westbury, 1969; Joyce and Rack, 1969; Rack and Westbury, 1974) on cat soleus, which, in addition to their broad scope, have the advantage that they employed distributed assynchronous stimulation of the ventral roots innervating soleus, thus producing smooth contractions at submaximal stimulation. This method approximates more closely than is possible with synchronous stimulation the natural activation of muscle *in vivo*. The mechanical behavior of muscle revealed by these studies was very complex indeed and included the following features:

1. Under isometric conditions, the steady state force resulting from constant stimulation was a monotonic, saturating function of stimulus rate. The rate of development of isometric force increased with the length at which the muscle was held.
2. At constant stimulation the steady state isometric force varied with length according to the well-known length-tension relation (Fig. 4-1) with a different length-tension curve for each level of constant stimulation.
3. In constant velocity releases, the force decreased continuously during the release and remained at all times below the isometric force appropriate to the instantaneous muscle length.
4. In isotonic (constant force) releases and stretches, the muscle velocity at any instant depended on the history of the motion, and, in general, there was no unique correspondence between force and velocity at a given length.
5. In constant velocity stretches, the muscle exhibited complex "yielding" phenomena in which the force decreased, often below its isometric value at the current muscle length. This is illustrated in the length-tension Figure 4-1. At the highest (tetanic stimulation) rate there was little yielding, whereas at the lower (more usual physiological) stimulation rates the muscle force decreased markedly after an initial sharp increase.
6. When the muscle was subjected to periodic length variations, the mean force dropped below the isometric value at the mean length and the force perturbation exhibited harmonic distor-

tion, these nonlinear effects becoming more prominent with increasing stretch amplitude and decreasing stimulation rate.

7. The instantaneous stiffness of the muscle, defined as the ratio of force change to length change in a small rapid stretch or release from an isometric state, increased approximately in proportion to the isometric force.

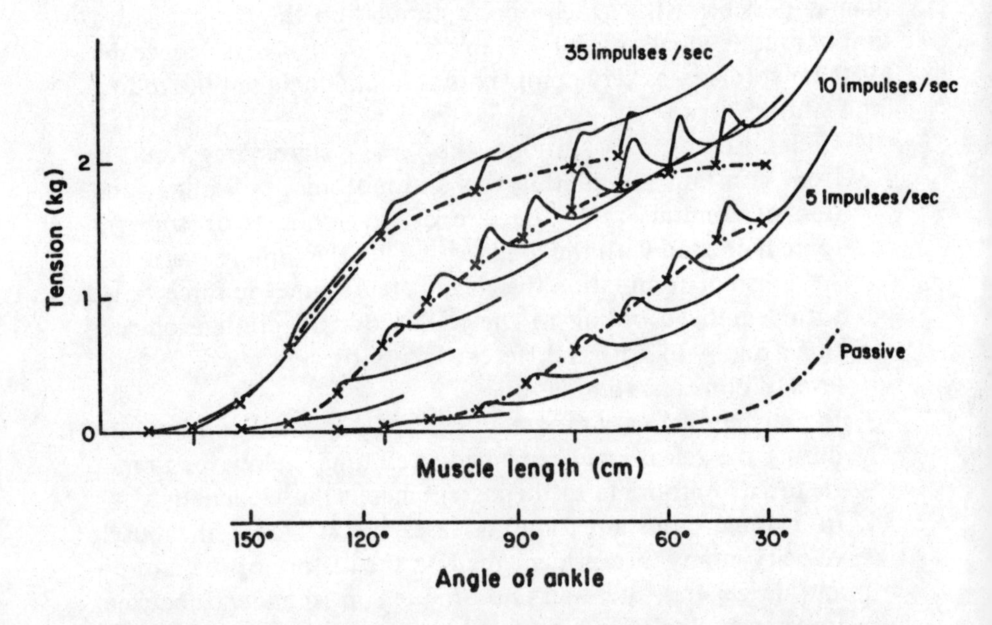

Figure 4-1. Tension response of isolated cat soleus to constant velocity (7.2 mm/sec) stretches at various stimulus rates. In each case the stretch started from the isometric state (interrupted curves) associated with a given muscle length and stimulus rate. From G.C. Joyce, P.M.H. Rack, and D.R. Westbury, The Mechanical Properties of Cat Soleus Muscle During Controlled Lengthening and Shortening Movements. *Journal of Physiology, 204*:461, 1969.

Item 4 raises questions about the validity of a force-velocity relation, except possibly for tetanized muscle near optimal length. A classic Contractile Element/Series Elastic Element model in which there is a unique correspondence between muscle force and Contractile Element velocity at each muscle length is inadequate to represent the yielding behavior illustrated in Figure 4-1 (see the discussion in the "Mathematical Models" section of this chapter), and a comparably simple, adequate quantitative model has yet to be proposed.

While the experiments of Rack et al. provided a more realistic model of natural excitation via asynchronous stimulation of ventral root filaments, they did not model the relation between mean firing rate and motor unit recruitment. Another problem not addressed in these experiments is the response of muscle to time-varying stimulation, but this has been examined by a number of other workers. Partridge measured the frequency response of both isometric (Patridge, 1965) and inertially loaded (Partridge, 1966) cat triceps surae to sinusoidally modulated stimulus pulse trains. Stein and his co-workers have carried out extensive frequency-response studies on isolated cat muscles (Mannard and Stein 1973; Bawa et al., 1976). The general conclusion from these and similar studies is that for sufficiently small length perturbations about a state of isometric contraction, muscle can be represented as a quasilinear system whose parameters depend on the isometric state. In particular, for isometric cat soleus, Mannard and Stein (1973) found that the muscle force perturbation δP was related to the stimulation (pulse rate) perturbation δE by the second-order equation

$$\delta \ddot{P} + 2\zeta\omega_n \delta \dot{P} + \omega_n^2 \delta P = \omega_n^2 G_0 \delta E \qquad (2)$$

The dot (·) represents time differentiation, and ω_n (natural frequency, ζ (damping ratio), and G_0 (low frequency gain) are system parameters characterizing the muscle. Note that these parameters are not constant, but vary with the mean stimulation, thus exhibiting the essentially nonlinear character of muscle. There is a close correlation between this small perturbation frequency response for isolated muscle and that measured for human skeletal muscle *in vivo*.

The results presented above are representative of the information currently available about the mechanical behavior of isolated muscle under controlled electrical stimulation. There have also been some studies involving the direct application of electrical stimulation to human skeletal muscle *in vivo*. These include experiments using transcutaneous stimulation by surface electrodes (Chrochetiere et al., 1967; Trncoczy et al., 1976) that have yielded some interesting observations on mechanical response, but these studies cannot be assumed representative of normal muscle function. Stein and his collaborators have used a more realistic method where a nerve supplying biceps brachii (Aaron and Stein, 1976) or soleus (Bawa and Stein, 1976) in human subjects has been stimulated by modulated pulse trains through subcutaneous needle electrodes. While still not natural, this method of stimulation approximates normal function better than surface electrodes. In these human studies, it was again found that the stimulation-force transfer function for isometric soleus had the character of a second-order low pass filter (eq. 2) and, further, that the system parameters varied systematically with muscle length. A typical frequency response obtained by Bawa and Stein (1976) is shown in Figure 4-2 for comparison with the corresponding response in voluntary movements of human subjects to be discussed in the latter part of this chapter. It should be noted that, in order to fit the phase response of the second order model to the observed phase data, both in the cat and human experiments, Stein et al. found it necessary to introduce a pure time delay, τ, of the order of 10 milliseconds between the stimulation and the force; that is, in equation two (2) the left-hand side is evaluated at time (t) while the right-hand side is evaluated at time $(t-\tau)$.

STEADY STATE BEHAVIOR

Since the work of Lippold and Inman et al. (1952) it has been accepted that quantitative relations exist between the measurable electrical activity of muscles *in vivo* and their mechanical response. However, until recently, clear correlations were available only in two cases: isometric and constant velocity/constant load ("isotonic") contractions. As these two cases are discussed in other parts of this book, only a brief account is included here for the sake of completeness. A review of this topic is also available in Bouisset (1973).

As shown by Lippold (1952), a monotonic relation can be demonstrated experimentally between the processed (in this case rectified and integrated) EMG from a muscle and the tension

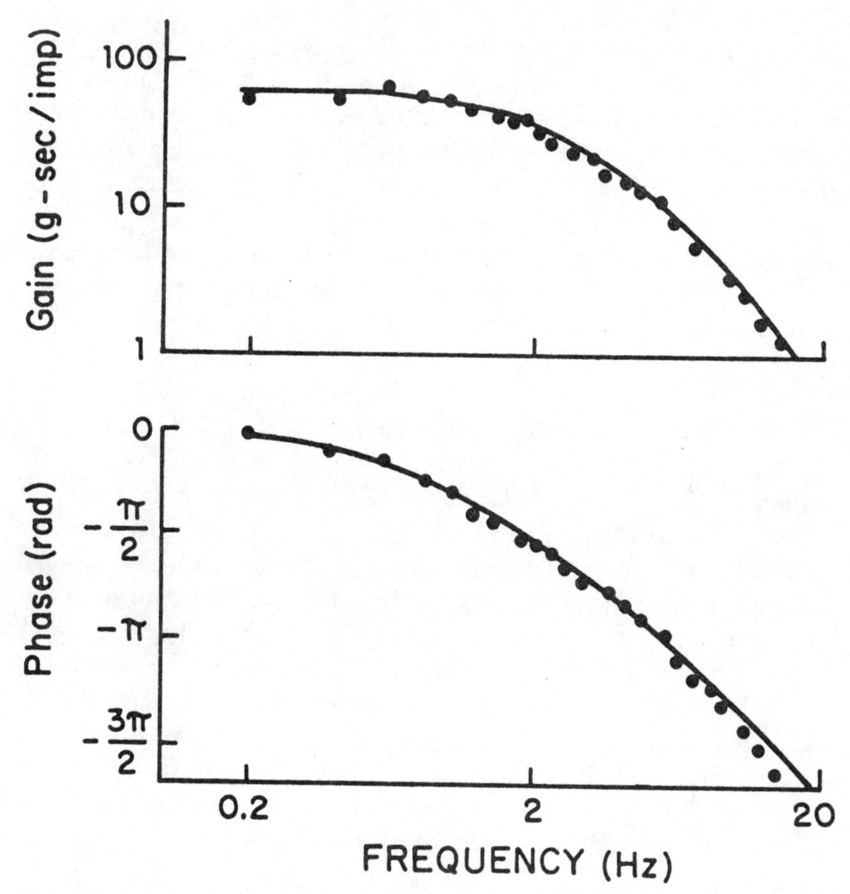

Figure 4-2. Frequency response of isometric human soleus under controlled stimulation of a nerve branch supplying this muscle. Adapted from P. Bawa and R.B. Stein, Frequency Response of the Human Soleus Muscle. *Journal of Neurophysiology, 39*:788, 1976.

generated by the muscle under isometric conditions. To be more precise, what Lippold and most subsequent investigators have shown is that there is a monotonic relation between the processed EMG from a representative (usually superficial) muscle of a group acting about a joint and the moment about that joint produced by that muscle group. This relation has been confirmed by many investigators for various muscles, electrode types, and EMG processing schemes. There has been considerable debate in the literature about whether the isometric force-EMG relations are linear or nonlinear, but much of this dispute seems unwarranted. The precise nature of the quantitative force-EMG relation may depend on the muscle, electrode type and placement, processing scheme, and fatigue; it may well be linear in one case and not in another. The important point is that for a given muscle, recording configuration, and processing scheme the isometric EMG is a monotonically increasing function of the force and thus an indicator of muscular contractile activity. Therefore, under isometric conditions (Vredenbregr and Rau, 1973), a functional relation of the form

$$P = g(E, x, t) \tag{3}$$

obtains, where E represents the EMG, and the dependence on the time (t) is included to account for fatigue effects. This time dependence is usually weak except at high loads and long durations. Over a limited range of forces, say from 0 to 50 percent of maximum voluntary effort, the processed EMG can usually be taken as proportional to the force with acceptable accuracy.

It is worth pointing out in this discussion of isometric force-EMG relations that these may vary greatly if the muscle is not truly isometric. This is due to the rapid variation near the isometric state of EMG with velocity at constant force. For example, Figure 4-3 shows the measured relations between processed (rectified, lightly filtered, and plainimetrically averaged) surface EMG from biceps and triceps as a function of the applied moment about the elbow joint for three conditions: (1) flexion at a very low velocity, (2) isometric, and (3) extension at a very low velocity; the angular velocities of the forearm were approximately 0.1 rad/sec. This figure shows that the EMG from a loaded muscle "shortening at zero velocity" may be considerably higher than that from the same muscle "lengthening at zero velocity," and both of these EMGs may differ from the true isometric EMG. This behavior is analogous to that of a mechanical system with dry (Coulomb) friction and may be related to the rapid increase in force per unit velocity when excised muscles are slowly stretched isotonically at constant stimulation (Katz, 1939).

There have been several investigations designed to measure the

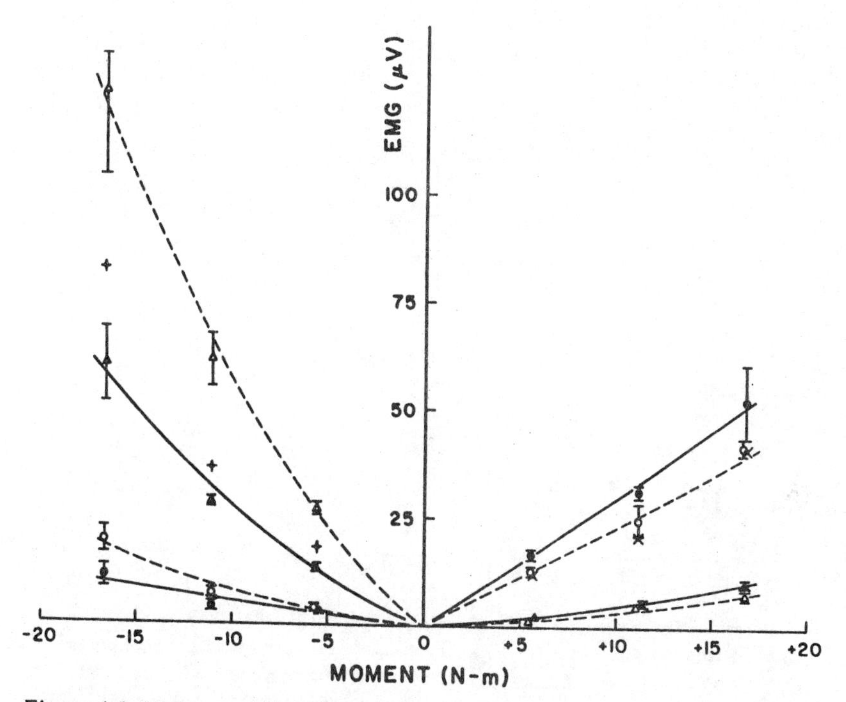

Figure 4-3. Mean rectified EMGs from biceps and triceps for a stationary, slowly flexing, and slowly extending forearm as a function of applied moment about the elbow joint. Circles = biceps and triangles = triceps; filled symbols = flexion and open symbols = extension. Crosses and pluses indicate isometric EMGs from biceps and triceps respectively. Velocities of flexion or extension were approximately 0.1 rad/sec. Positive and negative applied moments respectively tend to extend or flex the forearm. Each data point is the mean of five repeated trials for one subject. From G.I. Zahalak, J. Duffy, P.A. Steward and P.R. Paslay, The Apparent Internal Friction of Human Skeletal Muscle. In W.H. Boykin, Jr. (Ed.): *Workshop on the Biomechanics of Voluntary Human Motion*. Technical Report IWVHM-Roc-75, University of Florida, Gainesville, 1975.

force-velocity characteristics of normal human muscle *in vivo* both at maximal (Wilkie, 1950; Pertuzon and Bouisset, 1973) and submaximal (Bigland and Lippold, 1954; Komi, 1973; Zahalak et al., 1976) voluntary effort. In these experiments, an attempt was made to maintain constant simultaneously the load on a limb (applied in various ways) and the velocity (either the angular velocity

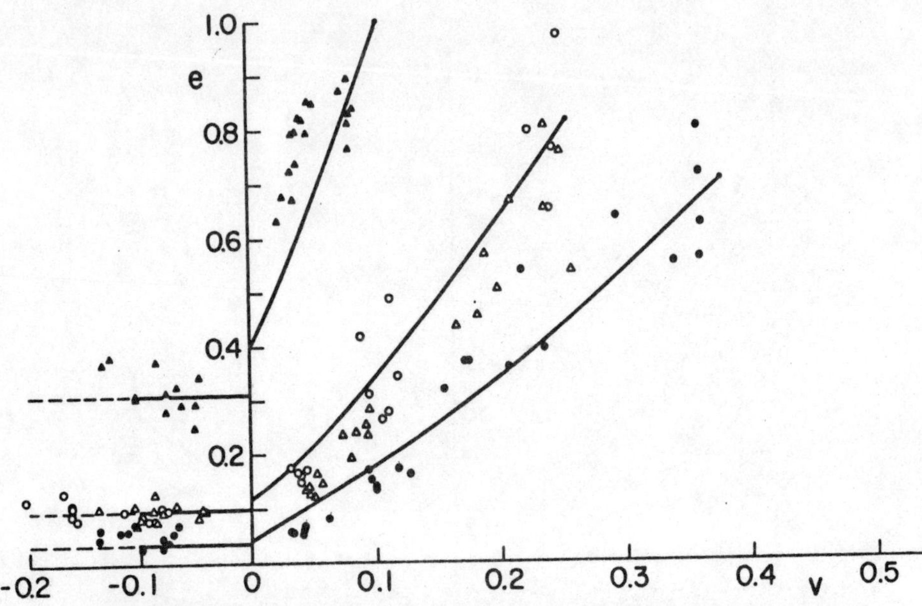

Figure 4-4. Force-velocity-EMG relations for the forearm flexors of one male athlete subject under constant load/constant velocity conditions. The velocity (v), biceps EMG (e), and load (p) have been normalized with respect to the maximum values recorded during the experiment. The filled circles, open circles and triangles, and filled triangles correspond to loads of approximately 6 percent, 16 percent, and 44 percent of maximum voluntary isometric effort. Flexion velocities are positive and extension velocities are negative. Adapted from G.I. Zahalak, J. Duffy, P.A. Steward, H.M. Litchman, R.H. Hawley, and P.R. Pasley, Partially Activated Human Skeletal Muscle: An Experimental Investigation of Force, Velocity, and EMG. *Journal of Applied Mechanics*, 98:81, 1976.

of the moving limb segment or the linear velocity of a point on the limb). Of course, whether these constraints resulted in isotonic constant velocity conditions in the participating muscles depended on whether the muscle forces were proportional to the external load and the muscle velocities were proportional to limb velocity; an assumption of proportionality would be indicated if the moment arms of the participating muscles and the applied moment remained substantially constant over the range of measurement. Bigland and Lippold (1954) showed, in the case of plantar flexion of the foot, that if the load is plotted as a function of velocity at constant EMG for shortening muscle one obtains a family of curves resembling the classic force-velocity hyperbolas, one for each EMG level.

Figure 4-4 shows typical results from a constant velocity test of the forearm flexors of a male athlete subject (Zahalak et al., 1976). In this experiment, the subject rotated his forearm in a sagittal plane at constant angular velocity under the action of a constant horizontal force applied at the wrist, and the EMG was measured at 90 degrees forearm flexion. This figure illustrates the marked difference in the behavior of a muscle shortening under load and that of a muscle lengthening under load: in the former case, the EMG increased with both force and velocity, while, in the latter case, it increases with force but appears substantially independent of velocity over the range of velocities examined. It was found that for velocities not exceeding half of the maximum velocity possible at each load (it is questionable whether steady state conditions could be assumed to prevail in this experiment at higher velocities) the data for shortening muscle could be fitted reasonably well by a relation in the form of Hill's equation

$$P_0(e) - p = (k_1 + k_{2p})v \tag{4}$$

The variables p_0, p, and v represent the (dimensionless) isometric force (in the limit as the *shortening* velocity goes to zero), force, and velocity respectively. Numerical values of the parameters k_1 and k_2 for six male athlete subjects were estimated by a least-squares identification procedure and are listed in Zahalak (1976).

STIFFNESS

In addition to the isometric and constant velocity responses, a number of investigators have measured the stiffness of limb musculature *in vivo*. One must exercise some caution in interpreting the results of stiffness measurements because various investigators mean different things by this term. Formally, stiffness may be defined as the change in force (or moment) per unit change in displacement (or rotation). The term *stiffness* has been applied to the following:

1. *Isometric stiffness*: the change in force per unit length along an isometric length-tension curve at constant stimulation.

2. *Passive stiffness*: the change in force per unit length associated with the displacement of a passive limb (in terms of the classical three-element model this is regarded as a measure of the parallel elasticity).
3. *Series stiffness*: the "instantaneous" change in force per unit change in length associated with the small displacement of a limb with actively contracting muscles, in the absence of reflex (this is a measure of the series elasticity in the classical model).
4. *Reflex stiffness*: the ratio of the force change to the length change during forced driving of a limb in the presence of reflex.

The isometric stiffness is simply the slope of the isometric tension length (or moment-rotation) curve at constant stimulation (or EMG). The passive stiffness has been estimated (Lestienne and Pertuzon, 1974) to be quite small for flexion-extension of the forearm (of the order of 1 Nm rad^{-1}). The series stiffness is an important mechanical property of muscle and has been estimated by a number of different methods for the forearm flexors and extensors. Wilkie (1950) used a rather indirect isometric method, involving prior knowledge of the force-velocity relation and an assumption of instantaneous and complete activation of the muscles, to conclude that the flexor (rotational) stiffness increased with muscle force and attained a value of the order of 200 Nm rad^{-1} at high levels of contraction on normal human subjects. A number of investigators have used quick-release techniques (Soechting et al., 1971; Zahalak et al., 1975; Goubel and Pertuzon, 1973; Cnockaert and Pertuzon, 1974) to estimate the series stiffness. Of course, quick-release tests are the traditional method for measuring this property in isolated muscle, but the interpretation of *in vivo* tests is complicated by the unavoidable limb inertia. Correcting for inertia and assuming, essentially, that for approximately 50 milliseconds following release the decrease in muscle force was attributable entirely to shortening of the series elastic element, Goubel and Pertuzon* (1973) obtained

*These authors report the series stiffness as the linear stiffness of an "equivalent flexor" which is a hypothetical muscle having the geometry of the biceps and the dynamic properties of the combined forearm flexors. To convert these values to the rotational stiffness (moment per unit rotation) of the forearm flexors, when the flexion angle is near 90 degrees they must be multiplied by the square of the moment arm of the biceps (approximately 0.044 m for adult males).

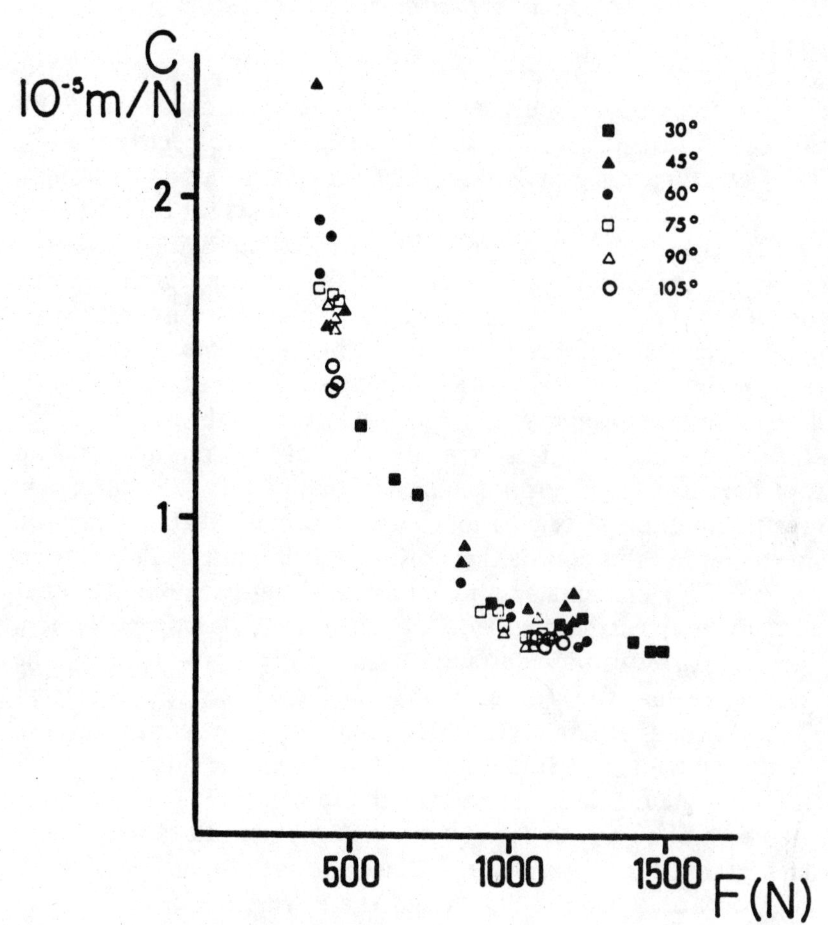

Figure 4-5. Series compliance (reciprocal stiffness) of the "equivalent flexor" vs. its force of contraction for one normal male subject, as determined by quick-release tests. The different symbols indicate various starting angles of flexion at the instant of release. From F. Goubel and E. Pertuzon, Evaluation de L'Elasticité du Muscle In Situ par une Méthode de Quick-Release. *Archives Internationales de Physiologie et de Biochimie*, 81:697, 1973.

the values of series compliance (reciprocal stiffness) shown in Figure 4-5. Further it was found that the stiffness data for five normal male subjects were well fitted by a power-law relation of the form

$$C = aF^b \quad (5)$$

where C is the compliance of the equivalent flexor in 10^{-5} m/N, F is the force of the equivalent flexor in N, and a and b are regression coefficients taking the values 394 and −0.9 respectively.

From Figure 4-5 it is clear that series stiffness is a strongly increasing function of muscle force, as is the case for isolated muscle. Cnockaert and Pertuzon (1974) carried out similar measurements for the elbow extensors and obtained similar results, except that the values of rotational stiffness at each level of isometric moment were considerably lower than for the flexors. It should be emphasized the stiffness measured in these quick-release tests is indeed a muscle property and can properly be labelled the series stiffness because the measurements were made over a time interval too short for significant mechanical reflex effects to occur. However the method of calculation takes no account of the nonelastic shortening of the contractile portion of the muscle; therefore, it probably underestimates the stiffness somewhat. As the series elasticity of isolated muscle is known to reside partly in the tendon and partly in the contractile tissue (Morgan, 1977), it is to be anticipated that the stiffness would vary separately with force and EMG, but this separate variation would not be evident in experiments involving a perturbation about an isometric state because in that state the EMG is a function of the force.

The reflex stiffness is not a muscle property but rather a transfer function that depends on intrinsic muscle properties, the characteristics of the reflex loops and the driving frequency. Discussion of this quantity is deferred to the section dealing with *in vivo* frequency response.

WHAT DOES THE EMG MEASURE?

Before describing the more general response of muscle *in vivo* under oscillatory conditions, it is necessary to consider what the EMG measures and how it is related to the mechanical response of muscle. Several authors have assumed explicitly or implicitly that the EMG should be a measure of the muscle force, but obviously this can only be true under quasistatic conditions when the EMG, force, and muscle length vary slowly with time. Constant velocity tests and the frequency-response tests to be discussed later in this chapter show clearly that EMG is not uniquely related to muscle force. What, then, does the EMG measure under general dynamic conditions? Some indications are provided by an examination of the microscopic mechanisms of contraction.

The EMG is known to be a weighted sum of the electrical disturbances produced by individual motor unit action potentials. The weighting coefficients depend on the size of a motor unit and its proximity to the recording electrodes. The intensity of the EMG as measured by its r.m.s. value or some other appropriate index increases with both the number of active motor units and their firing rates (Person and Libkind, 1970; DeLuca and VanDyk, 1975). According to the current physiological understanding of the contraction mechanism (Carlson and Wilkie, 1974; Huxley, 1974), the motor unit action potentials contributing to the EMG are associated with transient changes in the permeability of the sarcoplasmic reticulum to calcium ions. Increased electrical activity results in increased concentrations of calcium in the sarcoplasm, which, in turn, makes a larger proportion of binding sites on the actin filaments available for attachment to myosin. The proportion of actin sites available for binding may be taken as a measure of the "level of activation" of a muscle. Decreasing the electrical activity of a motor unit decreases the calcium permeability of the sarcoplasmic reticulum, resulting in a net uptake of calcium from the sarcoplasm by an active calcium pump and a consequent decrease in the level of activation. While these qualitative features of the excitation-contraction sequence are generally accepted, it has not yet been possible to develop them into reliable quantitative predictions of the dynamic relations between EMG and activation. For some examples of mathematical modeling in this area, one is referred to the work of Hatzel (1977; 1978).

In order to make progress in the absence of more detailed quantitative information one is forced to make some expedient, but with hope reasonable, assumptions. As the fraction of actin binding sites is not accessible to direct measurement, a practical measure of activation level can be taken as the force that would be developed under static conditions at a given length and stimulation level (EMG): this is essentially the A.V. Hill (1949) concept of the "active state." It must also be recognized that, since diffusion and chemical reactions intervene between changes in electrical activity and changes in activation level, the EMG will lead activation with certain characteristic time constants. Further, at a conduction speed of approximately 4 m/sec an action potential takes a small but appreciable time to travel from the innervation zone of a muscle to the tendons, (of the order of 10 milliseconds in human biceps brachii), and this time delay increases still further the lag of activation behind EMG.

Thus, in a broad sense, the intensity of the EMG may be regarded as a measure of the level of activation of the contractile tissue in a muscle, where the activation lags the EMG and leads the development of muscle force in time.

In addition to the theoretical problem of interpreting the EMG there is the practical problem of measuring and processing it. Space limitations preclude an extended discussion of this topic, but a few general comments are in order at this point. The raw EMG, that is, the myoelectric potential difference between two electrodes is a randomly fluctuating signal unsuitable for correlation with the mechanical response; it must be processed to extract a relatively smooth measure of a muscle's overall electrical activity. Many different analog and digital processing schemes have been used to produce measures such as the r.m.s. value, average rectified value, number of fluctuations per unit time, etc. These methods require the time-averaging of some property of the interference pattern over an interval that is large compared to the characteristic period of the EMG fluctuations (typically 10 to 20 milliseconds for surface EMGs). This presents no problem (except possibly that of fatigue) under static conditions where the EMG is a stationary random signal, and the various processing schemes are equivalent insofar as they all produce relatively smooth measures of electrical activity that vary monotonically with muscle force. However, under dynamic conditions when the mechanical parameters and the EMG amplitude may vary rapidly, time-averaging over sufficiently long intervals is no longer possible without grossly distorting the temporal information carried by the EMG. Thus it appears that in general the EMG should be regarded in an ensemble-average sense, and the intensity of electrical activity should be computed at fixed times by averaging over a collection of records from repeated trials under presumably identical conditions (again, fatigue effects may complicate the practical realization of this approach). An alternative method in the case of periodic signals is to average the EMG over many cycles at corresponding points in a cycle. Another possibility is to partially substitute instantaneous spatial averaging of several EMG signals, measured by electrode arrays placed over a muscle (Hogan, 1976; Emerson and Zahalak, 1981), for temporal averaging of a single signal, but this method appears to be limited by the relatively high correlations between the EMGs measured from a single muscle. There have been few studies of nonstationary EMGs employing ensemble-averaging techniques, and it is fair to say that the interpretation of such EMGs and their correlation with mechanical response is a very prominent issue in muscle mechanics.

DYNAMIC BEHAVIOR: PERTURBATIONS ABOUT AN ISOMETRIC STATE

Due to the formidable complexity of the mechanical behavior of muscle over its entire physiological range of motion, investigations of muscle dynamics under general time-varying conditions have been confined almost exclusively to the case of small perturbations in muscle length (generally on the order of 1 percent) about a steady isometric state. Under these conditions, it is possible to represent the observed mechanical response in terms of a quasilinear model, that is one in the form of a linear differential equation with parameters that vary with the isometric state. These results represent a first step toward the solution of the broader problem of the dynamic behavior of muscle during arbitrary motions.

Gottlieb and Agarwal (1971) investigated the dynamic relations between force and EMG for the ankle flexors and extensors under time-varying isometric conditions and found that their data could be represented by a second-order linear model of the form

$$\ddot{M} + (\tau_1^{-1} + \tau_2^{-1})\dot{M} + \tau_1^{-1}\tau_2^{-1}M = GE(t) \qquad (6)$$

where M is the foot torque, E is the EMG, G is a gain parameter, and τ_1 and τ_2 are time constants of the order of 100 milliseconds.

In this investigation the EMG was lightly filtered and used as an input to eq. six (6), which predicted reasonably good approximations to the measured foot torques. A more recent investigation (Crosby, 1978) indicates that the dynamic response of isometric triceps brachii is also well represented by a second-order linear model.

Frequency-response studies have considerably elucidated the dynamic response of muscle *in vivo*. These studies have employed voluntary oscillations (Coggshall and Bekey, 1970; Soechting and Roberts, 1975), forced oscillations (Nielson, 1972; Agarwal and Gottlieb, 1977), or both (Zahalak and Heyman, 1979; Cannon, 1980; Cannon and Zahalak, 1982). This section will review the results of one extensive representative series of such tests undertaken recently in the author's laboratory. A mathematical model for the perturbation response of a limb proposed in Zahalak and Heyman (1979) will be described in the next section along with a listing of the model parameter values measured in these experiments.

A typical experimental arrangement for a frequency-response test is shown in Figure 4-6. The subject sat immobile in a chair with his upper arm horizontal and his forearm fixed rigidly in a loading frame that permitted flexion and extension of the forearm in a sagittal plane about the elbow joint. In voluntary oscillation tests (VOTs), the subject oscillated his forearm at prescribed mean loads (at the wrist), frequencies, and force-perturbation amplitudes, while in the forced oscillation tests (FOTs) the subject maintained a constant level of contraction (by observing the output of a load cell at the wrist) while a small oscillatory displacement of constant amplitude was applied to the wrist. Normal adult males served as subjects in these experiments, and each subject underwent a full day of testing involving both VOTs and FOTs. The variables measured in each test were the force at the wrist, the angular position of the forearm, and EMGs from triceps, biceps, and brachioradialis.

To obtain a quantitative measure of electrical activity suitable for correlation with the mechanical response, the EMG was processed according to a scheme suggested by Neilson (1972). The EMG was first rectified and then low pass filtered with a 50 millisecond time constant. This smoothed the signal but produced

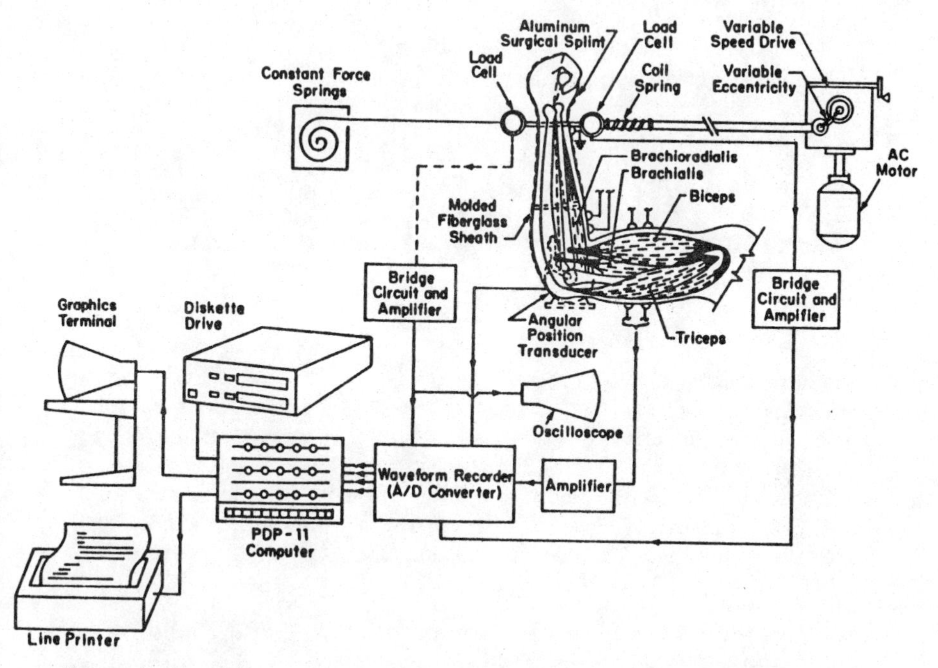

Figure 4-6. A typical experimental arrangement for a frequency-response test. From S.C. Cannon and G.I. Zahalak, The Mechanical Behavior of Active Human Skeletal Muscle in Small Oscillations. *Journal of Biomechanics, 15*: 111, 1982.

severe distortions of the phase and amplitude at all but the lowest frequencies. The distortion was compensated for by first approximating the filtered EMG by a sinusoid at the driving frequency and then correcting the phase and amplitude of this sinusoid by the known transfer characteristics of the smoothing filter. This procedure is illustrated in Figure 4-7, and it is clear that the resulting sinusoid provides a reasonable measure of the phase and amplitude of the muscle's electrical activity. It has been verified that, over the frequency range 0 to 15 Hertz this procedure yielded equivalent results to the method of Zahalak and Heyman (1979), which consisted of computing digitally a running r.m.s. based on 16 millisecond samples of raw EMG and then fitting a sinusoid to the resulting r.m.s. values.

The static force-EMG relations were measured several times

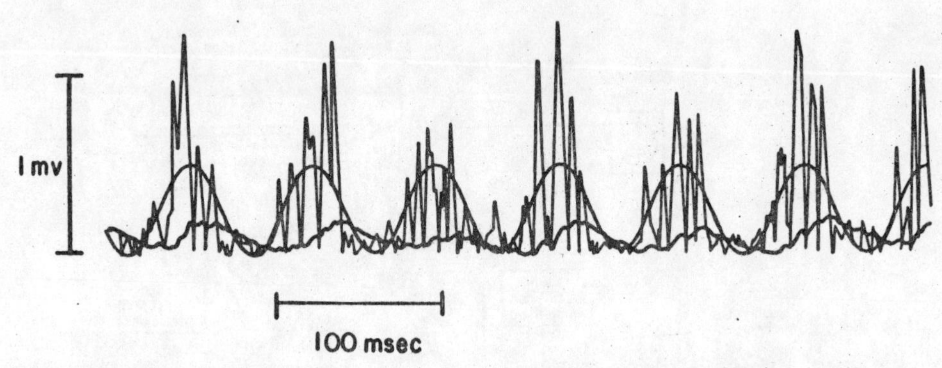

Figure 4-7. Results of EMG processing for a typical FOT record. The jagged curve is the rectified raw EMG, the wavy curve is the filtered rectified EMG, and the smooth curve is the least-squares sinusoidal approximation to the filtered signal, after it has been passed through a digital filter with characteristics inverse to those of the smoothing filter. From S.C. Cannon and G.I. Zahalak, The Mechanical Behavior of Active Human Skeletal Muscle in Small Oscillations. *Journal of Biomechanics, 15*:111, 1982.

throughout the day of testing for each subject and found to be reproducible and reasonably linear over a load range from 0 to about 60 percent of maximum voluntary effort (the range of loads used in this study). This indicates that fatigue effects did not produce significant distortions of data. The electrical activity of antagonists was always very low when either the flexors or extensors were loaded, and under isometric conditions the moment about the elbow could be related to the EMG of the loaded muscle group with sufficient accuracy by a simple proportionality factor

$$M = \Gamma E \qquad (7)$$

where M is the moment generated by the muscles and E is the processed EMG from biceps, brachioradialis, or triceps.

A typical response in an FOT is shown in Figure 4-8. This figure exhibits the amplitude ratio of forearm position perturbation to driving moment perturbation and the phase lag of angular position behind driving moment at three values of mean load on the triceps of one subject. Several interesting features are evident in this figure. The data segregate into three sets of response loci according to load. At each load the response shows resonance characteristics,

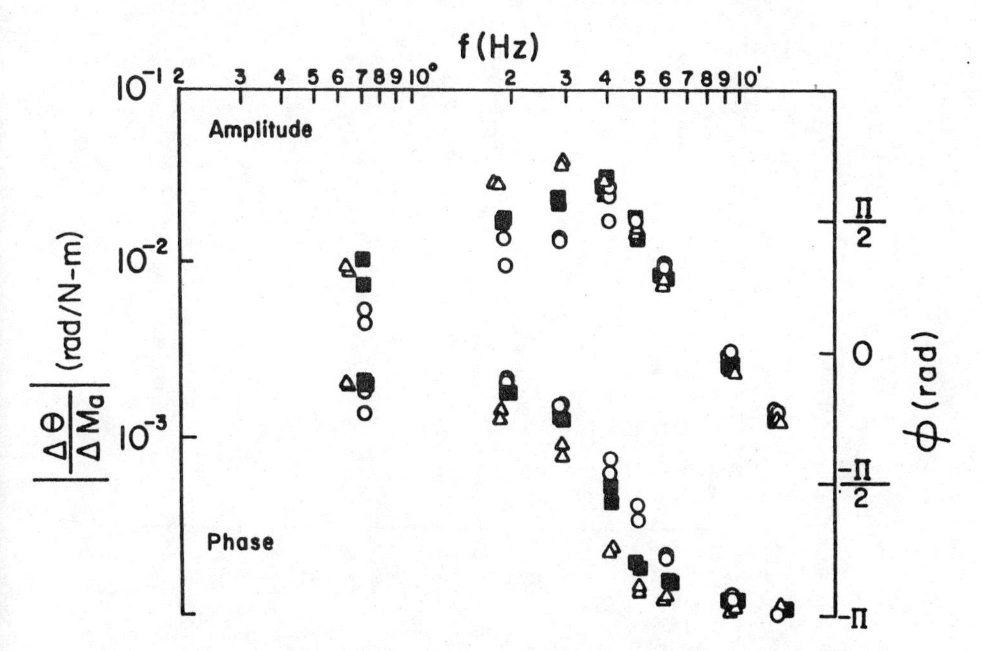

Figure 4-8. Bode plot of (flexion angle/applied moment) amplitude ratio (upper set of points) and phase lag of flexion angle behind moment (lower set of points) for one normal male adult subject in an FOT. Symbols indicate three different levels of mean applied moment tending to flex the forearm: open circles = 27 Nm, filled swaures = 13 Nm, open triangles = 4 Nm. The flexors were relaxed in these tests. From S.C. Cannon and G.I. Zahalak, The Mechanical Behavior of Active Human Skeletal Muscle in Small Oscillations. *Journal of Biomechanics, 15*:111, 1982.

that is an amplitude peak and a phase inversion from 0 to 180 degrees. At frequencies above 6 Hertz the amplitudes fall on a straight line of slope −2 indicating that the response is dominated by the forearm inertia. Further, both the amplitude and phase data, particularly the latter, show that the resonant frequency increases with increasing load. This indicates that the stiffness of the system increases with increasing muscular contraction, a conclusion that certainly agrees with the behavior of isolated muscle. In Zahalak and Heyman (1979) the stiffness was estimated from a measurement of the resonant frequency for four normal adult male subjects. Agarwal and Gottlieb (1977) have measured similar frequency-response characteristics for plantar flexion of the foot

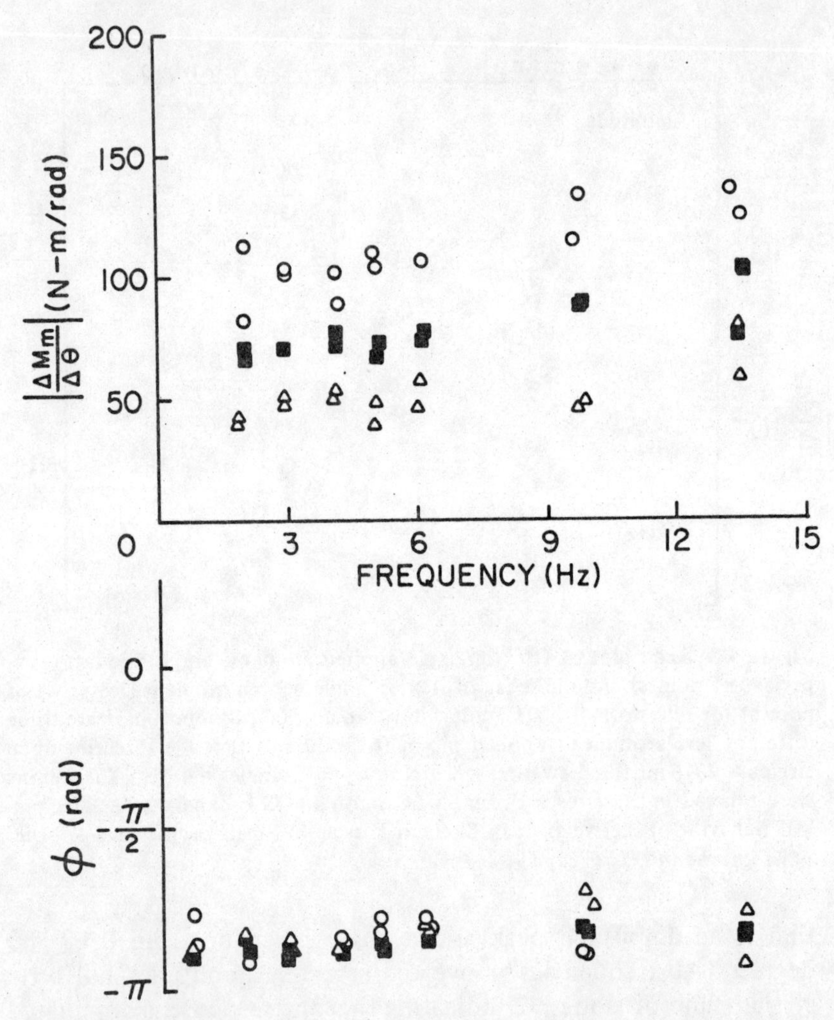

Figure 4-9. Amplitude ratio (muscle moment/flexion angle) and phase lead of muscle moment on flexion angle in an FOT. Data and symbols correspond to those shown in Fig. 4-8. From S.C. Cannon and G.I. Zahalak, The Mechanical Behavior of Active Human Skeletal Muscle in Small Oscillations. *Journal of Biomechanics, 15*:111, 1982.

in normal human subjects, using random rather than sinusoidal driving inputs.

Another, and perhaps more incisive, way to look at the data

of Figure 4-8 is to subtract the inertial and passive viscoelastic moments from the applied moment to obtain the moment due to the muscles (in this case primarily the triceps). Figure 4-9 exhibits the ratio of muscle moment perturbation, $\Delta\Theta$, and the phase lead of muscle moment on forearm position as a function of driving frequency (for the same data as Fig. 4-8). This transfer function is, in fact, the reflex stiffness mentioned in a previous section. At each load the amplitude ratio $|\Delta M_m/\Delta\Theta|$ is substantially constant, but increases with the load, whereas the phase remains approximately constant near $-\pi$.

Not all the measured amplitude responses were as flat as shown in Figure 4-9, but a majority were so over the frequency range 3 to 10 Hertz. This type of frequency response is that associated with an undamped elastic element–a pure spring. The stiffness may be estimated as the mean value of $|\Delta M_m/\Delta\Theta|$. The results of this estimate (over the frequency range 4 to 10 Hz) for the flexors and extensors of ten normal adult male subjects are shown in Figure 4-10. The stiffness is seen to be a strongly increasing function of contraction level. The quick release estimates of the muscular stiffness by Goubel and Pertuzon (1973) and Cnockaert and Pertuzon (1974) coincide approximately with the lower bounds of the frequency-response results exhibited in Figure 4-10, but, as noted previously, the former values are probably underestimates because the nonelastic shortening of the muscle during the recording interval is not accounted for in the stiffness calculation. In view of this consideration, it appears that the quick release and frequency-response estimates of muscular stiffness are in fair agreement. This last point is important because reflex modulations of muscular excitation do not strongly affect the quick release results, whereas such modulations are unavoidably present to some extent in the frequency-response tests. Nevertheless both the comparisons with quick release results and calculations of the magnitudes of reflex effects based on the mathematical model to be presented in the next section (Cannon, 1980) suggest that the stiffnesses presented in Figure 4-10 are serendipitously good estimates of the series stiffness, an intrinsic muscle property.

Reflex effects appeared in the FOTs as oscillatory perturbations in the EMG, and have been described in Cannon and Zahalak

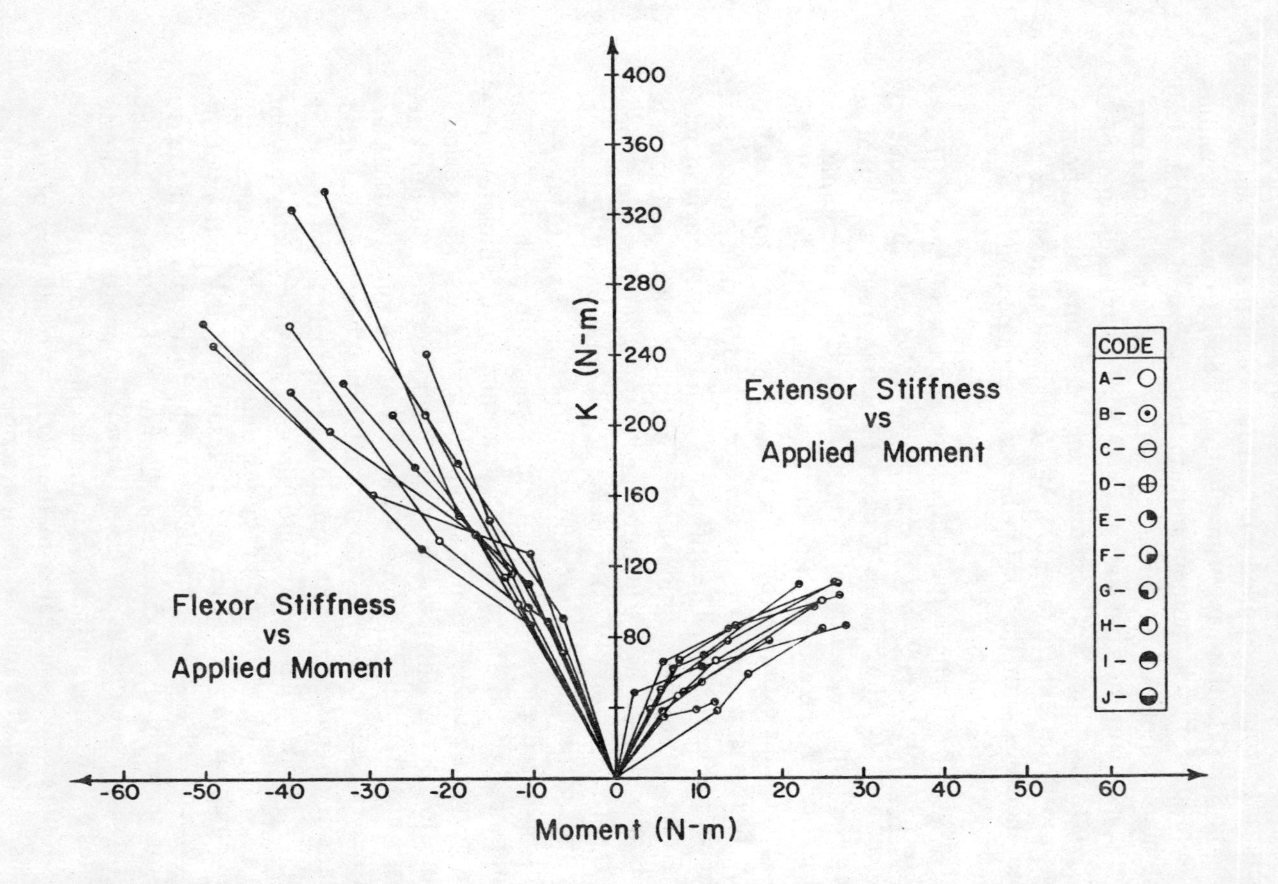

Figure 4-10. Muscular (rotational) stiffness for ten normal male adult subjects vs. mean applied moment, as measured by the mean value of the (muscle moment/flexion angle) amplitude ratio over the frequency range 4 to 10 Hz in the FOTs. From S.C. Cannon and G.I. Zahalak, The Mechanical Behavior of Active Human Skeletal Muscle in Small Oscilla-

(1981). It is beyond the scope of this chapter focusing on the mechanics of muscle to analyze this reflex activity in detail, but its prominent characteristics will be described to provide a more complete picture of the system response. As the reflex activity is presumably generated largely by stretch receptors within the loaded muscles, it was expected that the EMG would lag muscle stretch by a constant time delay associated with action potential propagation in the spinal reflex loop. A constant time delay would manifest itself as a linear decrease with frequency of the phase lead of the EMG on muscle stretch, and indeed such an approximately linear decrease of phase lead was observed for all subjects at frequencies above 6 Hertz. The reflex time delay could be estimated from the rate of decrease of phase lead with frequency and was found to have a value of about 25 milliseconds.

Mean phase variations of reflex EMG for ten subjects are displayed in Figure 4-11 where the frequency scale has been normalized by the estimated reflex conduction time delay for each subject. It is seen that the phase advance of reflex EMG on muscle stretch exceeds 90 degrees at approximately 5 Hertz (corresponding to $\omega\hat{\tau}\sim0.8$ in Fig. 4-11) and approaches the linear decrease characteristic of a pure time delay at higher frequencies. The phase variation shown in Figure 4-11 is that which would be produced by a sensor located in the muscle that generated a feedback signal proportional to stretch, stretch rate, and stretch acceleration: these are indeed the transfer characteristics of the muscle spindles for very small stretches (Matthews and Stein, 1969). In general, the reflex EMG amplitude increased with frequency at a given load and with load at a given frequency over the range tested. This is illustrated in Figure 4-12 for the triceps of one subject. The reflex EMG response tended to saturate at loads below 50 percent of maximum voluntary effort. It is evident that the total limb behavior, even in this apparently simple experiment, is governed by a complex interplay of limb inertia, stiffness, and other intrinsic mechanical properties of the muscles, as well as reflex feedback, and the latter must be taken into account in analyzing the frequency response.

In the second type of frequency-response test, the voluntary oscillation test (VOT), each subject voluntarily oscillated his fore-

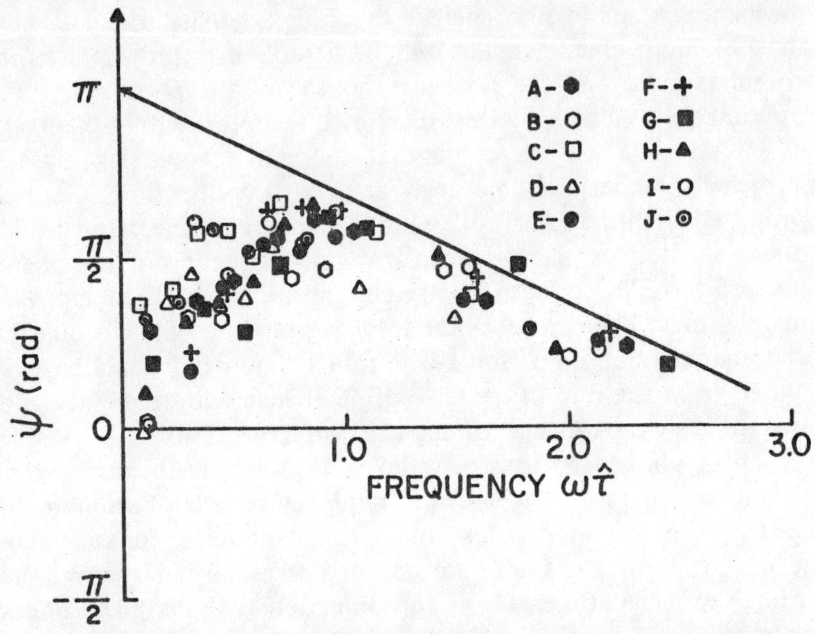

Figure 4-11. Mean phase advance of reflex EMG on muscle stretch as a function of driving frequency for ten normal adult male subjects in the FOTs. The (radian) driving frequency ω has been normalized separately for each subject by multiplying it by the estimated reflex time delay $\hat{\tau}$. From S.C. Cannon and G.I. Zahalak, Reflex Feedback in Small Perturbations of a Limb. *A.S.M.E. Biomechanics Symposium, AMD-Vol. 43*:117, 1981.

arm against an undamped elastic resistance at the wrist. The subjects attempted to control the mean load at the wrist, the load perturbation, and the frequency of oscillation. At higher frequencies co-contraction invariably occurred, whether desired or not, which led to some increase in the total muscular stiffness: this increase in stiffness was accounted for in the analysis of the VOT data (Cannon and Zahalak, 1982). To a good approximation, the flexor and extensor EMGs were 180 degrees out of phase at all mean loads, load perturbations, and frequencies.

The major results of these tests are displayed in Figure 4-13, which shows the amplitude ratio of muscle moment to a weighted sum of flexor and extensor EMGs, and the phase lag of muscle moment behind EMG. This weighted sum, which may be termed

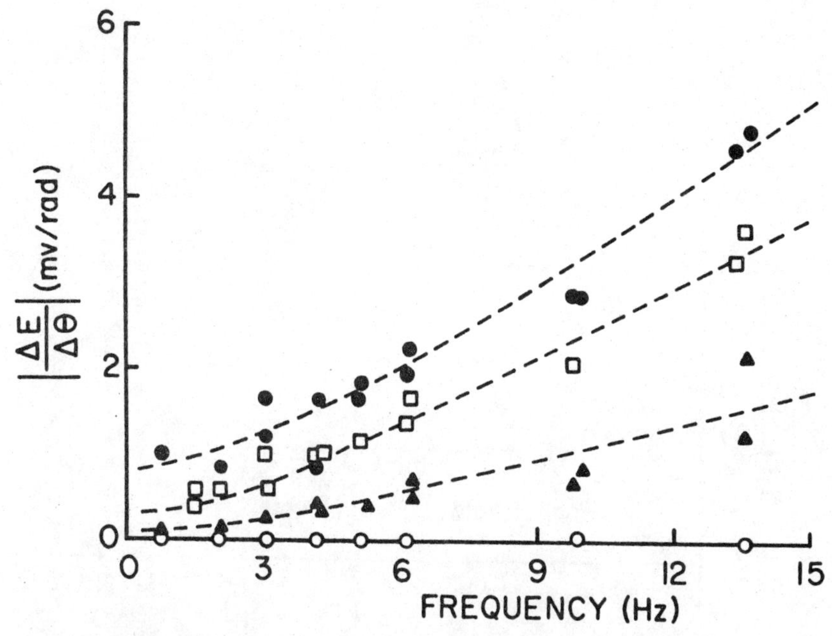

Figure 4-12. Amplitude ratio of the extensor reflex EMG to the flexion angle vs. frequency for one normal adult male subject in a FOT. The symbols indicate four different levels of applied moment: filled circles = 26 Nm, open squares = 14 Nm, filled triangles = 7 Nm, and open circles = 0 Nm (relaxed subject) (Cannon, 1980).

the "isometric moment perturbation amplitude," is defined as $(\Gamma_f \Delta E_f + \Gamma_e \Delta E_e)$ where ΔE_e and ΔE_f are respectively the perturbation amplitudes of the extensor and flexor EMGs, and Γ_e and Γ_f are the corresponding slopes of the isometric moment-EMG curves: it is a variable that occurs naturally in an analysis of the VOT via the mathematical model to be presented in the next section. Physically, it represents the amplitude of the muscle moment perturbation that would result if the flexor and extensor EMGs were 180 degrees out of phase and each muscle group generated a moment in accordance with its isometric moment-EMG curve; it is expected that the muscle moment perturbation amplitude should approach the "isometric moment perturbation amplitude" at low frequencies of oscillation. Further, the oscillation frequency has been normalized with respect to the natural frequency $\omega_n = \sqrt{K/I}$ for each subject, where K is the muscle stiff-

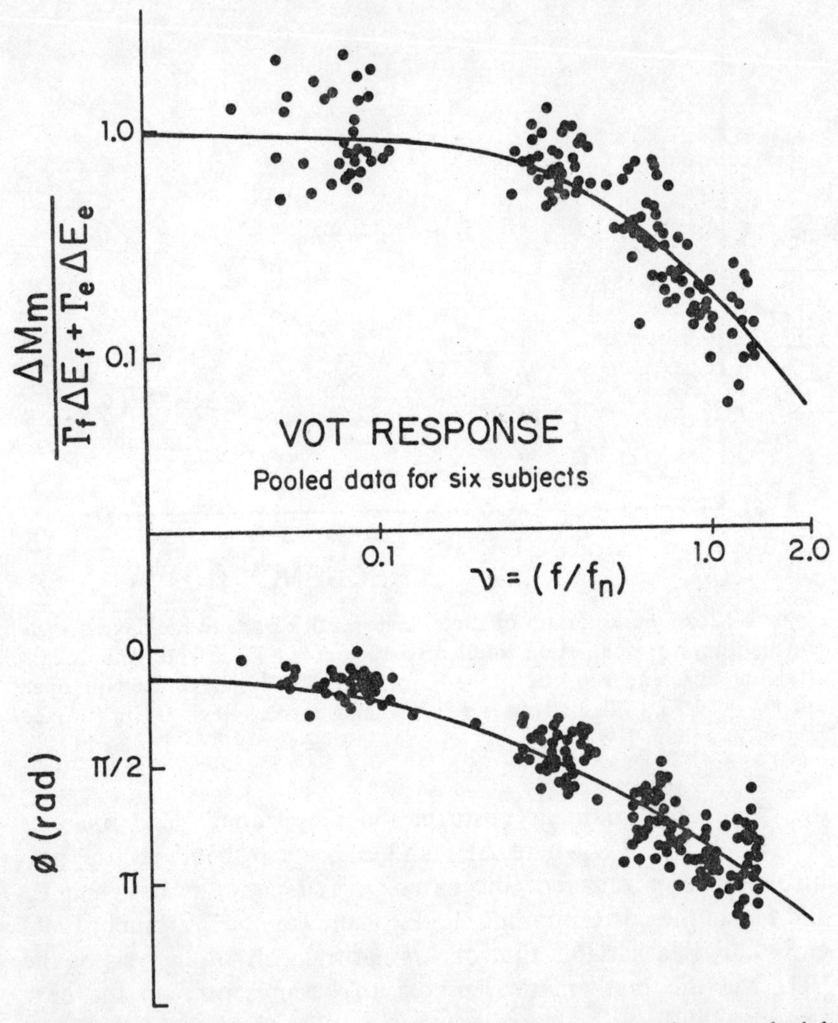

Figure 4-13. Bode plot showing the VOT frequency-response for normal adult male subjects. Pooled data for six subjects. The upper graph shows the amplitude ratio of the muscle moment to the EMG normalized as $\Gamma_f \Delta E_f + \Gamma_e \Delta E_e$ (see text). The lower graph shows the phase lag of muscle moment behind (positive) flexor EMG of (negative) extensor EMG. The frequency has been normalized with respect to the natural frequency $f_n = \sqrt{K/I/2\pi}$ where the stiffness K has been estimated for each test run from the measured mean EMGs via Figure 4-5. The smooth curves showing the model response (eq. 17) have been drawn using the following average values of the parameters

ness estimated from the measured mean EMGs via Figure 4-10 at each oscillation frequency, and I is the moment of inertia of the forearm; the physical range of frequencies represented by the data in Figure 4-13 is approximately 0.5 to 10 Hertz.

Figure 4-13 displays features of muscle dynamics that are consistent with the results of several other investigations of essentially isometric muscle, both isolated and *in vivo* (Partridge, 1965; Bawa et al., 1976; Aaron and Stein, 1976; Bawa and Stein, 1976; Soechting and Roberts, 1975), and should be compared with the frequency responses of electrically stimulated human muscle *in vivo* (Fig. 4-2). The amplitude of the muscle moment per unit EMG is seen to roll off very rapidly as the frequency increases, while the phase lag of muscle moment behind EMG reaches, and may exceed (*see* Fig. 7 of Zahalak and Heyman, 1979), 180 degrees at 10 Hertz, the highest frequencies attained in these experiments. These are the characteristics of a low-pass filter, where stimulation (EMG) is regarded as the input to the filter, and muscle moment is regarded as the output.

Quantitative analysis of the amplitude response indicates that a quasilinear second-order system provides a good representation, but this is not quite adequate for the phase response, as the phase of a second-order system cannot (at finite frequencies) reach 180 degrees. Formally, the data of Figure 4-13 can be made compatible with a second-order response if one assumes a small time delay of the order of 10 msec between arrival of stimulation at the muscle and the triggering of the processes that mobilize the contractile machinery (*see also* Aaron and Stein, 1976; Bawa and Stein, 1976; Zahalak and Heyman, 1979); one physical basis for such a time delay may be the fact that a conduction time of this order of magnitude is required for action potentials to travel from the innervation zone to the tendons of several human limb muscles. The presence of reflex feedback causes no particular problem in the analysis of the VOT because one measures the EMGs that are the sums of voluntary and reflex activity—one essentially opens the reflex

---

$(\tau_1, \tau_2, \tau, \gamma)$ = (0.05 sec, 0.05 sec, 0.0015 sec, 2.45 sec.). From S.C. Cannon and G.I. Zahalak, The Mechanical Behavior of Active Human Skeletal Muscle in Small Oscillations. *Journal of Biomechanics, 15*:111, 1982.
←

loops for analysis by monitoring the input to the muscle "plant."

The data presented in this section show that the dynamic behavior of muscle *in vivo* is very complex even when restricted to small motions about an isometric steady sate, involving the limb inertia, the intrinsic mechanical properties of muscle, and reflex feedback loops.

MATHEMATICAL MODELS

In this section, some mathematical models for musculoskeletal dynamics *in vivo* are discussed. No adequate models exist for arbitrary muscle motion over the entire physiological range, and none will be presented here. Most phenomenological models for muscle mechanics have been based on A.V. Hill's conceptual representation of muscle as consisting of a series elastic element (SE) in series with a contractile element (CE) (Hill, 1938). The extension of the SE and the velocity of the CE were assumed to be determined by the muscle force. The mathematical statement of this model is

$$\dot{x} = k^{-1}\dot{P} - V(P) \qquad (8)$$

where x is the muscle length, P is the force, k is the stiffness of the SE, V is the velocity of shortening of the CE as a function of force, and the superposed dots denote derivatives with respect to time. A model of this form has been applied to calculate the mechanical response of muscle, both isolated (Katz, 1939; Ritchie and Wilkie, 1958) and *in vivo* (Wilkie, 1950), under conditions of constant stimulation and no lengthening of the CE. Various modifications of equation eight (8) are possible to bring it into closer conformity with the experimental facts. First, the velocity of contraction under isotonic conditions depends on the level of activation, as measured by the isometric force P_0 corresponding to a given constant stimulation, in addition to P (Joyce and Rack, 1969; Mashima et al., 1972). Next, it is now generally accepted that a significant portion of the series elasticity resides anatomically in the contractile tissue (Jewell and Wilkie, 1958; Morgan, 1977) and, therefore, depends on activation level. Further, the mechanical properties of muscle, in particular P_0, are known to depend on muscle length (Gordon et al., 1966; Abbott and Wilkie, 1953). To account for these complications one could try to generalize equation 8 as

$$\dot{x} = k^{-1}(P, P_0, x)\dot{P} - V(P, P_0, x) \tag{9}$$

Equation nine may be adequate to describe muscle response in movements where no significant lengthening takes place; it has not been tested sufficiently by experiments. It is clear that no generalized Hill model of this form, in which there is a unique correspondence between muscle force and the velocity of the contractile element at constant length and activation, is adequate to describe muscle response during stretch. To see this, one needs only to consider that in a series of constant velocity stretches at constant stimulation $\dot{x}$ and P_0 are constant and $\dot{P} = (dP/dx)\dot{x}$ so that equation 9 may be written as

$$\frac{dP}{dx} = k(P, P_0, x)\left[1 + \frac{V(P, P_0, x)}{\dot{x}}\right] \tag{10}$$

This equation states that there exists a unique slope for the tension-length trajectories at each point of the P–x phase plane. Inspection of Figure 4-1 shows that this conclusion contradicts the experimental facts: trajectories starting at neighboring points on an isometric tension-length curve will intersect, so that the trajectory slope is not unique at the point of intersection. One could try to avoid this difficulty by postulating that V is a multivalued function of P, but such salvage exercises seem to be beyond the point of diminishing returns. It appears that the classic CE/SE model should be abandoned in seeking a mathematical representation for muscle under conditions where it both shortens and lengthens more than a small fraction of its mean length under load.

If consideration is restricted to small motions about an isometric steady state, it is possible to formulate a quasilinear model that appears to give an adequate representation of the experimentally observed response of muscle, at least as measured in forced and voluntary oscillation tests. This model will be developed in detail. In Zahalak and Heyman (1979), a Hill model in the form of equation 9 (with the x-dependence suppressed) was used as a point of departure for deriving dynamic equations describing the small perturbation response of the musculoskeletal system of a limb. Here the model will simply be postulated *ab initio*, relying for justification on the fact that it provides a reasonably good representation of both the FOT and VOT experimental responses. For an individual muscle, the mathematical statement of the model consists of two linear differential equations

$$\delta\dot{P} + \tau_1^{-1}\delta P = k\delta\dot{x} + \tau_1^{-1}\delta P_0$$

$$\delta\dot{P}_0 + \tau_2^{-1}\delta P_0 = g\tau_2^{-1}\delta E(t-\tau) \qquad (11)$$

In these equations P represents the muscle force, x the length, P_0 the isometric force (used as a measure of activation), and the symbol δ denotes a perturbation (from a steady isometric state) in the variable with which it is associated. The variable E is a measure of stimulation which may be identified with the EMG from a muscle *in vivo*. The first equation models the viscoelasticlike behavior of muscle at constant activation: k is the series elastic stiffness and $k\tau_1$ is the rate of change of force per unit velocity under isotonic conditions, so that τ_1 may be termed a viscoelastic time constant. The second equation models excitation-contraction dynamics: τ_2 is a contraction time constant associated with the rate of development of isometric force, and g is simply the proportionality factor between force and stimulation under isometric steady state conditions. A small pure time delay (τ) is included to model the spread of excitation along the muscle fibers at a finite rate and any other pure time lags that may exist in the system. A block diagram representation of the model as a two-input one-output system is given in Figure 4-14. It must be emphasized that equations 11 are to be interpreted as an *incremental* model where the parameters k, τ_1, τ_2, and g may be in general functions of the steady isometric state about which the perturbations take place.

To compare the predictions of equations 11 for muscle *in vivo* one must use them to synthesize a model for a limb, and this requires certain additional assumptions. For the sake of concreteness, a model for small perturbations of the forearm about a state

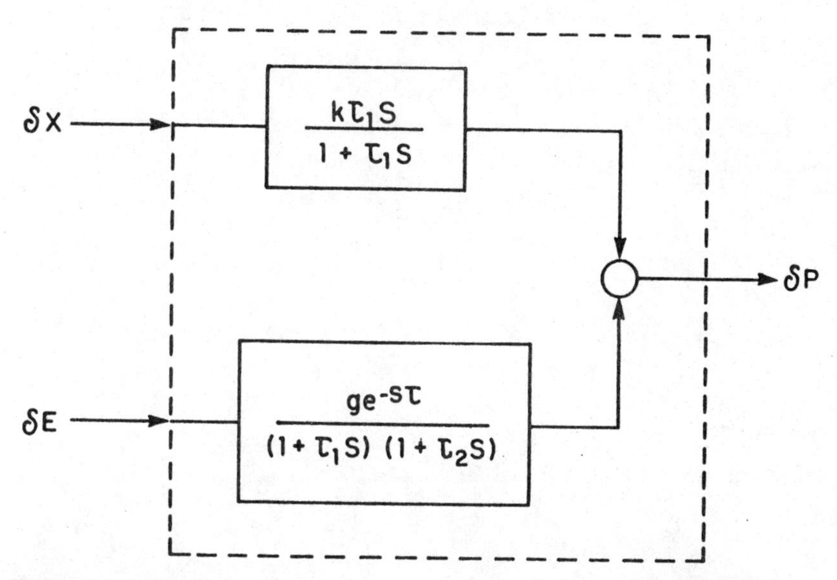

Figure 4-14. Block diagram of muscle model for perturbations about a steady state of isometric contraction. From S.C. Cannon and G.I. Zahalak, The Mechanical Behavior of Active Human Skeletal Muscle in Small Oscillations. *Journal of Biomechanics, 15*:111, 1982.

of steady isometric contraction at 90 degrees flexion will be developed. The moment M_i about the elbow joint contributed by the i-th muscle is equal to a_iP_i where a_i is the moment arm of that muscle: a_i is a function of the angle of flexion θ. Thus the muscle moment perturbation is equal to $\delta M_i = a_i\delta P_i + P_i\delta a_i$. It is easy to show that $a_i = \pm (dx_i/d\theta)$, where the positive sign applies to the extensors and the negative sign applies to the flexors. Further, if $a_i(\theta)$ is close to a stationary value, as will be true for the major flexors and extensors when $\theta = 90°$, $\delta a_i = (da_i/d\theta)\delta\theta$ will be a small quantity and will be a neglected leaving $\delta M_i = a_i\delta P$ approximately. If use is made of these facts after multiplying both sides of eqs. 11 by a_i, one obtains the dynamic equations for the moment contributed by a particular muscle about the elbow joint.

$$\begin{aligned} \delta\dot{M}_i + \tau_{1i}^{-1}\delta M_i &= n_iK_i\delta\dot{\theta} + \tau_{1i}^{-1}\delta M_{oi} \\ \delta\dot{M}_{oi} + \tau_{2i}^{-1}\delta M_{oi} &= \tau_{2i}^{-1}\Gamma_i\delta E(t-\tau_i) \end{aligned} \qquad (12)$$

where $K_i = a_i^2 k_i$ is the rotational stiffness of the muscle, $\Gamma_i = a_i g_i$, $\delta M_{oi} = a_i \delta P_{oi}$, and $n_i = -1$ for the flexors and $+1$ for the extensors. These two equations may be combined into the single equation

$$L_i \delta M_i = n_i K_i (\delta \ddot{\theta} + \tau_{2i}^{-1} \delta \dot{\theta}) + \tau_{1i}^{-1} \tau_{2i}^{-1} a_i g_i \delta E_i (t - \tau_i) \qquad (13)$$

where L_i is the second-order linear differential operator

$$L_i = \frac{d^2}{dt^2} + (\tau_{1i}^{-1} + \tau_{2i}^{-1}) \frac{d}{dt} + \tau_{1i}^{-1} \tau_{2i}^{-1}.$$

The equation of motion of the forearm is

$$I \delta \ddot{\theta} = \delta M - \sum_i n_i \delta M_i \qquad (14)$$

where I is the moment of inertia of the forearm about the elbow joint and δM is the moment perturbation of nonmuscular origin.

As they stand eqs. 13 and 14 would require a knowledge of the externally applied moment, the stimulation (EMG) of each muscle, as well as a complete set of stiffnesses, time constants, etc. in order to determine the motion of the forearm. While this presents no particular computational problem, it is clearly intractable from an experimental point of view. Progress can be made if one accepts two additional assumptions that do not seem unreasonable in first approximation and which are no more restrictive than the assumptions already inherent in the model:

1. The time constants τ_{1i}, τ_{2i} and the time delay τ_i have approximately the same values for all participating muscles.
2. The perturbations in stimulation (EMG), whether of voluntary or reflex origin, are proportional in synergistic muscles.

Then, as shown in Zahalak and Heyman (1979), eqs. 13 and 14 reduce to a single fourth-order system equation for the small-perturbation dynamics of the forearm

$$\left[\frac{d^2}{dt^2} + (\tau_1^{-1} + \tau_2^{-1}) \frac{d}{dt} + \tau_1^{-1} \tau_2^{-1}\right] (I \delta \ddot{\theta} - \delta M) + K(\delta \ddot{\theta} + \tau_2^{-1} \delta \dot{\theta})$$

$$= \tau_2^{-1} \tau_1^{-1} \left[\Gamma_f \delta E_f (t - \tau) - \Gamma_e \delta E_e (t - \tau)\right]. \qquad (15)$$

In this equation, $K = \sum_i K_i$ is the sum of the rotational stiffness of all the active contributing muscles (both flexors and extensors), and Γ_f and Γ_e are linear combinations of the Γ_i for the flexors and extensors respectively. The perturbations δE_f and δE_e are respectively those of the EMG from a representative flexor (say, the biceps or brachioradialis) and a representative extensor (the triceps). Equation 15 contains only directly *measurable* variables δM, $\delta\theta$, δE_e, δE_f and can therefore be used to predict the system performance under various experimental conditions for comparison with the observed response. The externally applied moment δM is the sum of an applied driving moment δM_a and the (usually small) passive viscoelastic resistance of the forearm-loading apparatus combination.

$$\delta M = \delta M_a - K_o \delta\theta - C_o \delta\dot{\theta}$$

Under strictly isometric conditions $\delta\theta$ is identically zero and the resultant muscle moment δM_m is equal to δM; if only one of the muscle groups is active then eq. 15 reduces to the form proposed by Gottlieb and Agarwal, eq. 6, with the modification that E is evaluated at $(t - \tau)$ rather than t (this modification affects only the phase of the response at higher frequencies).

In the forced oscillation tests at most one synergistic muscle group is active (say, the extensors) and one assumes $\delta\theta = \Delta\theta_e{}^{i\omega t}$, $\delta E_e = \Delta E_e e^{i(\omega t + \Psi)}$, $\delta E_f = 0$, and $\delta M_a = \Delta M_a e^{i(\omega t + \phi)}$. Then it can be shown that the frequency-response is given by

$$\left|\frac{K\Delta\theta}{\Delta M_a}\right| e^{-i\phi} \tag{16}$$

$$= \left[\beta_o{}^2 - \nu^2 + i2\beta_o{}^2\zeta_o\nu + \frac{i2\zeta_1\nu}{1 + i2\zeta_1\nu} + \frac{a e^{i(\psi - \tau^*\nu)}}{(1 + i2\zeta_1\nu)(1 + i2\zeta_2\nu)}\right]^{-1}$$

where $\nu = \omega/\omega_n$, $\omega_n = \sqrt{K/I}$, $\zeta_o = \tau_o\omega_n/2$, $\zeta_1 = \tau_1\omega_n/2$, $\zeta_2 = \omega_n\tau_2/2$, $\beta_o{}^2 = K_o/K$, $\tau^* = \omega_n\tau$, and $a = (\Gamma_e/K)(\Delta E_e/\Delta\theta)$.

The quantities $(\Delta E_e/\Delta\theta)$ and ψ are obtained as functions of frequency in a forced oscillation test from measurements of the reflex EMG.

To simulate a voluntary oscillation test we assume $\delta\theta = \Delta\theta_e{}^{i\omega t}$, $\delta E_f = \Delta E_f e^{i(\omega t + \phi)}$, and $\delta E_e = \Delta E_e e^{i(\omega t + \phi - \pi)}$. In this test, $\delta M_a = -\hat{K}\delta\theta$ where $\hat{K}$ is the rotational stiffness of the elastic resistance at the wrist and the passive viscoelastic resistance $-(K_o\delta\theta + C_o\delta\dot{\theta})$ can be neglected. Then, straightforward mathematical manipulation yields the frequency-response function

$$\left|\frac{\Delta M_m}{\Gamma_f \Delta E_f + \Gamma_e \Delta E_e}\right| e^{-i\phi} = (1 + i2\zeta_2\nu)^{-1} \left[(1 + i2\zeta_1\nu) + \frac{i2\zeta_1\nu}{\gamma^2 - \nu^2}\right]^{-1} e^{-i\tau^*\nu} \quad (17)$$

where $\Delta M_m = (\hat{K} - I\omega^2)\Delta\theta$ is the resultant muscle moment and $\gamma^2 = \hat{K}/K$.

The parameters of eq. 15 have been evaluated by least-squares estimation procedures (Cannon and Zahalak, 1982) for ten normal adult male subjects from the results of the frequency response studies described in the preceding section, and are listed in Table 4-II. Some additional characteristics of the subjects are listed in

Table 4-I

CHARACTERISTICS OF SUBJECTS

SUBJECT	HEIGHT (M)	WEIGHT (Kg)	MOMENT ARM (M) (1)	MOMENT OF INERTIA (Kg - m²) (2)	$(M_f)_{max}$ (N-m) (3)	$(M_e)_{max}$ (N-m) (3)
A	1.77	80.7	0.254	0.108	62	44
B	1.79	80.7	0.254	0.115	49	40
C	1.75	72.6	0.279	0.117	61	45
D	1.83	74.8	0.270	0.122	61	48
E	1.75	61.2	0.241	0.082	47	32
F	1.83	71.7	0.254	0.105	59	42
G	1.88	74.8	0.292	0.126	31	38
H	1.84	71.2	0.254	0.097	65	50
I	1.73	74.8	0.254	0.117	58	39
J	1.75	79.4	0.267	0.097	46	45

(1) Distance from the elbow joint to the wrist.

(2) Determined from the high-frequency amplitude response $\Delta\Theta/\Delta M_a$ in the FOTS. These are the combined moments of inertia of the subjects' forearms and the sheath.

(3) Isometric flexor and extensor moments at maximum voluntary effort.

From S. C. Cannon and G. I. Zahalak, The Mechanical Behavior of Active Human Skeletal Muscle in Small Oscillations. *Journal of Biomechanics* (in press).

Table 4-II

SUBJECT	(1) K_o (N-m)	(1) C_o N-m-sec	(2) $(K/M)_e$	(2) $(K/M)_f$	(3) Γ_{TR} N-m/mv	(3) Γ_{BC} N-m/mv	(3) Γ_{BR} N-m/mv	(4) τ_1 msec	(4) τ_2 msec	(4) τ msec
A	1.5	0.48	4.3	6.5	102	105	126	44	48	11
B	9.0	0.94	4.5	9.2	154	240	138	39	42	17
C	5.3	0.11	4.5	5.4	98	97	98	60	66	11
D	6.9	0.92	3.0	5.6	66	91	83	*	*	*
E	4.8	0.81	3.6	8.5	44	68	61	36	41	28
F	3.9	0.68	4.6	5.8	101	111	199	49	53	21
G	6.2	0.82	3.4	7.8	68	72	94	39	41	16
H	7.6	0.58	4.4	7.1	87	79	81	39	41	17
I	4.6	1.01	5.0	8.5	126	114	111	56	64	17
J	7.7	0.81	4.9	10.7	131	108	198	40	57	14

* There were insufficient high frequency data in the VOT results for this subject to identify the time constants accurately.
(1) Passive viscoelastic parameters.
(2) Ratios of extensor and flexor stiffness to the mean muscle moment. These are the slopes of regression lines through the origin of Figure 4-10.
(3) Ratios of the moment to the EMG under isometric steady-state conditions for the triceps, biceps, and brachioradialis respectively.
(4) Values of the time constants and time delay determined from the VOT data by least-squares estimations procedures.
From S. C. Cannon and G. I. Zahalak, The Mechanical Behavior of Active Human Skeletal Muscle in Small Oscillations. *Journal of Biomechanics* (in press).

Table 4-I. A measure of the accuracy of the model is provided by Figure 4-13 and Figure 4-15. In the former figure, the theoretical response of eq. 17 is compared to the pooled measured responses for six subjects: the theoretical curves are drawn using average values of the parameters for these subjects. Most of the scatter in this diagram reflects intersubject variability, and the data for individual subjects were much less variable. Figure 4-15 evaluates the predictions of eq. 16 in a forced oscillation test for one subject at two values of applied load, using the *measured* reflex EMG perturbations as an input. The agreement between the model and the experimental data appears to be adequate. It should be noted that eq. 15 does not give good predictions for both VOT and FOT responses if the reflex EMG is neglected.

As eq. 15 is capable of predicting the observed response in two very different situations—forced oscillations and voluntary oscillations—it can be provisionally accepted as a minimal adequate model for small-perturbation limb dynamics *in vivo*.

CLOSURE

While some progress has been made since the pioneer work of

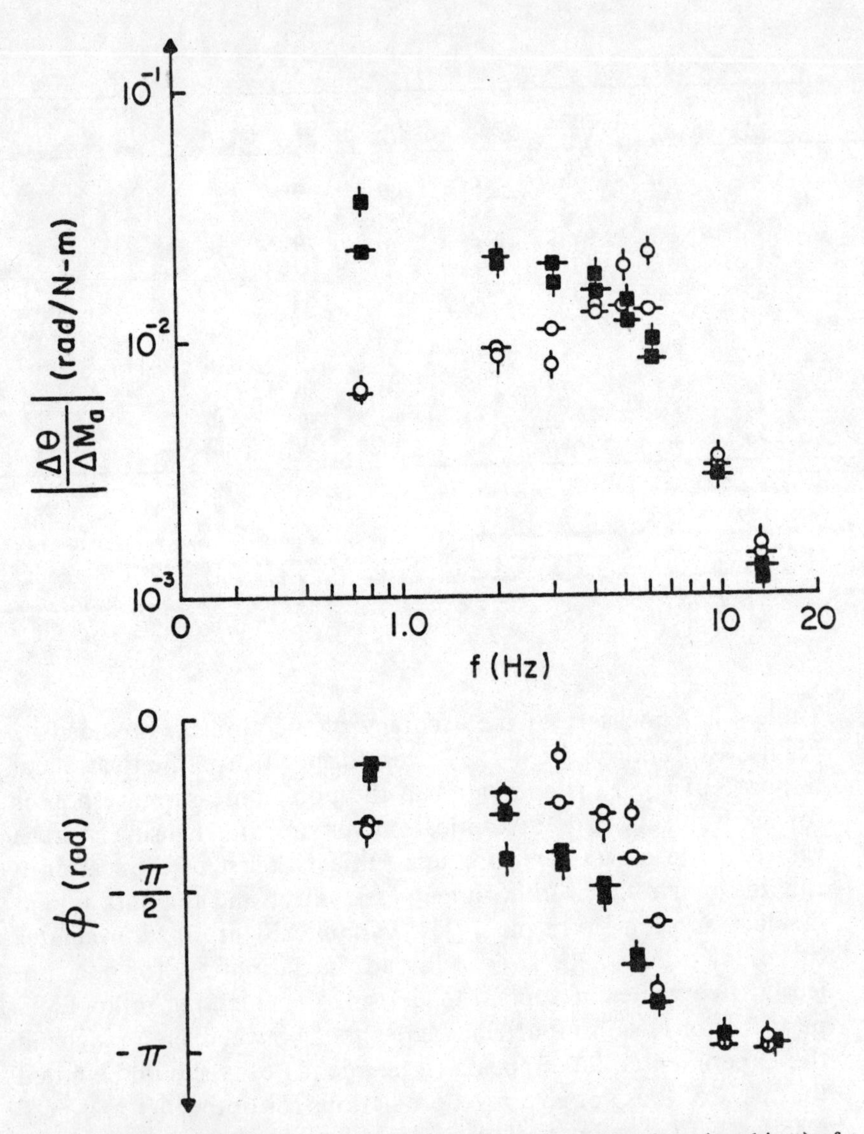

Figure 4-15. Comparison for the extensors (one normal adult male subject) of the frequency response measured in an FOT with that predicted by the model (eq. 16), using the parameters of Table 4-II and the *measured* amplitudes and phases of the reflex EMGs as inputs to the model. Symbols indicate two different levels of mean applied moment: open circles = 26 Nm and filled squares = 7.4 Nm. Horizontal marks indicate measured data and vertical marks indicate model predictions, (each point is the mean of two tests). From S.C. Cannon and G.I. Zahalak, The Mechanical Behavior of Active Human Skeletal Muscle in Small Oscillations. *Journal of Biomechanics, 15*:111, 1982.

thirty years ago when it appeared that "*no* quantitative relation between EMG and tension exists when a muscle is allowed to change in length" (Inman et al., 1952), the quantitative characterization of human skeletal muscle mechanics *in vivo* remains a very difficult problem due to the complexities of intrinsic muscle behavior, coactivation, reflex, and the processing and interpretation of time-varying EMGs. Significant quantitative correlations are available only for isometric, constant velocity constant load, and small-displacement oscillatory conditions. In the last case, the muscle behavior seems reasonably well represented by a second-order quasilinear model whose parameters, particularly the stiffness, are functions of the steady isometric state about which the oscillations take place. For unrestricted motion over the entire physiological range the quantitative description of muscle behavior *in vivo* is an open question and very much a contemporary issue in muscle mechanics.

REFERENCES

Aaron, S.L. and Stein, R.B. (1976) Comparison of an EMG-controlled prosthesis and the normal human biceps brachii muscle. *Am. J. Physical Med.*, 55:1

Abbott, B.C. and Wilkie, D.R. (1953) The relation between the velocity of shortening and the tension-length curve of skeletal muscle. *J. Physiology.*, 120:214

Agarwal. G.C. and Gottlieb, G.L. (1977) Compliance of the human ankle joint. *J. Biomechanical Engineering*, 99:166

Bawa, P., Mannard, A., and Stein, R.B. (1976) Predictions and experimental tests of a viscoelastic muscle model using elastic and inertial loads. *Biological Cybernetics*, 22:139

Bawa, P. and Stein, R.B. (1976) Frequency response of human soleus muscle. *J. Neurophysiology*, 39:788

Bigland, B. and Lippold, O.C.J. (1954) The relation between force, velocity and integrated electrical activity in human muscles. *J. Physiology*, 123:214

Bouisset, S. (1973) EMG and muscle force in normal motor activities. In *New Developments in Electromyography and Clinical Neurophysiology*. J.E. Desmedt, ed., S. Karger, Basel. 1:547

Cannon, S.C. (1980) *Mechanics and Reflex Feedback in Small*

Perturbations of a Limb: Measurements and Models. M.S. Thesis, Washington University, St. Louis

Cannon, S.C. and Zahalak, G.I. (1981) Reflex feedback in small perturbations of a limb. *A.S.M.E. Biomechanics Symposium*, AMD-Vol. 43:117

Cannon, S.C. and Zahalak, G.I. (1982) The mechanical behavior of active human skeletal muscle in small oscillations. *J. Biomechanics, 15*:111

Carlson, F.D. and Wilkie, D.R. (1974) *Muscle Physiology*. Prentice Hall, Englewood Cliffs, N.J.

Cnockaert, J.C. and Pertuzon, E. (1974) Sur la gémetrie musculoskeletique du triceps brachii. *European J. Applied Physiology*, 32:149

Coggshall, J.C. and Bekey, G.A. (1970) EMG-force dynamics in human skeletal muscle. *Medical and Biological Engineering*, 8:265

Crochetiere, W.J., Vodovnik, L., and Reswick, J.B. (1967) Electrical stimulation of skeletal muscle–A study of muscle as an actuator. *Medical and Biological Engineering*, 5:111

Crosby, P.A. (1978) Use of surface electromyogram as a measure of dynamic force in human limb muscles. *Medical & Biological Engineering & Computing*, 16:519

De Luca, C.J. and VanDyk, E.J. (1975) Derivation of some parameters of myoelectric signals recorded during sustained constant force isometric contractions. *Biophysical Journal*, 15: 1167

Emerson, N.D. and Zahalak, G.I. (1981) Longitudinal electrode arrays for electromyography. *Medical & Biological Engineering & Computing* (in press)

Gordon, A.M., Huxley, A.F., and Julian, F.J. (1966) The variation of isometric tension and sarcomere length in vertebrate muscle fibers. *J. Physiology*, 184:170

Gottlieb, G.L. and Agawal, G.C. (1971) Dynamic relationship between isometric muscle tension and the electromyogram in man. *J. Applied Physiology*, 30:345

Goubel, F. and Pertuzon, E. (1973) Evaluation de l'élasticité du muscle in situ par une méthode de quick-release. *Archives Internationales de Physiologie et de Biochimie*, 81:697

Hatze, H. (1977) A myocybernetic control model of skeletal muscle. *Biological Cybernetics*, 25:103

Hatze, H. (1978) A general myocybernetic control model of skeletal muscle. *Biological Cybernetics*, 28:143

Hill, A.V. (1938) The heat of shortening and the dynamic constants of muscle. *Proceedings of the Royal Society*, London, Ser. B, 126:136

Hill, A.V. (1949) The abrupt transition from rest to activity in muscle. *Proceedings of the Royal Society*, Ser. B, 136:399

Hogan, N.J. (1976) *Myoelectric Prosthesis Control: Optimal Estimation Applied to EMG and the Cybernetic Considerations for its Use in a Man-Machine Interface*. Ph.D. thesis, Massachusetts Institute of Technology, Cambridge

Huxley, A.F. (1974) Review lecture: Muscular contraction. *J. Physiology*, 243:1

Inman, V.T., Ralston, H.J., Saunders, J. B de C.M., Feinstein, B. and Wright, E.W., Jr. (1952) Relations of human electromyogram to muscular tension. *Electroencephalography and Clinical Neurophysiology*, 4:187

Jewell, B.R. and Wilkie, D.R. (1958) An analysis of the mechanical components in frog's striated muscle. *J. Physiology*, 143: 515

Joyce, G.C. and Rack, P.M.H. (1969) Isotonic lengthening and shortening movements of cat soleus muscle. *J. Physiology*, 204:475

Joyce, G.C., Rack, P.M.H., and Westbury, D.R. (1969) The mechanical properties of cat soleus muscle during controlled lengthening and shortening movements. *J. Physiology*, 204: 461

Katz, B. (1939) The relation between force and speed in muscular contraction. *J. Physiology*, 96:45

Komi, P.V. (1973) Relationship between muscle tension, EMG and velocity of contraction under concentric and eccentric work. *New Developments in Electromyography and Clinical Neurophysiology*. J.E. Desmedt, ed., S. Karger, Basel. 1:596

Lestienne, F., and Pertuzon, E. (1974) Determination, in situ, de l' élasticité du muscle humain inactive. *European J. Applied Physiology*, 32:159

Lippold, O.C.J. (1952) The relation between integrated action potentials in a human muscle and its isometric tension. *J. Physiology*, 117:492

Mannard, A. and Stein, R.B. (1973) Determination of the frequency response of isometric soleus muscle in the cat using random nerve stimulation. *J. Physiology*, 229:275

Mashima, H., Akazawa, K., Kushima, H., and Fujii, K. (1972) The force-load-velocity relation and the viscous-like force in the frog skeletal muscle. *Japanese J. Physiology*, 22:103

Matthews, P.B.C. and Stein, R.B. (1969) The sensitivity of muscle spindle afferents to small sinusoidal changes in length. *J. Physiology*, 200:723

Morgan, D.L. (1977) Separation of active and passive components of short-range stiffness of muscle. *Am J. Physiology*, 232:C45

Neilson, P.D. (1972) Frequency-response characteristics of the tonic stretch reflexes of biceps brachii muscle in intact man. *Medical and Biological Engineering*, 10:460

Partridge, L.D. (1965) Modification of neural output signals by muscles: A frequency-response study. *J. Applied Physiology*, 20:150

Partridge, L.D. (1966) Signal-handling characteristics of load-moving skeletal muscle. *Am J. Physiology*, 210:1170

Person, R.S. and Libkind, M.S. (1970) Simulation of electromyograms showing interference patterns. *Electroencephalography and Clinical Neurophysiology*, 28:625

Pertuzon, E. and Bouisset, S. (1973) Instantaneous force-velocity relation in human muscle. *Biomechanics III*. 3rd International Seminar, Rome, 1971, S. Karger, Basel. 8:230

Rack, P.M.H. and Westbury, D.R. (1969) The effect of length and stimulus rate on tension in the isometric cat soleus muscle. *J. Physiology*, 204:443

Rack, P.M.H., and Westbury, D.R. (1974) The short range stiffness of active mammalian muscle and its effect on mechanical properties. *J. Physiology*, 240:331

Ritchie, J.M. and Wilkie, D.R. (1958) The dynamics of muscular contraction. *Journal of Physiology*, 143:104

Soechting, J.F. and Roberts, W.J. (1975) Transfer characteristics between EMG activity and muscle tension under isometric conditions in man. *J. Physiology*, Paris. 70:779

Soechting, J.F., Stewart, P.A., Hawley, R.H., Paslay, P.R., and Duffy, J. (1971) Evaluation of neuro-muscular parameters describing human reflex motion. *J. Dynamic Systems, Measurement, and Control*, 93:221

Stein, R.B. (1974) Peripheral control of movement. *Physiological Reviews*, 54:215

Tennant, J.A. (1971) The dynamic characteristics of human skeletal muscle modeled from surface stimulation. *Technical Report NASA CR–1691*. Stanford University

Trncoczy, A., Bajd, T., and Malezic, M. (1976) A dynamic model of the ankle joint under functional electrical stimulation in free movement and isometric conditions. *J. Biomechanics*, 9:509

Vredenbregt, J. and Rau, G. (1973) Surface electromyography in relation to force, muscle length and endurance. *New Developments in Electromyography and Clinical Neurophysiology*, J.E. Desmedt, ed., S. Karger, Basel. 1:607

Wilkie, D.R. (1950) The relation between force and velocity in human muscle. *J. Physiology*, 110:249

Zahalak, G.I., Duffy, J., Steward, P.A., Litchman, H.M., Hawley, R.H., and Paslay, P.R. (1976) Partially activated human skeletal muscle: An experimental investigation of force, velocity and EMG. *J. Applied Mechanics*, 98:81

Zahalak, G.I., Duffy, J. Stewart, P.A., and Paslay, P.R. (1975) The apparent internal friction of human skeletal muscle. *Workshop on the Biomechanics of Voluntary Human Motion*. W.H. Boykin Jr., ed. Technical Report IWVHM - Proc - 75, University of Florida, Gainesville. p. 21

Zahalak, G.I., El-Hindawy, K., Steward, P.A., Hawley, R.H., and Duffy, J. (1975) Measurement of the series elasticity of human muscle *in vivo. Proceedings of the ASME Biomechanics Symposium*, AMD-Vol. 10:83

Zahalak, G.I. and Heyman, S.J. (1979) A quantitative evaluation of the frequency-response characteristics of active human skeletal muscle *in vivo. J. Biomechanical Engineering*, 101: 28

Chapter 5

ELECTRICAL STIMULATION OF PARALYZED MUSCLE

A Practical Application of Muscle Mechanics

J.S. PETROFSKY and C.A. PHILLIPS

HISTORICAL BACKGROUND

UNLIKE simple movements, such as opening the eyelids or lifting a finger, movements such as walking or standing require the use of many muscle groups. To complicate matters further, these muscle groups contract and relax out of phase with one another, making the control of muscle movement a complex problem. In the healthy individual, co-ordination of movement is accomplished by the interaction between joint and muscle receptors, tendon receptors, vestibular receptors, and integration centers in the spinal cord and the brain. However, when the normal neural pathways are disrupted due to spinal injury, paralysis below the level of the injury results. Although the central command is no longer present to the alpha motor neuron pool below the level of the injury, if the cell bodies of these neurons have been undamaged, the motor units remain healthy and unharmed. How-

---

ever, the muscle fibers decrease in size due to disuse atrophy as is often seen following tenotomy or immobilization of the muscles in a cast. Thus, although healthy motor neurons and muscles are present, the individual remains paralyzed, often for life, due to the absence of central command.

Through the first World War, most types of spinal injury did not pose a long-term problem to the health community, since, up until this time, most cases of spinal cord injury resulted in death. For example, of the incidents of spinal injury from stab and gunshot wounds in the American and British armies in World War I, the mortality rate was 80 percent and 47 and 65 percent respectively for these two armies (Vellacott and Webb-Johnson, 1919; Cushing, 1927). The reason for this high incidence of death has been attributed to the lack of antibiotics and the inability of the medical community to deal with shock (Burke and Murray, 1975; Guttman, 1976). With the discovery of penicillin and the sulfonamides and a better understanding of the mechanisms of spinal and cardiovascular shock, the incidence of death from spinal injury has dropped dramatically (Sunderlund, 1968; Guttman, 1976).

In addition to the increase in the rate of recovery immediately following spinal injury, there has been a dramatic increase in the longevity of these patients as well. During World War I, of the few patients who survived spinal injury, the mortality rate was 80 per cent in the first three years. However, due to the current advances in medicine, especially due to chemotherapy, the problems of spasticity, urinary tract infections, bed sores, hypotension, paralytic ileus, renal failure, and other disorders associated with spinal injury can be managed quite well. With the present level of proficiency in clinical medicine, then, patients who have incurred some sort of spinal injury may expect to live a normal life span (Cox and Grubb, 1974, Burke and Murray, 1975; Guttman, 1976; Smart and Sanders, 1976).

Although the United States is not currently involved in a war, the number of veterans with incurable spinal cord injuries from World War II, the Korean War, and the Viet Nam War is alarming. What is more alarming, however, is the high incidence of spinal cord injuries that occur today without a major war. These injuries occur from gunshot and stab wounds, compression and stretch injuries (as occur in sports), disease-related lesions of the spinal cord, and auto accidents. The incidence of incurable spinal injury from auto accidents alone averages between 2500 and 3500 cases per year in the United States (Smart and Sanders, 1976). Although these numbers are small compared to the number of cases of cancer reported each year (annual incidence is over 1.5 million cases) (Smart and Sanders, 1976), there are two striking features of these spinal injuries. First, the average patient who incurs a spinal injury is between thirty-one and thirty-six years of age (Smart and Sanders, 1976). Second, as a result of the young age and the low mortality rate for these patients, it has been estimated that the total lifetime cost of maintaining a spinal patient including the initial hospitalization was estimated at $350,000 as far back as 1960 (Burke, 1960) and, with the current trend in the price of medical care, the current figure probably exceeds a million dollars.

With the current longevity of these patients, it has become increasingly important to provide a means of mobility to allow them to move in their environments. For the quadriplegic, this is indeed a difficult task, especially since they may require a respirator for life support if the lesion or injury has occurred in or above the cervical spinal cord. However, for the paraplegic, the wheelchair has been the most commonly used means of mobility. With the aid of their wheelchairs and specially designed cars with hand controls, these patients are able to move with some degree of freedom in their environment. However, in spite of legislation to make buildings more accessible to the wheelchair patients, their lives are still restricted, and, due to loss of mobility and the paralysis of their bodies, they often suffer a marked change in their psychological profiles (Burke and Murray, 1975).

Several alternate forms of mobility have also been tested. One of these is the active exoskeleton. This type of device has been under study in several labs (Scott, 1968; Vukobratovic et al., 1974; Townsend and Lepofsky, 1976) and involves the placement of the patient in a metal frame work of hydraulic levers and servo motors that can be used to move the device. It is the aim of this research to develop these "vertical wheelchairs" into mobile supportive devices within which the patient can move around or even climb steps. The limiting factors of these devices lie in their size and high power utilization (Reswick and Vodovnik, 1967; Scott, 1968; Vukobratovic et al., 1974).

Another approach to the restoration of mobility is by replacing the lost central command to the alpha motor neurons with direct electrical stimulation of the muscle. Although electrical stimulation of paralyzed muscle has been used since 1840 to reduce the spasticity associated with spinal injuries (Guttman, 1976), it was not until the early 1960s that studies began to be published concerning the restoration of movement by electrical stimulation Liberson et al., 1962; Long and Masciacrelli, 1963; Trenkoczy et al., 1976; Zealer and Dedo, 1977).

However, direct electrical stimulation can only be used in cases where the cell bodies of the alpha motor neurons remain unaffected by the injury. With the exception of the immediate area around the injury, this is normally the case when the injury results from a bullet or stab wound to a restricted area of the spinal cord or due to a tumor. Under these conditions, although the interneurons connecting the higher and the lower levels of the spinal cord are destroyed, the cell bodies of the lower motor neurons are uninjured (Guttman, 1976). As a result, the alpha motor neurons and their associated muscle fibers remain healthy (Sunderland, 1968; Guttman, 1976). However, since these motor neurons lose their normal pattern of central nervous traffic, the muscles atrophy from disuse and the fast twitch motor units show a reduction in their contraction speed and increase in their oxidative capacity (Vrobova, 1963; Sunderland, 1968; Brown, 1973; Dubowitz and Brooke, 1974; Van der Meulen et al., 1974). When stimulated electrically, then, these muscles respond in a manner typical of slow twitch muscle at first, but, with repeated electrical stimulation, will begin to hypertrophy and take on the biochemical and physiological appearance of normal mixed muscle (Vrobova, 1963; Sunderland, 1968; Cherepakhin et al., 1977). In contrast, if the injury has destroyed the cell bodies of the motor neurons, the muscle suffers irreversible denervation atrophy (Guttman, 1976; Dubowitz and Brooke, 1974).

In the tetraplegic, for example, if the injury has occurred high enough in the spinal cord, although the motor neurons in the phrenic nerves are not injured, there is no voluntary respiration since central transmission has been interrupted (Guttman, 1976). To restore a diaphragmatic breathing pattern, several investigators have employed either direct electrical stimulation of the phrenic nerves (Sarnoff et al., 1948a; 1948b; Glenn et al., 1964; Van Heeckeren and Glenn, 1966; Stemmer et al., 1967; Judson and Glenn, 1968; Glenn et al., 1970; 1973) or intravenous or percutaneous phrenic nerve stimulation (Daggett et al., 1966; 1970). This form of treatment has also proven useful to enhance the depth of respiration in chronic hypoventilation (Stemmer et al., 1967; Glenn et al., 1970; 1973).

Another problem area for both paraplegics and tetraplegics lies in the loss of control of voluntary bladder emission (Guttman, 1976). To resolve this problem, several groups of investigators have attempted to apply electrical stimulation to the bladder to restore voluntary movement (Timm and Bradley, 1969; 1971; 1973; Nashold et al., 1972; Nauman and Milner, 1978). In the case of these muscles, the electrodes have been apllied directly to the bladder wall. However, the use of this procedure has been limited by surgical complications in attaching the wires and maintaining a good electrode contact to the muscles of the bladder (Guttman, 1976).

Similar attempts have been made to provide proportional control to the skeletal muscle used to move the appendages. This has ranged from electrical stimulation of both agonist and antagonist muscles to restore stability to the joints for postural control of partly paralyzed patients to the suggested use of complete electrical stimulation to restore both hand and leg movement (Crochetiere et al., 1967; Vodovnik et al., 1967; McNeal et al., 1969; Milner et al., 1970; Kiwerski, 1973; Rebersek and Vodovnik, 1973; Peckham, 1976; Petrofsky et al., 1976; Petrofsky, 1978; 1979a; 1979b; 1980; Solomonow et al., 1978). To restore the mobility of these muscles, it has been suggested that the muscles be stimulated by electrodes placed on the skin (Crochetiere et al., 1967; Milner et al., 1970). However, this form of stimulation requires large voltages and currents that can burn and irritate the skin (Scott, 1968). For this reason, other investigators, e.g. Peckham (1976), have suggested using wire intramuscular electrodes to stimulate the muscle. Although the stimulation currents and voltages are still high, this offers an improvement over the use of surface electrodes. However, due to the movement of the muscles during contraction, Peckham has reported that the electrodes are hard to keep in place (Peckham, 1976). A different approach has also been suggested by several other investigators (Petrofsky et al., 1976; Petrofsky, 1978; 1979a; 1979b; Solomonow et al., 1978) involving the stimulation of the motor nerve itself. This technique offers the advantage of using still lower stimulation currents and voltages (stimulation voltages necessary to recruit all the motor units in the muscle typically average less than 1 volt) (Petrofsky, 1978).

DESIGN CRITERIA

If electrical stimulation is to be applied to paralyzed muscle at least four basic criteria (Fig. 5-1) must be met. First, to enable a patient to stand in place or hold objects, a potential stimulation system must be able to exert fine control over the tension developed by paralyzed muscle during isometric contractions. However, although good proportional control is an important design objective and has been achieved in some of the studies listed above, it is equally important to stimulate the muscle in a manner that will not rapidly fatigue the muscle (a design criteria rarely satisfied in any of these same studies). This is particularly important in the design of systems to potentially enable a patient to stand, in that, without a low degree of fatigability of his muscles

Isometric Exercise
1. proportional control of tension
2. fatigue

Dynamic Exercise
1. velocity control
2. fatigue

Figure 5-1. Four criteria for re-establishing movement in paralyzed muscle.

the patient would rapidly fall down. Another design criterion is to develop a system that enables a paralyzed patient to move his muscles in a co-ordinated fashion, i.e. the system must be able to precisely control the velocity of movement of the paralyzed muscles. Furthermore, as stated above, this control technique must be able to exert such control while inducing only minimal muscle fatigue.

CONTROL OF MUSCLE DURING ISOMETRIC CONTRACTIONS

Our first objective was to develop a proportional control technique for muscle during isometric contractions.

From the time of Sherrington (1894), electrical stimulation of muscle has typically employed the use of a single pair of synchronously firing electrodes either applied directly to the muscle or to its motor nerve (indirect stimulation). To develop smooth contractions in the muscle with this form of stimulation requires frequencies of stimulation typically as high as 100 Hertz, whereas the stimulation frequency which is necessary to tetanize the muscle fully can be as high as 300 Hertz (Brown and Burns, 1949). These frequencies are far in excess of the normal physiological frequency range associated with voluntary motor unit recruitment. During voluntary activity, the frequency of firing of motor neurons ranges from 5 to 15 Hertz during weak isometric contractions (Bigland and Lippold, 1954, Edwards and Lippold, 1956; Milner-Brown and Stein, 1975) to typically as high as 50 Hertz during powerful isometric contractions (Milner-Brown and Stein, 1975), although brief periods where frequencies of motor unit discharge as high as 70 Hertz have been noted (Marsden, Meadows and Merton, 1971). Further, unlike voluntary tensions that can be maintained steadily for long periods of time (Lind, 1959; Rohmert, 1968), stimulation at high frequencies leads to the rapid onset of fatigue, the tension falling within seconds to only a fraction of the original value: presumably this occurs due to neuromuscular blockade (Brown and Burns, 1949). For these reasons, synchronous stimulation appears to be a poor means of reactivating paralyzed muscle.

During voluntary activity, however, motor units are rarely recruited synchronously (Milner-Brown and Stein, 1975). Instead, the normal pattern of motor unit activation is asynchronous in nature. Curiously, it was not until the late 1960s that any physiological studies of the effect of asynchronous stimulation on the physiological properties of muscle were examined (Rack and Westbury, 1969). By dividing the ventral roots of the spinal cord innervating the soleus muscle into three equally sized populations and then stimulating these in turn, Rack and Westbury (1969) found that the muscles were able to develop smooth contractions at stimulation frequencies of 5 Hertz, whereas synchronous stimulation at this frequency resulted in only a series of unfused muscle twitches.

The soleus muscle differs from most muscles in that it is composed of only slow twitch motor units. Therefore, in our own work, we examined the plantaris and medial gastrocnemius muscles (largely fast twitch muscles) as well as the soleus of the cat and found that they too could be tetanized by sequential stimulation. We also found that submaximal tensions could be maintained smoothly for a length of time similar to that seen during voluntary isometric contractions in man (Petrofsky and Lind, 1979; Lind and Petrofsky, 1979: Petrofsky, 1978; 1979a). During these contractions, no neuromuscular blockade was found throughout the length of time the contractions were sustained.

However, division of the ventral roots of the spinal cord is far too severe a procedure to be used in paraplegics. For this reason, Peckham (1976) tried to obtain the advantage of sequential stimulation by injecting three electrodes directly into the muscle. Timm and Bradley (1969, 1971, 1973) and Nauman and Milner (1978) also recognized the advantage of sequential stimulation and tried to use it to stimulate the paralyzed bladder.

Although muscle control was realized at lower stimulation frequencies (as cited previously), due to difficulties with the electrode placement on moving muscle, these techniques have proven to be difficult to use.

However, in our lab, we have developed a technique to stimulate muscle sequentially through its motor nerve (Petrofsky, 1978). To accomplish this end, a sleeve electrode is inserted over the peripheral motor nerve innervating a muscle. Inside the sleeve electrode are three platinum contact wires that run for a distance of 1 mm parallel to the motor nerve. Each of the three electrodes is alternately selected as the active electrode while the other electrodes serve as the reference. In this manner, groups of motor units are alternately stimulated to produce the beneficial effects of sequential stimulation. To vary the tension developed in the muscle, the stimulation voltage is increased to stimulate more motor units. Ideally, the electrode would divide the nerve into three equal populations of motor neurons (Fig. 5-2). However, with sufficient stimulation voltage to "recruit" all the motor units, there is double and triple stimulation of a small percent of the motor units. This can be reduced by shaping the electric stimulation fields in a triangular pattern by adding three additional reference electrodes as shown in Panel A of Figure 5-3. To stimulate the muscle, two electrodes are always "floating" while one becomes active and two others are used as references.

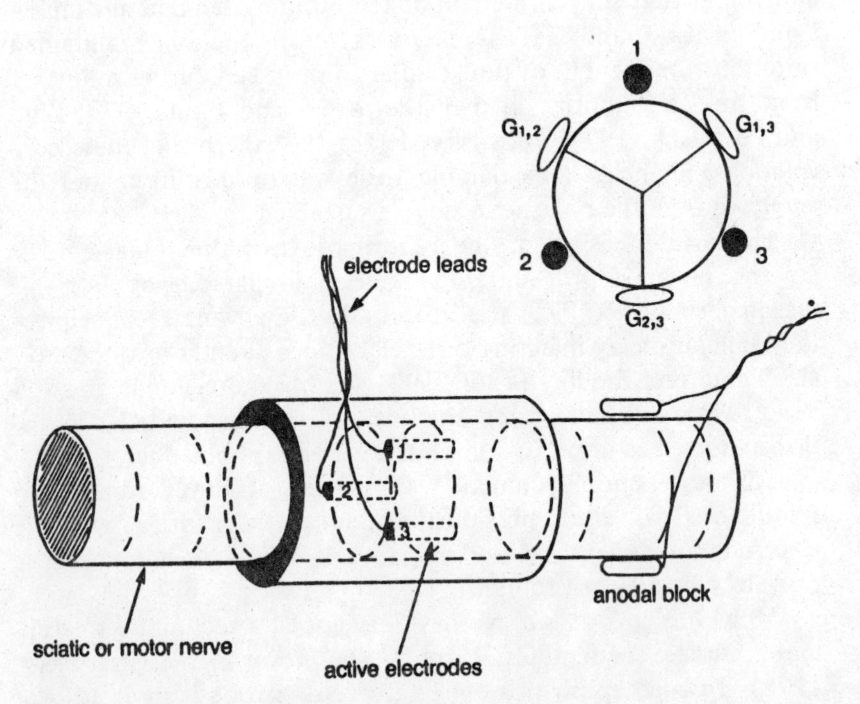

Figure 5-2. A representation of the sleeve electrode.

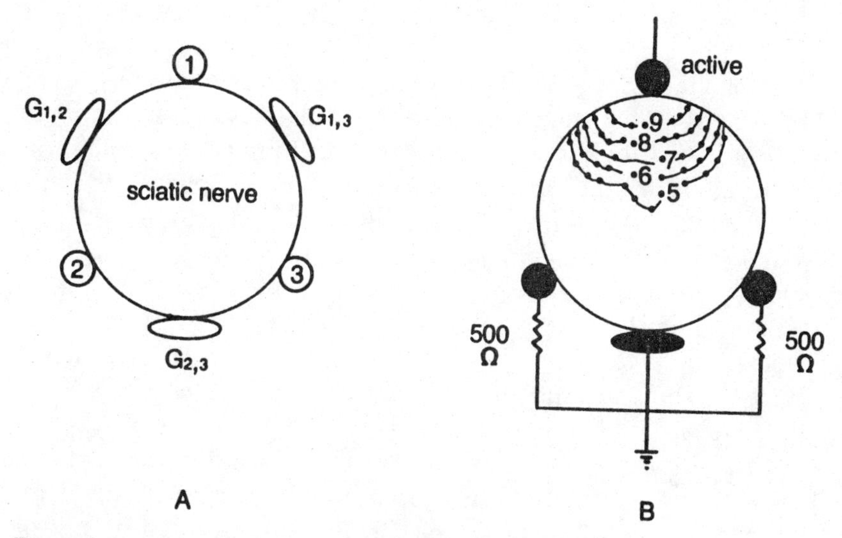

Figure 5-3. A cross-section of the sequential electrode. From J.S. Petrofsky, Sequential Motor Unit Stimulation Through Peripheral Motor Nerves. *Medical and Biological Engineering and Computing, 17*:87-93, 1979.

A resistor network is switched electronically between the three stimulating electrodes such that as each electrode becomes active, the other two are connected to ground through the network. At the same time, the reference electrode between these two electrodes is grounded directly (*see* Fig. 5-3, Panel B). As each of the three original electrodes becomes active, the network is switched appropriately around the motor nerve.

When the stimulating current fields were measured with a glass micro-electrode, they matched those desired. (Fig. 5-3, Panel A shows the fields generated during the stimulation of one of the three electrodes.) Using this electrode, the muscles were able to develop smooth contractions at stimulation frequencies of 10 Hertz or less and were tetanized at stimulation frequencies of only 35 Hertz, as had been reported for sequential stimulation through the spinal cord (Petrofsky, 1978). When recordings were made from individual motor units within the muscle to check for double firing of the motor neurons, it was found that even when the stimulation voltage was increased to the level necessary to stimulate all the motor units in the muscle; multiple stimulation of motor neurons was always restricted to less than 5 percent of the motor units.

From the above, then, it would appear that sequential motor unit stimulation through peripheral motor nerves can mimic the normal asynchronous stimulation pattern found for skeletal muscle. In addition, the force of isometric contractions can be varied by adjusting the stimulation voltage to the electrode since this would result in an increase in the intensity of the stimulation field (Petrofsky, 1978; 1979a). It is also necessary to be able to sustain isometric contractions. Although initially a tension can be achieved by a combination of stimulation voltage and frequency, as contractions are maintained, motor units will fatigue and the stimulator must compensate for this.

In man, during brief isometric contractions, tension is graded by recruiting the motor units in an orderly fashion by size from the smallest motor units to the largest motor units respectively (slow twitch motor units), at less than about half of our strength (Bigland and Lippold, 1954; Olsen et al., 1968; Milner-Brown and Stein, 1975). Once a motor unit is recruited, its frequency of firing (range 5 to 10 Hz) remains almost constant until all of the motor units within the muscle are active. Due to the recruitment of motor units alone, a muscle can develop as much as one half of its maximum isometric strength (Milner-Brown and Stein, 1975).

To develop the remaining half of a muscle's strength, the frequency of firing of the motor units is increased by the central nervous system to frequencies as high as 40 to 50 Hertz (Bigland and Lippold, 1954; Milner-Brown and Stein, 1975). This simple pattern of neural activity is believed to occur during sustained isometric contractions (Bigland and Lippold, 1954). Under these circumstances, a target tension is first reached, by a combination of recruitment and firing frequency as described above. However, as the contraction is sustained, motor units begin to fatigue . Tension is maintained by first recruiting more motor units and, finally, after all motor units are recruited, by increasing the frequency of firing of the alpha motor-neuron pool.

To see if this same pattern could be induced by electrical stimulation, a microprocessor controlled stimulator was designed (Fig. 5-4) and tested on the medial gastrocnemius muscle of the

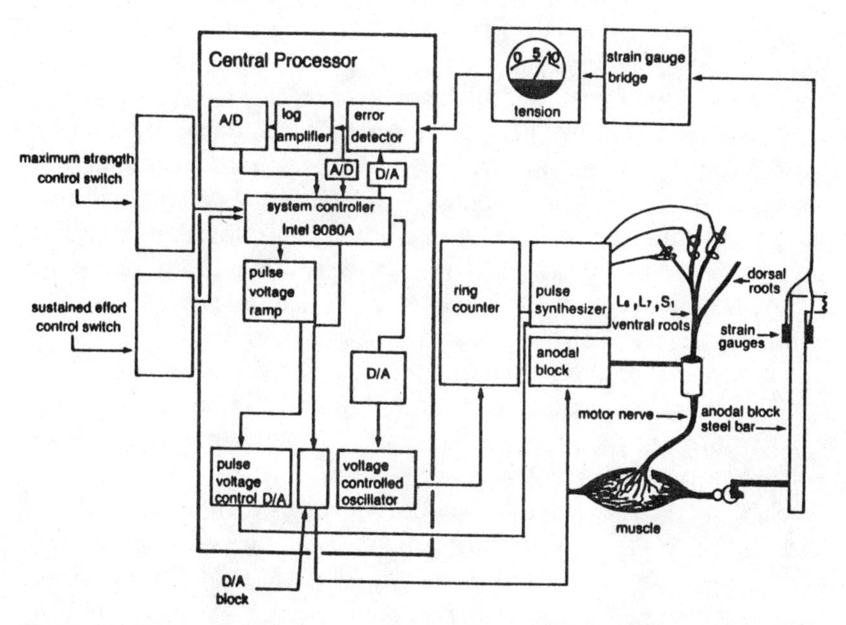

Figure 5-4. Schematic of the microprocessor controlled stimulator.

cat, since this muscle has a similar fiber composition (about 50% fast twitch and 50% slow twitch motor units) as is found in most muscles in man (Ariano et al., 1973: Dubowitz and Brooke, 1974).

The first problem was to enable the computer to set the recruitment order of the motor units by size from the smallest (slowest) to the largest (fastest) respectively. This was accomplished by adding an anodal block electrode to the sequential electrode array just proximal to the muscle. The voltage to the electrode array was increased until the maximum tetanic tension of the muscle was reached. The block voltage was then increased until the muscle was fully relaxed (all motor units were blocked). Since fast twitch motor units are usually more susceptible to anodal block than slow twitch units (Petrofsky, 1979a), removing the block had the effect of recruiting the motor units by size from the smallest to the largest motor units respectively. The pattern of stimulation used during the sustained contractions was then based on that reported for voluntary activity in man. To achieve a given tension, recruitment was increased by the computer; the stimulation frequency was held constant at 10 Hertz. Once the desired tension was reached, recruitment was increased to maintain the tension as motor units began to fatigue (sensed by a tension input into the computer). Once all motor units were recruited, the computer increased the frequency of stimulation to all the motor units to maintain the tension. When this was no longer effective, the contractions were terminated. The length of time the tension was sustained was called the endurance time.

To test the program, each of four medial gastrocnemius muscles in each of four cats was stimulated as described above to enable the muscles to develop brief and sustained isometric contractions at tensions ranging between 10 and 100 per cent of each muscle's maximum tetanic strength (determined by tetanic stimulation of all of the motor units). During 3 sec contractions the muscles were able to sustain smooth isometric contractions at all tensions examined. The tension recorded from a typical muscle is shown in Figure 5-5. Further, the program was able to cause the muscle to maintain smoothly these tensions to fatigue.

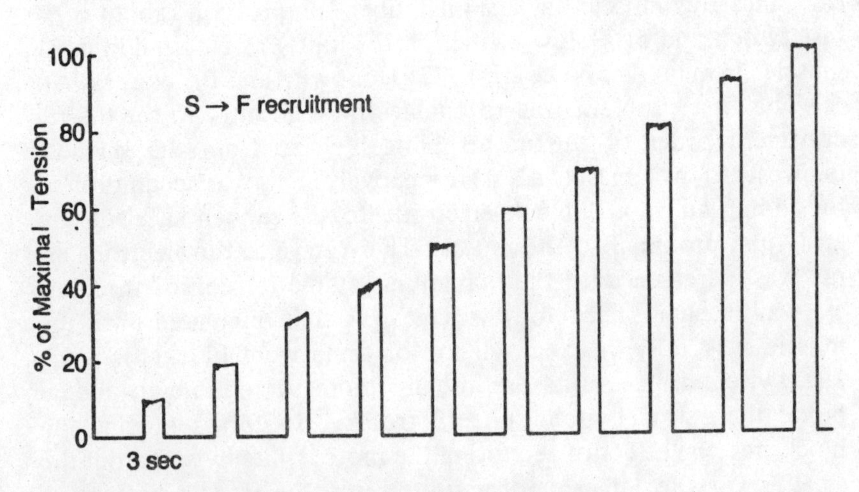

Figure 5-5. The tension developed during brief isometric contractions (with recruitment proceeding from the slowest to fastest units).

The average endurance of all four muscles is shown in Figure 5-6 with the respective standard deviations. As a basis for compari-

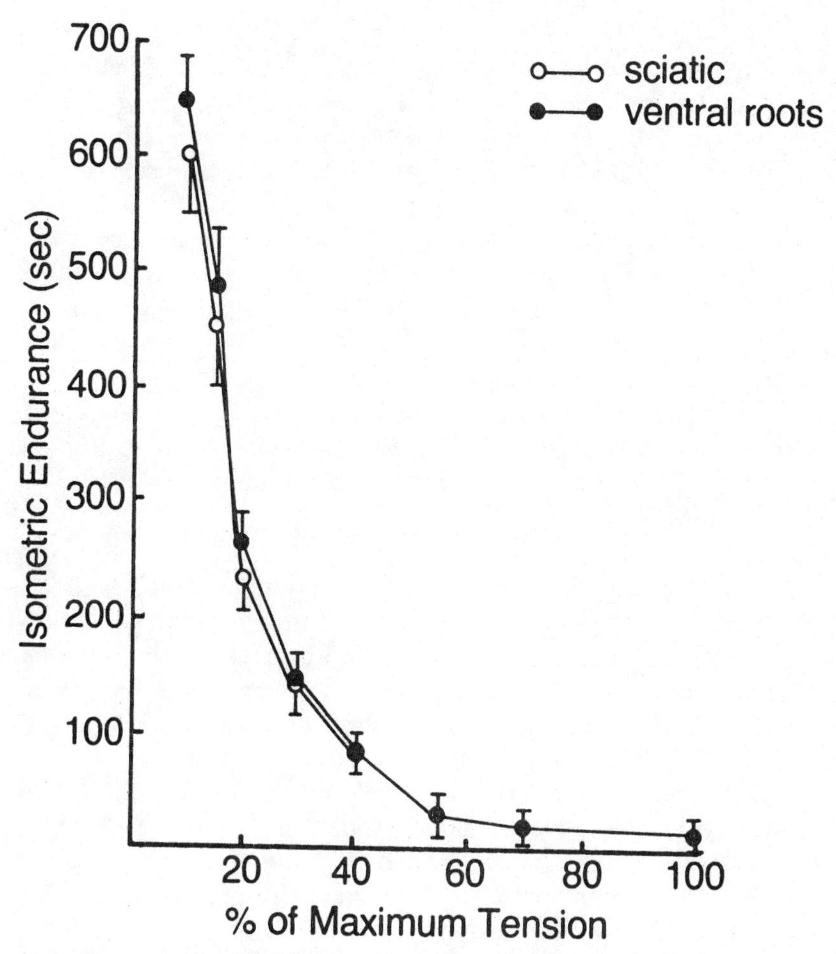

Figure 5-6. Average isometric endurance in the medial gastrocnemius muscle of 4 cats (○) compared to four human subjects (●). From J.S. Petrofsky, Control of the Recruitment and Firing Frequencies of Motor Units in Electrically Stimulated Muscles in the Cat. *Medical and Biological Engineering and Computing, 16*:302-308, 1978.

son, the isometric endurance of the cats (○) was compared to that of the handgrip muscles of four human subjects (●). The similarity between the endurance in the cat mixed muscle (gastrocnemius) and the handgrip muscle in man was striking. In both man and the cat, the muscles fatigued rapidly when the tension exerted was greater than about 10 percent of the muscle's maximum strength. Of particular interest was the fact that in both man and cat, the muscles were able to sustain isometric tensions for indefinite periods of time with no sign of fatigue when the tension exerted by the muscles was less than 10 percent of the muscle's maximum strength. In man, this phenomenon allows us to maintain the tension necessary to control posture for long periods of time without muscle fatigue. Since the same phenomenon is also found in the cat during this form of electrical stimulation, it appears that this technique will be useful for postural control as well.

This technique apparently satisfied both design criteria for isometric contractions in that fine control of movement in paralyzed muscles can be achieved with low fatigability.

Reverse Recruitment

One limitation to the use of the previously mentioned technique must be cited here. To set the recruitment order of the motor units from the slowest to fastest motor units respectively, i.e. forward recruitment, it was necessary to use an anodal block electrode. While this electrode was excellent for the acute experiments cited previously, it can be anticipated that, for chronic use of the electrodes, the constant DC potential applied by these electrodes across the motor nerve to induce anodal block may cause some pathological damage to the nerve. We, therefore, conducted a final series of experiments to determine the importance of forward recruitment of motor units.

The experiments were similar to those described previously. First, following the program cited above, each of four medial gastrocnemius muscles in each of four cats were stimulated to sustain brief and fatiguing isometric contractions at tensions between 10 and 100 percent of each muscle's maximum strength. Here, however, two series of experiments were performed. In the first, the recruitment order was set from the slowest to fastest motor units by use of the anodal block electrode. In the second series of experiments, although the experimental procedures were the same, motor unit recruitment was achieved by increasing the stimulation voltage to the sleeve electrode rather than recruiting all of the motor units and then blocking the impulses. Since large motor units are more susceptible to electrical stimulation than are the small motor units, the recruitment order was therefore the opposite, i.e. reverse recruitment. The results of these experiments are shown in Figures 5-7 and 5-8.

During the brief isometric contractions the tension developed by the muscles at low isometric tensions for recruitment from the largest to the smallest motor units (Fig. 5-7) was not as precisely controlled as that obtained during similar contractions when the recruitment order was set from the smallest to largest motor units (Fig. 5-3). However, for contractions at tension greater than about 50 percent of the muscle's maximum strength, the tensions were controlled equally as well with both recruitment orders.

The endurance for sustained isometric contractions at tensions between 10 and 100 percent of each muscle's maximum strength are shown in Figure 5-8 for recruitment from the smallest to largest motor units (open symbols) and recruitment from the largest to smallest motor units (closed symbols), respectively. Each point in these figures shows the average results of the four cats plus and minus the respective standard deviations. For tensions greater than 40 percent of the maximum strength of the muscles, there was no difference between the endurance of the cats where recruitment

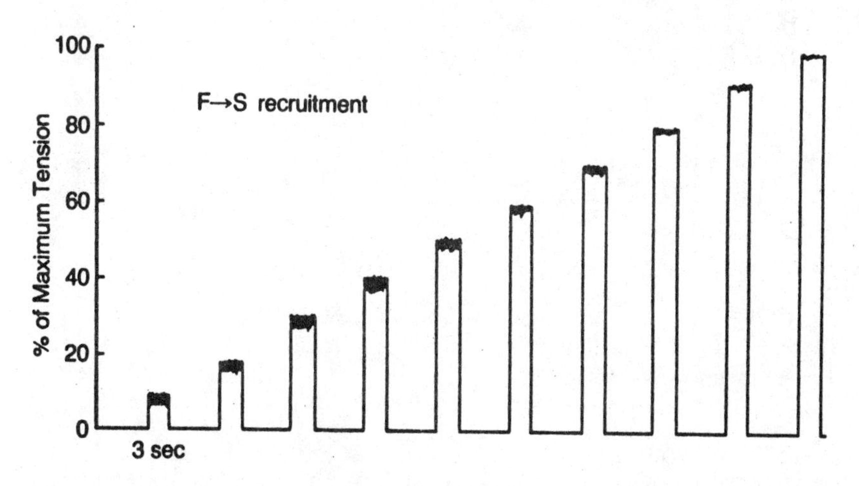

Figure 5-7. The tension developed during brief isometric contractions with recruitment proceeding from the fastest to slowest motor units.

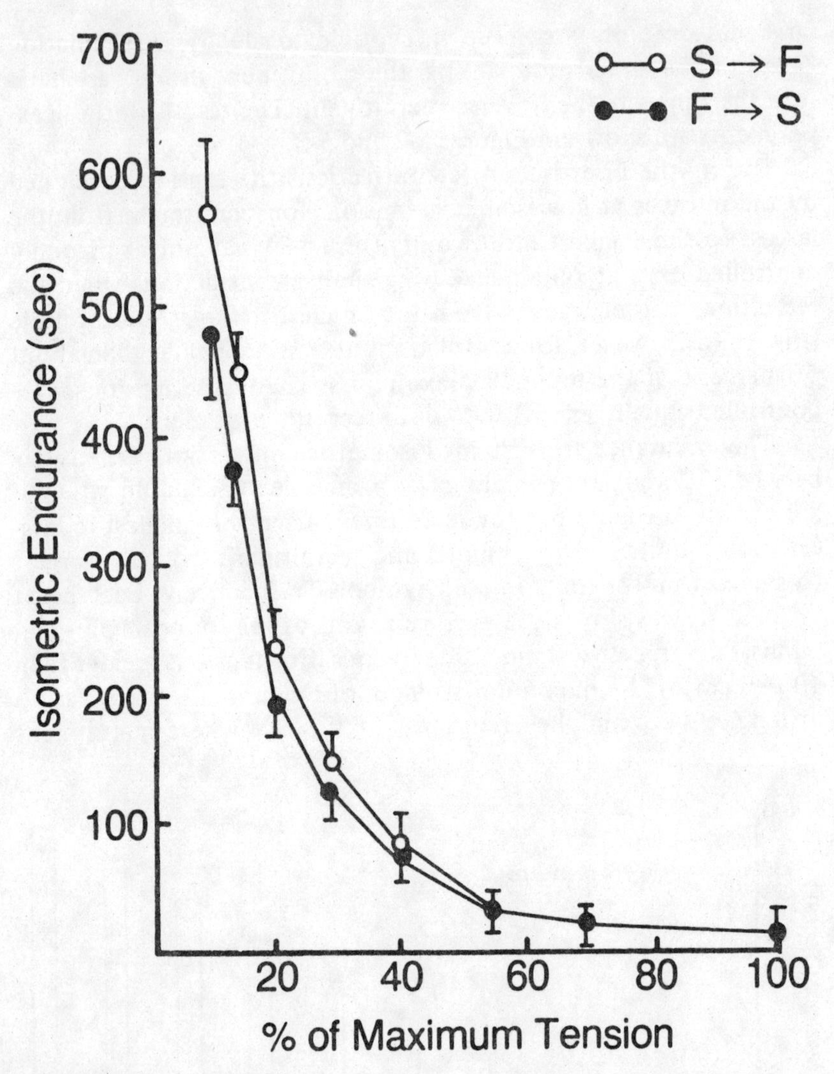

Figure 5-8. Isometric endurance in four cat muscles with recruitment proceeding from the smallest to largest (open symbols) and largest to smallest (closed symbols) motor units.

proceeded from the smallest to the largest or the largest to the smallest motor units. However, for contraction at tension less than 40 percent of the muscle's strength, the endurance was progressively less as the tension exerted during the contraction was reduced. Apparently then, for postural contractions and low strength contractions, the body achieves a greater degree of control and a longer endurance with recruitment proceeding from the smallest to the largest motor units. However, although the controllability of the muscle and the endurance were lower for these tensions, the difference seen between these two methods of stimulation was small compared to the advantage that may be reached by not using the anodal block electrodes, i.e. pathological damage, making this perhaps the ideal method of stimulation. Further work in this area is needed before this decision can be made.

CONTROL OF MUSCLE DURING DYNAMIC EXERCISE

Although this initial series of experiments showed the feasibility of using a microprocessor controlled stimulator to maintain isometric contractions, it still remained to be determined if a computer controlled stimulator could control movement during dynamic exercise as well. Therefore, an additional series of experiments was conducted to assess the feasibility of using a microprocessor to provide constant velocity contractions in the medial gastrocnemius muscle of the cat. Some of these experiments have been reported elsewhere (Petrofsky and Phillips, 1979). This was accomplished in two phases. First, the relationship between motor unit recruitment and force and velocity in the medial gastrocnemius muscle was established. From this data base, an appropriate computer program was written to control movement in these muscles. Since the anodal block electrode may be detrimental to the nerves during chronic stimulation, these experiments were repeated with recruitment proceeding in both directions. In both cases, the frequency of stimulation was kept constant at 35 Hertz.

The experimental setup differed here from that described previously. Here an aluminum clamp was placed around the foot about half the distance between the calcaneus and the end of the toes (Fig. 5-9). A steel chain connected the clamp to a variable load. Interposed between the load and the clamp was a magnetic lock, which was used to hold the muscle at a constant length prior to the isotonic contractions. The lock was mounted on a steel isometric strain gauge transducer bar. To measure isometric tension in the muscle, the lock was turned on throughout the duration of

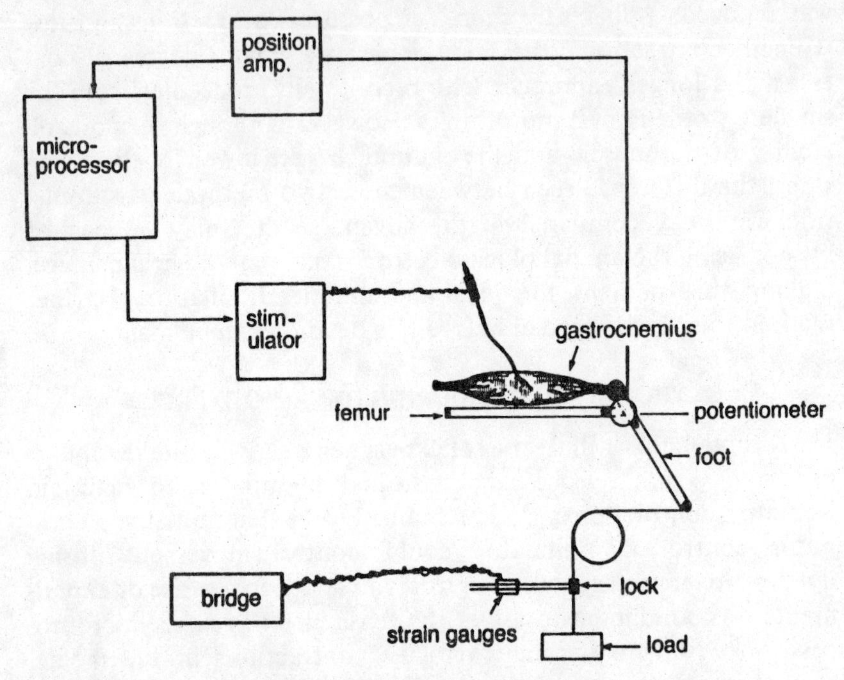

Figure 5-9. Schematic representation of the experimental apparatus.

stimulation. To allow the muscle to contract isotonically, the lock was turned off at the onset of stimulation.

The position of the ankle joint was sensed by a precision potentiometer attached to the leg and foot bones. This potentiometer formed part of a Wheatstone bridge, the output of which was digitized by one of the analog-to-digital channels of the microprocessor. The output of the bridge was also continuously differentiated before being digitized by a second analog-to-digital converter channel. This differentiated output was used as the velocity of the contraction of the muscle.

These experiments were conducted in two phases. First, it was necessary to establish the data base that would be used for the computer program. A series of experiments was therefore conducted to find out (1) how quickly the muscle would respond and obtain its maximum speed of shortening following the onset of stimulation and if this would vary with different degrees of activation of the muscle and (2) the nature of the force-velocity relationship with differing degrees of activation of a mixed muscle.

Next, a program was written and tested based on this data and in a final series of experiments the program was tested by assessing the ability of the muscle to sustain constant velocity contractions.

P_{mx} AND THE VELOCITY OF SHORTENING AT DIFFERENT LEVELS OF ACTIVATION

The results of the determinations of the maximum isometric tension (P_{mx}) and the velocity of shortening at different levels of load are shown in Figure 5-10 for recruitment from the slowest to fastest motor units.

Each point in this figure shows the average results of the medial gastrocnemius muscle in four different cats. As the clamp voltage was reduced to each of three levels and finally turned off entirely, there was an increase in the maximum isometric tension of the muscles (P_{mx}), as shown in this figure. The muscles were then allowed to shorten, and loads were applied that were less than P_{mx} to produce the force-velocity diagrams shown in this figure. V_{mx} has been estimated in all cases by extrapolation of the curves. The curves relating the load to the velocity of contraction were of the general form as those predicted by Hill (1938).

As the number of active motor units was increased, there was a progressive increase in the V_{mx} as well as the P_{mx} (reported above). Thus, for a progressive increase in the recruitment of the motor units from the slowest to the fastest in the muscle, a family of curves is established, each of which follows the general form described by Hill (1938).

From the raw data shown in this figure, a and b constants, described by Hill (1938), were calculated to best fit the data on a digital computer.

To quantify the degree of recruitment for the four different curves, the recruitment was expressed in terms of the isometric tension generated by the muscle at the four different stimulation levels. In these terms, curve a represented 100 percent activation while b, c, and d represented 73, 41, and 23 percent activation

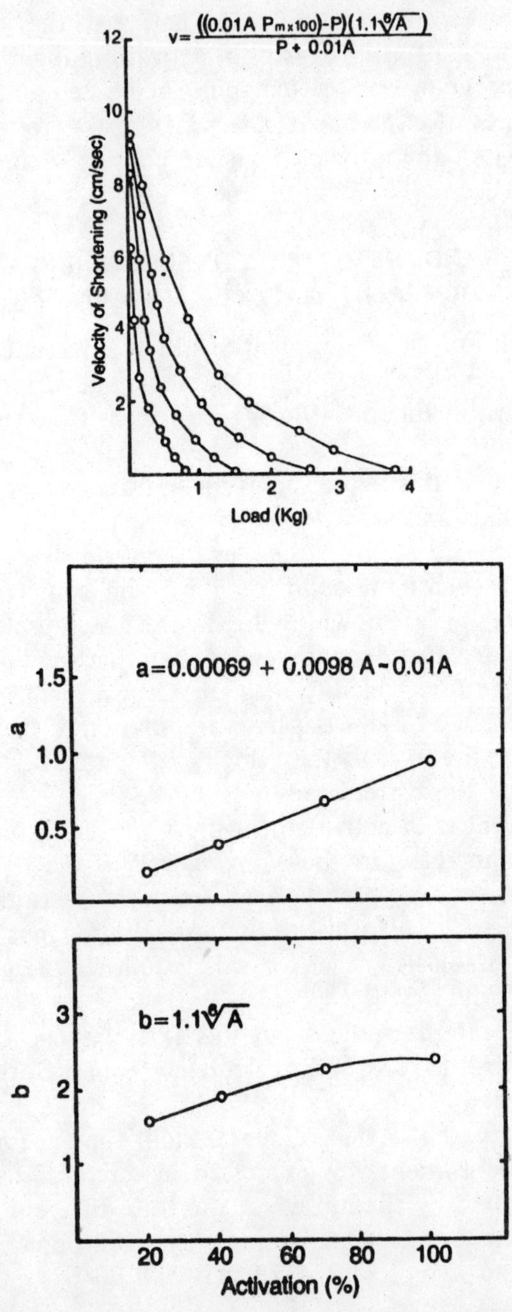

Figure 5-10. Force-velocity diagrams of the medial gastrocnemius at four levels of recruitment.

respectively. The relationship between the Hill co-efficient a and the degree of activation in the muscle (A) was found to be almost linear and could be expressed by the equation

$$a = 0.01A \tag{1}$$

where a is the a co-efficient of the Hill equation and A is the percent of activation of the muscle expressed as (100) P/P_{mx} for isometric contractions at a given degree of recruitment.

The relationship between the b co-efficient of the Hill equation and A was not linear; however, it could be approximated by the equation

$$b = 1.1 \sqrt[6]{A} \tag{2}$$

Using these equations, the a and b co-efficients can be replaced in the original Hill equation. However, P_{mx} is also a function of the degree of activation and can be approximated by the expression

$$P_{mx} = 0.01A\ P_{mx100} \tag{3}$$

where P_{mx100} is the maximum isometric tension when all motor units are recruited, P_{mx} is the maximum isometric tension at any level of activation, and A is the percent of the muscle that is activated.

Substituting back into the original Hill equation, to calculate the velocity of contraction for this muscle, it is only necessary to know the maximum isometric tension, the current load on the muscle, and the degree of recruitment:

$$v = \frac{(0.01A\ P_{mx100}) - P(1.1\ A^{(1/6)})}{P + (0.01A)} \tag{4}$$

These same experiments were repeated on the medial gastrocnemius muscles of four additional cats to determine a similar equation for recruited proceeding from the fastest to slowest motor units (no anodal block electrode).

The time from the onset of stimulation to the achievement of the peak velocity in the muscle was always less than 13 ms. How-

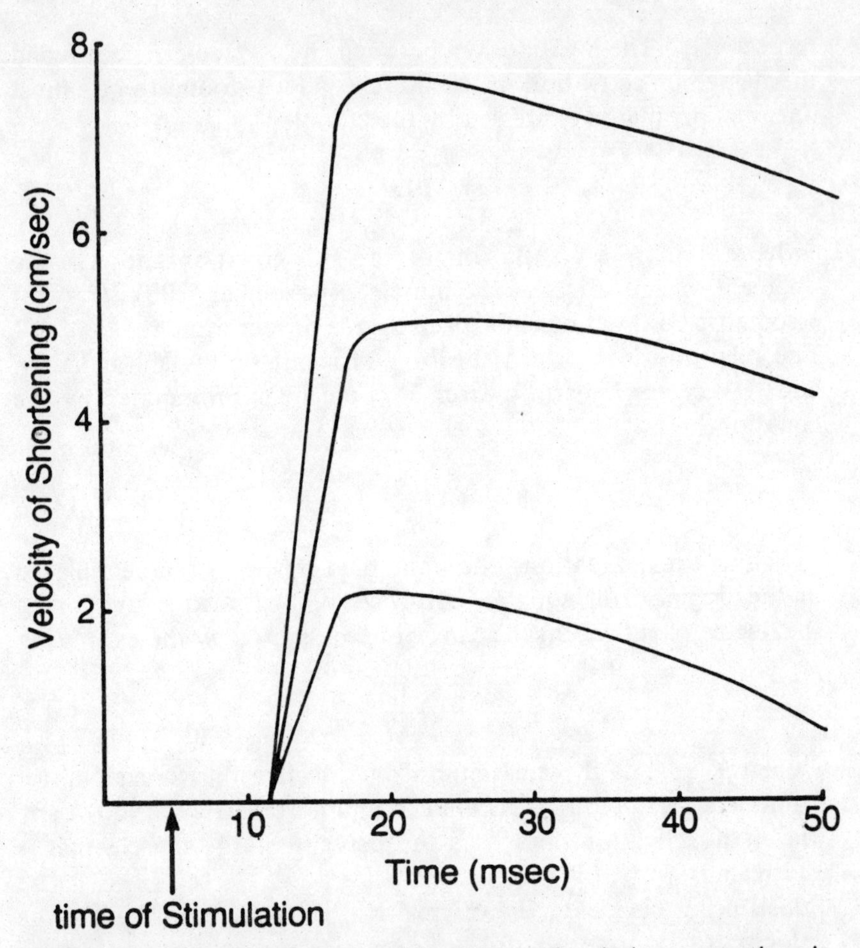

Figure 5-11. Velocity of the medial gastrocnemius during contractions induced at three different levels of recruitment. The vertical arrow on the abscissa shows the time of stimulation.

ever, although the time delay was always constant at a given degree of recruitment in the muscle, the time required to reach peak velocity was progressively shorter when more motor units were recruited. For example, Figure 5-11 shows a tracing of the time courses of activation in the medial gastrocnemius in one of the cats where muscle was contracting against a load of 0.5 kg.

With all the motor units recruited (upper curve), the muscle was able to develop a speed of shortening of about 8 cm/sec. The time required from the onset of stimulation (shown by the vertical arrow on the abscissa) to the peak velocity was about 9.5 ms. However, with only about half of the muscle activated (middle curve), the time required to reach peak tension was about 11 ms; whereas with only about one-fourth of the motor units activated, the peak velocity was just over 2 cm/sec and the time required for the muscle to reach its peak velocity was about 13 ms. In all cases, there was a gradual reduction in the velocity as the muscle continued to contract.

COMPUTER PROGRAM TO MAINTAIN CONSTANT VELOCITY

From this data, a computer program was developed to try to maintain constant velocity contraction in the medial gastrocnemius muscle. The assumption here was that the desired velocity of contraction would be known but the imposed load on the muscle would not; therefore, the computer program would have to compensate by altering patterns of recruitment for differing loads.

At the start of the program, all registers in the computer were cleared and the desired velocity of contraction was entered into memory. Next, the amplitude of the EMG during a brief maximal isometric contraction was determined and also stored in the memory. To begin the contraction, a ramped decrease in the clamp voltage was initiated on the motor nerve of the medial gastrocnemius. The rate of decrease of the voltage was set at 10 volts per second. With a stimulation frequency of 40 Hertz, the interpulse interval would be 25 ms. Since less than 15 ms is necessary for the muscle to reach its peak velocity, this allowed the computer to accurately track the effect on velocity of each increase in recruitment. Nearly all of the motor units would need to be active to obtain the desired velocity of contraction. Therefore, to set the recruitment more quickly at that which would be necessary for the desired velocity, the computer needed to be provided not only with the desired velocity but the load as well. For this reason, the load was estimated as follows.

At the end of the first firing of all three electrodes, the current velocity of contraction was assessed, and, from this value, the load imposed on the muscle was estimated from Equation 4.

An anodal block voltage was then applied to the muscle that would cause the muscle to develop the velocity desired. With each firing of the sequential electrode, the velocity of the contraction of the muscle was assessed. If the velocity of contraction exceeded that desired by the program, the recruitment was reduced. If the velocity of contraction was less than that desired by the program, recruitment was increased; in both cases, the voltage was changed at a rate of 10 volts per second. In addition, with each electrode rotation, the distance the muscle had contracted was assessed. If the desired shortening distance had been reached (9mm), the contraction was terminated.

CONSTANT VELOCITY CONTRACTIONS IN MUSCLE

The results of these experiments are shown in Figure 5-12. This figure shows the average velocity of contraction ± the S.D. achieved by four cat medial gastrocnemius muscles during contractions set to maintain velocities of 5, 10, and 15 mm/sec against a load of 1 kg. These experiments were repeated with each of two orders of recruitment: recruitment from the largest to smallest motor units (△) and recruitment from the smallest to largest motor units (○). The results of these experiments clearly showed that the computer program used in the present series of experiments was able to provide excellent control of the velocity of movement of cat skeletal muscle by either order of recruitment.

SUMMARY

It would seem, then, that it is feasible to use a microprocessor to control the movement of muscle for both postural control (isometric contraction) and during movement (dynamic contractions). In a human, externally placed joint position transducers could provide the necessary position information for the microprocessor to operate a program similar to the one described here. The stimulation electrodes could be Rf powered as well and implanted surgically as was done by Glenn et al. (1973). With stimulation electrodes Rf coupled through the skin, the chance of infection would be markedly reduced. The only factor lacking would be the control signal to direct the microprocessor. The surface EMG of the

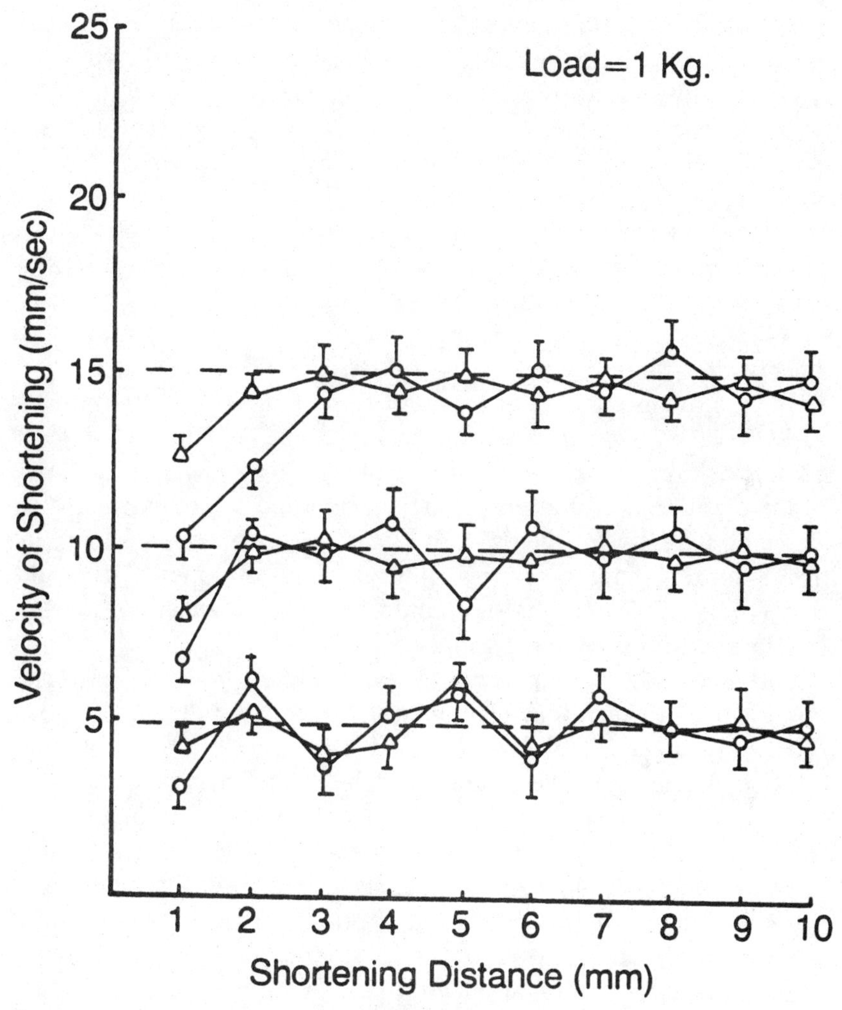

Figure 5-12. The velocity of contraction of the cat medial gastrocnemius muscle with recruitment in opposite orders (*see* text).

unparalyzed muscle may also provide this capacity. The EMG is a complex interference pattern arising from the summation of asynchronously firing motor units in active muscle. When the EMG is recorded through electrodes placed on the surface of the skin (surface EMG), it offers a noninvasive tool with which to observe the electrical activity in the underlying muscle. During brief isometric contractions, most authors agree that there is a linear relationship between the tension exerted and either the RMS or average amplitude of the surface EMG (Bigland and Lippold 1954, Milner-Brown and Stein 1975). On this basis, then, it would appear that the amplitude of the surface EMG would provide a good index of isometric tension.

After the loss of an appendage or paralysis, the muscles above the injury still contract in the same manner during attempted movement as if the limb were still functional (Abe, 1970; Hogan, 1974). The linear relationship between the EMG amplitude and tension has therefore been used by designers of prothesis control systems not only to provide proportional control over movement (Abe, 1970; Hogan, 1974; Aaron and Stein, 1976), but to lock joint position as well (DiSano et al., 1974) by using the EMG from the uninjured muscles. Further, microprocessors are presently being used to decode the EMG from groups of muscles to allow for a more complex control of prosthetic movement. It would seem that similar technology could be used as an input for the microprocessor controlled stimulator.

The control system used to electrically stimulate paralyzed muscle would have four elements: (1) sensors to detect the position and the velocity of motion of the joints, (2) a series of stimulators which have electrodes on the motor nerves innervating various leg muscles, (3) some sort of control signal to interpret the motion desired by the patient, and (4) a central processing unit to interpret the control signals and the sensor inputs and provide the proper sequence of motion to stand, walk, climb stairs, or whatever movement is desirable. The use of a microprocessor for such an application has a number of advantages over hardwired sequential circuits or programmable logic arrays. Because of the diversity in the types of injury and the degree of paralysis, the degree of control that the microprocessor might need to impose may vary from stimulating only the lower leg and foot muscles of a single leg, to alternate stimulation of all the muscles of both legs. In the former example, the control signal might come from the EMG recorded from the upper leg muscles, while, in the latter example, the control signal might arise from either a decoding of the electrical activity in the ab-

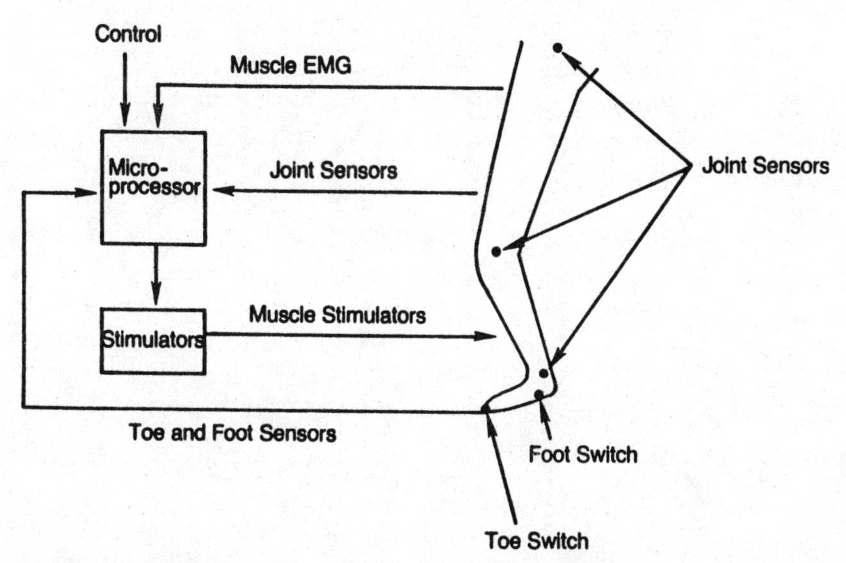

Figure 5-13. Diagrammatic representation of the anticipated microprocessor control system.

dominal or trunk muscles (since it has been shown that these muscles still maintain the same electrical rhythm during attempted walking as if the lower muscles were still functioning) or a combination of button control on a motion select keyboard worn on the belt of the patients and the muscle EMG. In both cases, at least three sources of sensory information will be needed by the microprocessor as shown in Figure 5-13.

First, the position of the joints and the rate of change in movement and position of the leg(s) will need to be known. Second, since walking may be on an incline or up or down stairs, contact sensors may be needed on the heel and toe of the foot to provide information about the height of the ground and obstacle position respectively. Finally, the computer will also need information concerning the number of motor units recruited in the various muscles that it is stimulating in order to set the level of the stimulation voltage appropriately and to sense the load being imposed on the muscles.

Obviously, the programming required by the computer will be complex, and (because of differences in the muscle involvement) although the basic hardware would be the same among individuals, the programming would be different. This provides no problem with the current generation of microprocessors, since it would only entail blowing a different PROM program for each patient. However, before PROM programs could be implemented, a variety of problems with respect to functional electrical stimulation of muscle must be resolved. These include the effects of muscle fatigue (Petrofsky, Weber, and Phillips, 1980), initial length (Petrofsky and Phillips, 1981), resting muscle tension (Phillips and Petrofsky, 1981), the oxidative capacity and functional performance of muscle (Petrofsky and Phillips, 1980), and the biomechanics of multi-muscle stimulation (Phillips and Petrofsky, 1980).

REFERENCES

Aaron, S.L. and Stein, R.B. (1976) Comparison of an EMG-controlled prosthesis and the normal biceps brachii muscle. *Am. J. Phy. Med.*, 55(1):1-14

Abe, H. (1970) Electromyographic study of the amputation stump of the lower extremity. *J. Jap. Orthop. Assoc.*, 44(5):351-364

Ariano, M., Armstrong, R.B., and Edgerton, V.R. (1973) Hindlimb muscle fiber populations of 5 animals. *J. Hist. Cytochem.*, 21:51-55

Bigland, B. and Lippold, O.C.J. (1954) Motor unit activity in the voluntary contraction of human muscle. *J. Physiol.*, 125:322

Brown, G.L. and Burns, B.D. (1949) Fatigue and neuromuscular block in mammalian skeletal muscle. *Proc. Roy. Soc. Ser. B.*, 136:182-196

Brown, W.F. (1973) Functional compensation of human motor units in health and disease. *Neurol. Sci.*, 20(2):199-209

Burke, D.C. and Murray, D.D. (1975) *Handbook of Spinal Cord Medicine*. London, Macmillan Press Limited

Burke, M.H. (1960) Survival of patients with injuries of the spinal cord. *J. Am. Med. Assoc.*, 172:121-124

Cherepakhin, M.A., Kakurin, L.I., Il'ina-Kakeuva, E.I., and Fedorenko, G.T. (1977) Evaluation of the effectiveness of electrostimulation of the muscles in preventing disorders related to prolonged limited motor activity in man. *Kosm. Biol. Aviakosm. Med.*, 11:64-68

Cox, W. and Grubb, R. (1974) Trauma to the central nervous system. In *Neurological Pathophysiology*. (Ed.) Eliasson, S., Prensky, A., and Hardin, W. New York, Oxford U. Press

Crochetière, W.T., Vodovnik, L., and Reswick, J.B. (1967) Electrical stimulation of skeletal muscle—A study of muscle as an actuator. *Med. Biol. Eng.*, 5:111-125

Cushing, H. (1927) The Med. Dept. United States Army in World War. Surgery II, Part I.:757

Daggett, W.M., Piccinini, J.C., and Austen, W.G. (1966) Intracaval electrophrenic stimulation. I. Experimental application during barbiturate intoxication, hemorrhage, and ganglionic blockade. *J. Thorac. Cardiovas. Surg.*, 51:676

Daggett, W.M., Piccinini, J.C., and Austen, W.G. (1970) Intravenous electrical stimulation of the phrenic nerve: A new technique for artificial respiration. *Surg. Forum*, 16:177

DiSano, L., Iannuzzi, M., Finkelstein, S., and Mason, C. (1974) An EMG-controlled knee locking prosthesis. *Proc. 2nd Annual New England Bio. Eng. Conference*, 185-191

Dubowitz, V. and Brooke, M. (1974) *Muscle Biopsy: A Modern Approach*. Philadelphia, W.B. Saunders

Edwards, R.G. and Lippold, O.C.J. (1956) The relation between force and integrated electrical activity in fatigued muscle. *J. Physiol.*, 132:677-681

Glenn, W., Hageman, J.H., Mauro, A., Eisenberg, L., Flanigan, S., and Harvard, B.M. (1964) Electrical stimulation of excitable tissue by radiofrequency transmission. *Ann. Surg.*, 160:338

Glenn, W., Holcomb, W., Gee, J., and Ranjit, R. (1970) Central hypoventilation: long-term ventilatory assistance by radiofrequency electrophrenic respirations. *Ann. Surg.*, 166:755-772

Glenn, W., Holcomb, W., Hogan, J., Matano, I., Gee, J., Motoyama, E., King, C., Poirier, R., and Forbes, G. (1973) Diaphramatic pacing by radiofrequency transmission in the treatment of chronic ventilatory insufficiency. *J. Thorac. Cardiovas. Surg.*, 66:505-571

Guttman, L. (1976) *Spinal Cord Injuries, Comprehensive Management and Research*. Oxford, Blackwell Scientific Publications

Hill, A.V. (1938) Heat of shortening and the dynamic constants of muscle. *Proc. R. Soc.*, 126:136-195

Hogan, N. (1974) An evaluation of EMG as a proportional control signal. *Proc. 2nd Annual New England Bio. Eng. Conference*, 179-183

Judson, J.P. and Glenn, W.W.L. (1968) Radiofrequency electrophrenic respiration. *JAMA*, 203:1033

Kiwerski, J. (1973) Effects of electrical stimulation of paralyzed muscles in quadriplegia. *Pol. Tyg. Kok.*, 28:158-159

Liberson, W.T., Holmquest, H.J., Scott, D., and Daw., M (1961) *Arch. Phys. Med. Rehabil.*, 42:101

Lind, A.R. (1959) Muscle fatigue and recovery from fatigue induced by sustained contractions. *J. Physiol.*, 147:162

Lind, A.R. and Petrofsky, J.S. (1979) The amplitude of the EMG during sustained isometric contractions. *Muscle and Nerve*, 2:257-264

Long, C., III, and Masciacrelli, V.D. (1963) An electrophysiologic splint for the hand. *Arch. Phys. Med. Rehabil.*, 44:499

Marsden, C., Meadows, J., and Merton, P. (1971) Isolated single motor units in human muscle and their rate of discharge during maximal voluntary effort. *J. Physiol.*, 217:12-13

McNeal, D.R., Wilemon, W.K., Mooney, V., Boggs, R., and Tumaki, J. (1969) The effect of peripheral nerve implanted electrical stimulation on motor control in stroke patients. *World Congr. Neurological Sciences*, New York

Milner, M., Quanbury, A.O., and Basmajian, J.V. (1970) Surface electric stimulation of lower limb. *Arch. Phys. Med. Rehabil.*, 51:540-545

Milner-Brown, H.S. and Stein, R.B. (1975) The relation between the surface electromyogram and muscular force. *J. Physiol.*, 246:549

Nashold, B.S., Jr., Friendman, H., Glenn, J.F., Barry, S.W., and Avery, R. (1972) Electromicturition in paraplegia. Implantation of a spinal neuroprosthesis. *Arch. Surg.*, 104:195-202

Nauman, S. and Milner, M. (1978) A sequential stimulator for electrical restoration of the micturition reflex. *IEEE Trans. Biomed. Eng.*, 25:307-311

Olsen, C., Carpenter, D., and Henneman, E. (1968) Orderly recruitment of muscle action potentials. *Arch. Neurol.*, 19:591-597

Peckham, P.H. (1976) Control of contraction strength of electrically stimulated muscle by pulse width and frequency modulation. *Eng. in Med. and Biol.*, 18:116

Petrofsky, J.S. (1978) Control of the recruitment and firing frequencies of motor units in electrically stimulated muscles in the cat. *Med. Biol. Eng.*, 16:302-308

Petrofsky, J.S. (1979a) Sequential motor unit stimulation through peripheral motor nerves. *Med. Biol. Eng. Comput.*, 17:87-93

Petrofsky, J.S. (1979b) A digital-analogue hybrid 3 channel sequential stimulator. *Med. Biol. Eng. Comput.*, 17:421-424

Petrofsky, J.S. (1980) Computer analysis of the surface EMG during isometric exercise. *Comput. Biol. Med.*, 10:83-95

Petrofsky, J.S., LaDonne, D., Rinehart, J., and Lind, A.R. (1976) Isometric strength and endurance during the menstrual cycle in healthy young women. *Europ. J. Appl. Physiol.*, 35:1-10

Petrofsky, J.S. and Lind, A.R. (1979) Isometric endurance in fast and slow muscles in the cat. *Am. J. Physiol.*, 236:c185-c191

Petrofsky, J.S. and Phillips, C.A. (1979) Constant velocity contractions in skeletal muscle by sequential stimulation of muscle efferents. *Med. Biol. Eng. Comput.*, 17:583-592

Petrofsky, J.S. and Phillips, C.A. (1980) The influence of recruitment order and fibre composition on the force-velocity relationship and fatigability of skeletal muscles in cat. *Med. and Biol. Eng. and Comput.*, 18:381-390

Petrofsky, J.S. and Phillips, C.A. (1981) The influence of temperature, initial length and electrical activity on the force-velocity relationship of the medial gastrocnemius muscle of the cat. *J. Biomechanics*, 14:297-306

Petrofsky, J.S., Weber, C. and Phillips, C.A. (1980) Mechanical and electrical correlates of isometric muscle fatigue in skeletal muscle in the cat. *European J. Physiol.*, 387:33-38

Phillips, C.A. and Petrofsky, J.S. (1980) Velocity of contraction of skeletal muscle as a function of activation and fiber composition: A mathematical model. *J. Biomechanics*, 13:549-558

Phillips, C.A. and Petrofsky, J.S. (1981) The passive elastic force-velocity relationship of cat skeletal muscle: Influence upon the maximal contractile element velocity. *J. Biomechanics*, 14: 399-403

Rack, P.M.H. and Westbury, D.R. (1969) The effects of stimulus rate on tension in isometric contraction of the cat soleus muscle. *J. Physiol.*, 204:443-460

Rebersek, S. and Vodovnik, L. (1973) Proportionally controlled functional electrical stimulation of upper extremity. *Arch. Phys. Med. Rehab.*, 54:378

Reswick, J.B. and Vodovnik, L. (1967) External power in prosthetics and orthotics: An overview. *Artif. Limbs*, 11:5-21

Rohmert, W. (1968) Rechts-links Vergleich bei achtjahvgin Kindern. *Int. Z. Agnew. Physiol.*, 26:363

Sarnoff, S.L., Hardenburgh, E., and Whittenberger, J.L. (1948a) Electrophrenic respiration. *Am. J. Physiol.*, 155:1

Sarnoff, S.J., Hardenbergh, E., and Whittenberger, J.L. (1948b) Electrophrenic respiration. *Science*, 108:482

Scott, R.N. (1968) Myoelectrical control systems. In *Advances in Biomedical Engineering and Medical Physics*. Vol. 2. (Ed.) Levine, S.N. New York, John Wiley & Sons, Inc.

Sherrington, C.S. (1894) On the anatomical constitution of nerves of skeletal muscles; with remarks on recurrent fibres in the ventral spinal nerve root. *J. Physiol.*, 17:211-258

Smart, C.N. and Sanders, C.R. (1976) *The Costs of Motor Vehicle Related Spinal Cord Injuries*. Washington, D.C., Insurance Institute for Highway Safety

Solomonow, M., Foster, J., Eldred, E., and Lyman, J. (1978) Proportional control of paralyzed muscle. *Fed. Proc.*, 37:215

Stemmer, E., Crawford, D., List, J., Heker, B., and Conolly, J. (1967) Diaphragmatic pacing in the treatment of hypoventilation syndrome. *J. Cardiovas. Surg.*, 173:649-688

Sunderland, S. (1968) *Nerves and Nerve Injuries*. Baltimore, The Williams and Wilkins Company.

Timm, G.W. and Bradley, W.E. (1969) Electrostimulation of the urinary detrusor to effect contraction and evacuation. *Invest. Urol.*, 6:562-568

Timm, G.W. and Bradley, W.E. (1971) Electromechanical restoration of the micturition reflex. *Biomed. Eng.*, 18:274-280

Timm, G.W. and Bradley, W.E. (1973) Technologic and biologic considerations in neuro-urologic prosthesis development. *Biomed. Eng.*, 20:208-212

Townsend, M.A. and Lepofsky, R.J. (1976) Powered walking machine prosthesis for paraplegics. *Med. Biol. Eng. Comput.*, 14: 436-443

Trnkoczy, A., Bajd, T., and Maleizi, C.M. (1976) A dynamic model of the ankle joint under functional electrical stimulation in free movement and isometric conditions. *J. Biomech.*, 9:509-519

Van der Meulen, J.P., Peckham, P.H., and Mortimer, J.T. (1974) Trophic functions of the neuron. 3. Mechanisms of neurotrophic interactions. Use and disuse of muscle. *Ann. N.Y. Acad. Sci.*, 228:177-189

Van Heeckeren, D.W. and Glenn, W.W.L. (1966) Electrophrenic respiration by radiofrequency induction. *J. Thorac. Cardiovas. Surg.*, 52:655

Vellacott, P.N. and Webb-Johnson, A.E. (1919) Spinal injuries. *Lancet*, i: 173

Vodovnik, L., Crochetière, W.S., and Reswick, J.B. (1967) Control of a skeletal joint by electrical stimulation of antagonists. *Med. Biol. Eng. Comput.*, 5:97-109

Vrbova, G. (1963) The effect of motorneurone activity on the speed of contraction of striated muscle. *J. Physiol.*, 169:513-526

Vukobratovic, M., Hristic, D., and Stojilkovic, Z. (1974) Development of active anthropormorphic exoskeletons. *Med. Biol. Eng. Comput.*, 12:66-80

Zealer, D.L. and Dedo, H.H. (1977) Control of paralyzed axial muscles by electrical stimulation. *Trans. Am. Acad. Ophthalmol. Otolaryngol.*, 84:310

Chapter 6

DIAGNOSTIC CARDIAC MECHANICS

Clinical Application of the Biomechanics of Cardiac Function

C.A. PHILLIPS, D.N. GHISTA, and J.S. PETROFSKY

THE analysis of the functional diagnostic aspects of the left ventricle (LV) during systole has seen some significant advances in rigorous mechanics terms over the past decade with respect to thermodynamics-energetics, contractile stress-strain, myocardial power, and motion of fluid (wall motion) in the LV chamber.

Proceeding from micro- to macromechanics as well as from cause to effect phenomena resulting from left ventricular contraction during systole, then the aspects that need to be quantitated and physiologically interpreted are (1) the myocardial energetics (chemical power generated), (2) the myocardial contractile stress versus shortening strain characteristics, (3) myocardial contractile power, and (4) the resulting pressure distribution and motion of the fluid (wall motion) in the left ventricular chamber resulting in an effective ejection fraction and cardiac output. No single work has addressed itself to these aspects; however, the current literature, when surveyed in general, provides a great deal of insight into the mechanical, physiological, and clinical aspects of LV pumping performance.

MYOCARDIAL ENERGETICS

The chemo-mechanical energy conversion process was first

described for skeletal muscle using the principles of irreversible thermodynamics by Caplan (1966). He assumed that the entire muscle is a self-regulated linear energy converter (with a variable chemical affinity), and his analysis resulted in Hill's hyperbolic relationship between force and shortening velocity. Subsequently, Bornhorst and Minardi (1970a, 1970b) proposed an alternate theory in which each individual cross-bridge operated as a linear energy converter (with a constant chemical affinity for the driving chemical reaction). Powell (1971) compared Caplan's modification of his original model (1968a, 1968b), which assumes that the number of cross-bridges remains constant (while the chemical affinity varies), with that of Bornhorst and Minardi (1970a). His study lent support to Bornhorst and Minardi's hypothesis, which assumes that the number of active cross-bridges varies while the local affinity is constant and equal to the overall affinity for the driving chemical reaction.

Following these developments in applying irreversible thermodynamics to describe skeletal muscle performance, it was logical that the approach be extended to cardiac muscle. Bloomfield et al. (1971, 1972) have utilized irreversible thermodynamics to describe the chemo-mechanical conversion process in heart muscle. Assuming 100 percent conversion efficiency, these investigators showed that, during isovolumic systole, any change in pressure is exactly equal to a change in the free energy of the left ventricle. This simply means that as the pressure changes, so does the ability for the ventricle to perform the work of displacing blood into the artery. Although their theoretical development was valid only for the isovolumic phase of systole, these investigators proceeded to extrapolate this equivalence to the entire systolic cycle and develop power terms. Stein and Sabbah (1975) have also defined an isovolumic systolic power term using fluid mechanics instead of irreversible thermodynamics and found similar results, though the exact equality between the two methods is still in question (Moreno, 1976; Stein et al., 1976).

Phillips et al. (1979b) recently derived equations for the instantaneous chemical power (Fig. 6-1) generated by the myocardium from work performed throughout the cardiac cycle. Their equations allow prediction of the chemical energetics in terms of two mechanical variables (stress and velocity) and three constants: two equation constants (k and K) and a reference velocity (V_R). These investigators assigned values to the first two constants, so that when the stress history and velocity history throughout a cardiac cycle is determined and the characteristic reference velocity is defined, the instantaneous chemical power and work (Fig. 6-2) can be solved for that cycle.

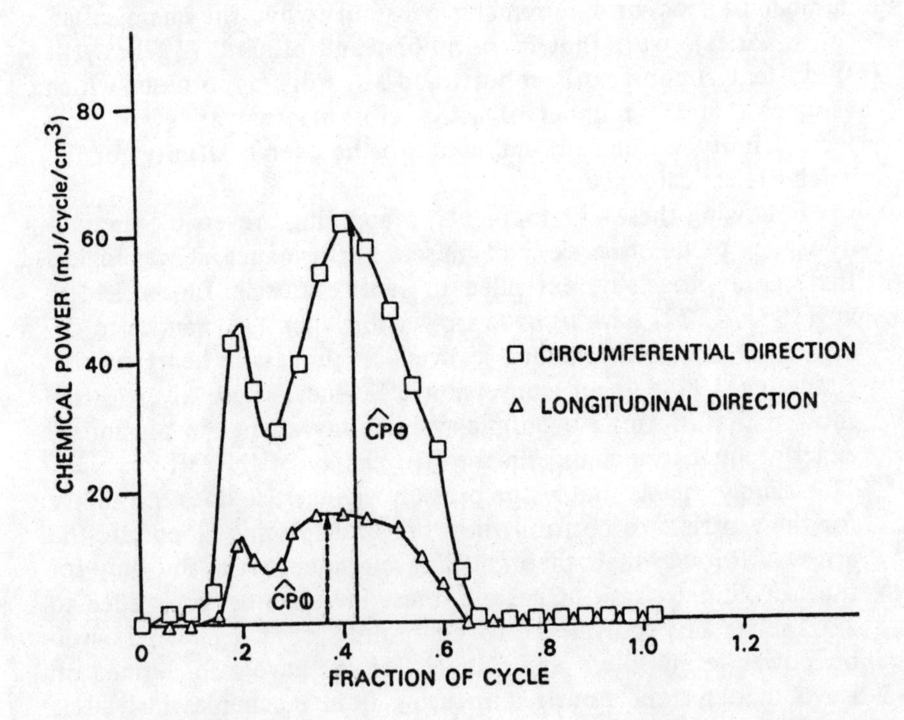

Figure 6-1. Chemical power in the circumferential direction (□) and longitudinal direction (△) of the left ventricular during a cardiac cycle. The solid vertical arrow denotes the peak circumferential chemical power (CP_θ) while the broken arrow denotes the peak longitudinal chemical power (CP_ϕ). From C.A. Phillips, E.S. Grood, W.S. Scott, and J.S. Petrofsky, Cardiac Chemical Power:2. *Medical and Biological Engineering and Computing,* *17*:503-509, 1979.

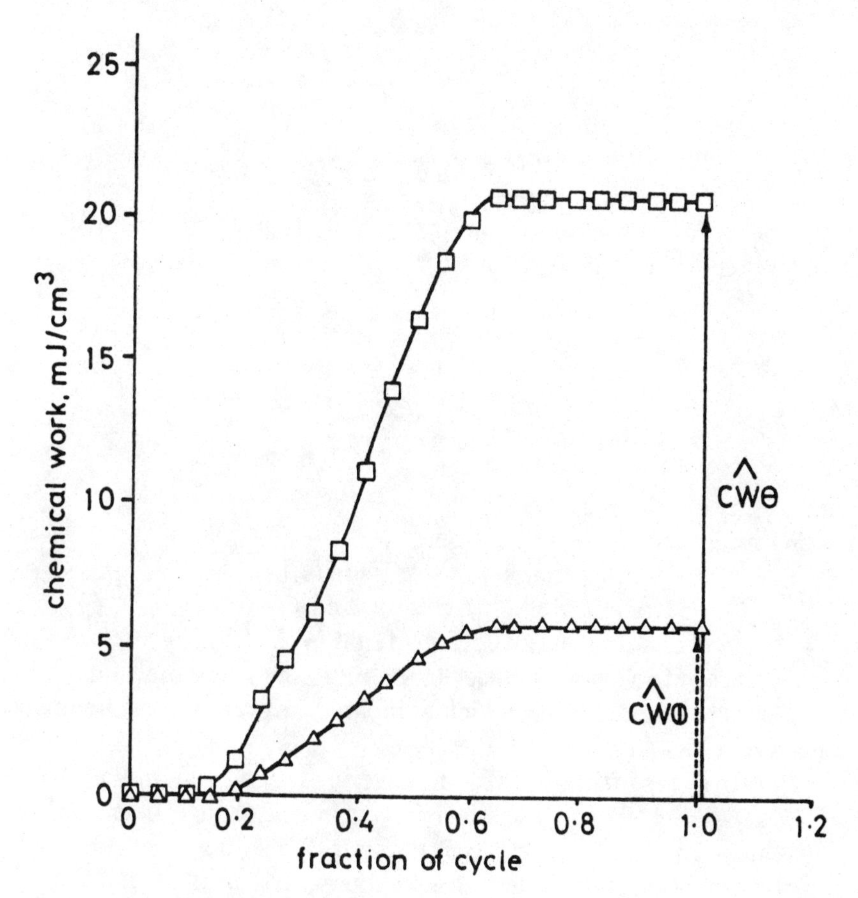

Figure 6-2. Chemical work in the circumferential direction (□) and longitudinal direction (△) of the left ventricle during a cardiac cycle. The solid vertical arrow denotes endcycle circumferential chemical work (CW_θ) and the broken vertical arrow denotes end-cycle longitudinal chemical work (CW_ϕ). From C.A. Phillips, E.S. Grood, W.S. Scott, and J.S. Petrofsky, Cardiac Chemical Power:2. *Medical and Biological Engineering and Computing, 17*:503-509, 1979.

The equations for chemical power (CP), chemical work (CW), and efficiency (η) (Fig. 6-3) were defined by Phillips et al. (1979c) for both the circumferential (subscript θ) and longitudinal (subscript ϕ) axes of the LV:

$$CP_{\theta} = \frac{(\sigma_{CF\theta})(V_{CE\theta})(V_{max\theta}/V_{CE\theta} + 15)}{15(1 - V_{CE\theta}/V_{max\theta})} \quad (1)$$

$$\eta_{\theta} = \frac{15(1 - V_{CE\theta}/V_{max\theta})}{(V_{max\theta}/V_{CE\theta} + 15)} \quad (2)$$

and CW_{θ} is the time integral of CP_{θ}.

$$CP_{\phi} = \frac{(\sigma_{CF\phi})(V_{CE\phi})(V_{max\phi}/V_{CE\phi} + 15)}{15(1 - V_{CE\phi}/V_{max\phi})} \quad (3)$$

$$\eta_{\phi} = \frac{15(1 - V_{CE\phi}/V_{max\phi})}{(V_{max\phi}/V_{CE\phi} + 15)} \quad (4)$$

and CW_{ϕ} is the time integral of CP_{ϕ}.

With respect to these equations:

σ_{CF} = contractile filament stress (*vide infra*),
V_{CE} = contractile element velocity, and
V_{max} = maximum (unloaded shortening velocity).

Average efficiency for eleven hemodynamically normal patients was just under 0.50, which compared favorably with the upper limit set by Curtin et al. (1974).

For five compensated volume overload patients studied by Phillips et al. (1979c), chemical power, work, and efficiency is not significantly different from normals for either the circumferential or longitudinal directions. Apparently, part of the compensation process to a volume overload involves the maintenance chemical power and work per cubic centimeter at near normal values.

For six decompensated volume overload patients studied by Phillips et al. (1979c), there is a significant decrease in average circumferential efficiency, combined with a significant increase in the peak circumferential power and work. Apparently, in the decompensated volume overload ventricle, the peak mechanical power output (computed as peak chemical power times average efficiency) is maintained at or above normal levels at the expense of increased chemical power generation. This implies a reduced efficiency in coupling the chemical power generated to the mechanical power output.

For the two compensated pressure overload patients studied by Phillips et al. (1979c), circumferential chemical power and

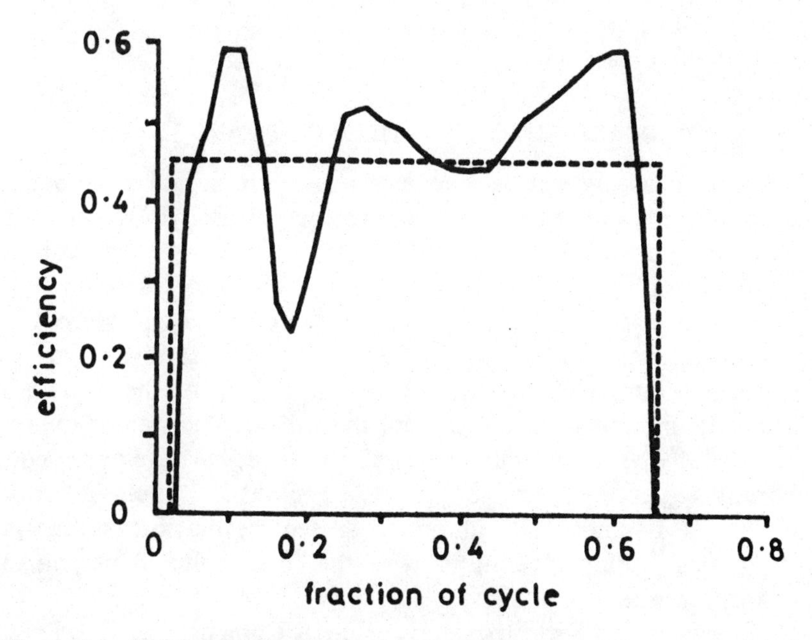

Figure 6-3. The solid line denotes instantaneous efficiency for the circumferential direction of the left ventricle during a fraction of the cardiac cycle, and the broken line denotes the average efficiency during that portion of the cardiac cycle. From C.A. Phillips, E.S. Grood, W.S. Scott, and J.S. Petrofsky, Cardiac Chemical Power:2. *Medical and Biological Engineering and Computing, 17*:503-509, 1979.

work are individually higher and circumferential efficiency is individually lower as compared to the eleven normals. Because these ventricles are clinically compensated, the low circumferential efficiency is an ominous finding. Since much of the chemical power is being dissipated due to inefficient chemo-mechanical coupling, there is essentially no safety margin. Should either of these two patients become clinically decompensated, a rapid loss of left ventricular mechanical power output might be expected.

In conclusion, chemical power, work, and efficiency appear to provide additional information on the energetic state of the normal and abnormal left ventricle. Left ventricular energetics represent one more item of information that can be integrated with other clinical, laboratory, and analytical information in order to allow a more complete understanding of ventricular function in health and disease.

MYOCARDIAL CONTRACTILE STRESS AND STRAIN

A typical myocardial medium's element should be modeled as a hybrid composite shell element comprised of three components: an active or contractile component of preferentially oriented contractile filaments, in parallel and series with two sets of passive elastic components of the sarcolemma and connective tissue (Ghista, 1974; Fung, 1970; 1971). The total incremental stress resultants (N^T) of the left ventricular model shell would be obtained in terms of the (previously indicated) instantaneous incremental distributed chamber wall pressure, by developing equilibrium equations for the appropriate geometrical thick-shell model loaded symmetrically about an axis of revolution (simulating the chord joining the midpoint of the aortic valve plane and the ventricular apex).

For a hybrid orthotropic myocardial element, we would then need to put down the relations between (1) the total incremental strains (ϵ^T) and the total incremental stress resultants (N^T), (2) the total incremental strains and the shell surface displacements, (3) the total incremental strains (ϵ^T) and the incremental strains in the contractile element (ϵ^E) and in the elastic matrix components (ϵ^S, ϵ^P) in the fiber direction, and (4) the total incremental stress resultants in the contractile and elastic components. We would then need to solve the above four sets of equations (along with the boundary condition to simulate the ventricular shell being supported along the aortic valve ring) to obtain the distributions of the directions of the fibers and the magnitudes of the contractile stress and strain for each imaged frame instant in terms of (or normalized with respect to) the computed chamber wall pressure and the monitored wall displacements.

Such a comprehensive model is unavailable at the present time. However, it may be useful for the readers to at least note the state of the art works contributory to or suggestive of the previously indicated approach. To this end, we could take note of the following works and employ them as references for various aspects of the comprehensive approach plan: Mirsky's ellipsoidal thick shell model (1969) for the stress state in the ventricular medium, Wong's cardiac contraction simulation model (1971), Panda and Natarajan's finite element layered shell left ventricular model (1977), and Glantz's (1977) three-component model with viscoelastic parallel elements. We will now discuss in some detail the attempts of Grood, Phillips, and Mates (1979).

Grood et al. (1979) have adopted a two-component myocardial element in which the contractile component is in parallel with the elastic matrix component (Fig. 6-4). They put down the force equilibrium equations for the element relating the total stress in the element to the stresses in the contractile and passive elastic components. The elastic matrix stress is derivable from a strain energy function W. By invoking the incompressibility criterion $\lambda_x \lambda_y \lambda_z = 1$, they obtain a relation for the component stresses in the fiber direction (n):

$$\sigma_n^T = \gamma_E \sigma_n^E + \sigma_n^c \tag{5}$$

which is adopted to determine the following relationship for the circumferential midwall contractile stress at the equator of an ellipsoidal shaped left ventricular model (where the fibers are oriented circumferentially):

$$\sigma_\theta^c = \sigma_\theta^T - \gamma_E(\sigma_r^T + \lambda_\theta \frac{\delta W}{\delta \lambda_\theta}) \tag{6}$$

wherein γ_E is the fraction of a wall cross-sectional area that is occupied by the elastic matrix (taken to be equal to 0.93), and λ_θ is the extension ratio. From the above relationship, Grood et al. (1979) empirically determine the value of σ_θ^C by (1) determining the circumferential wall stress σ_θ^T from a thick-walled ellipsoidal model of Falsetti (1970), (2) assuming the radial wall stress σ_r^{TT} to be equal to $-P/2$, (3) employing equation 6 for the diastolic phase (when σ_θ^c is zero) for evaluating the strain energy function term

$$\gamma_E \lambda_\theta \frac{\delta W}{\delta \lambda_\theta} = \sigma_\theta - \gamma_E \sigma_{rr} \tag{7}$$

wherein the values of the right hand side terms are conveniently adopted as their average diastolic values, equal to half their end-diastolic values.

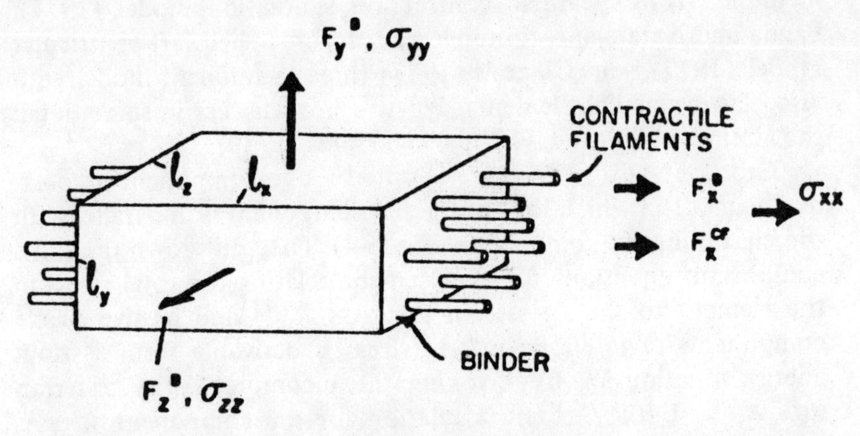

Figure 6-4. A model of heart muscle. One-dimensional active contractile filaments are embedded in a passive three-dimensional binder. The binder is composed primarily of the intracellular water, but also includes the cell organelles, sarcolemma, connective tissue and interstitial fluid. The normal stress in the x-direction is due to forces acting in both the contractile filaments and the binder. In the y and z directions the normal stress is supported by the binder alone. From E.S. Grood, C.A. Phillips, and R.E. Mates, Contractile Filament Stress in the Left Ventricle and Its Relationship to Wall Stress. *Journal of Biomechanical Engineering, 101*:225-231, 1979.

Phillips and Grood (1978) calculated both the Lagrangian stress (σ) and contractile filament stress (σ_{CF}) for both the circumferential (subscript θ) and longitudinal (subscript ϕ) directions in 39 patients:

$$\sigma_\theta = (\sigma_\theta)^* \text{ (Area Ratio)}_\theta \quad (8)$$

$$\sigma_\phi = (\sigma_\phi)^* \text{ (Area Ratio)}_\phi \quad (9)$$

$$\sigma_{CF\theta} = \sigma_\theta + (P/2) \text{ (Area Ratio)}_\theta \quad (10)$$

$$\sigma_{CF\phi} = \sigma_\phi + (P/2) \text{ (Area Ratio)}_\phi \quad (11)$$

where:

$(\sigma_\theta)^*, (\sigma_\phi)^*$ = an uncorrected circumferential (longitudinal) left ventricular wall stress (Falsetti et al., 1970),

$(\text{Area Ratio})_{\theta,\phi}$ = a normalization term (*vide infra*),

P/2 = an approximation of the radial stress component (Grood, et al., 1979).

Contractile filament stress was 24 percent higher than the Lagrangian stress in the circumferential direction on the average for the thirty-nine patients (Fig. 6-5).

Although the above analyses (Grood et al., 1979; Phillips and Grood, 1978) assume a uniform stress distribution across the wall, recently, investigators have used thick-walled theories to predict nonuniform stress distributions. Wong and Rautaharju (1968) analyzed the stress distributions assuming the heart is a passive, isotropic system. Their analysis applied the above approximations, except the geometry of the heart was modeled as a thick ellipsoidal shell with equal semiminor axes. This model predicted wall stresses that were highest at the endocardium and lowest at the epicardium. Streeter et al. (1970) developed a similar model but considered the curvature and orientation of the muscle fibers. His model predicted stresses that were highest at the midwall and lowest at the endocardium and epicardium. This method is limited in that empirical fiber curvature data is unavailable for the entire cardiac cycle. Mirsky (1970) employed a thick-walled, prolate spheroid of constant wall thickness as his model. He also included relations for various degrees of anisotropy and nonhomogeneity of the myocardium. He assumed the circumferential modulus of elasticity was parabolic in nature so that his predicted wall stress distribution corresponds with Streeter et al. (1970) predictions based on fiber orientation. Hanna (1973) used a ten-layered spherical myocardium as his model. The model, based on the force-length-velocity dependence of muscle, predicted the greatest stress at the epicardium and the lowest at the endocardium. Janz and Waldron (1975) showed that nonhomogeneity of the myocardium is required in order to calculate a constant fiber stress through the myocardial wall. Their model required that the myocardial elastic modulus be lowest at the endocardium and highest at the epicar-

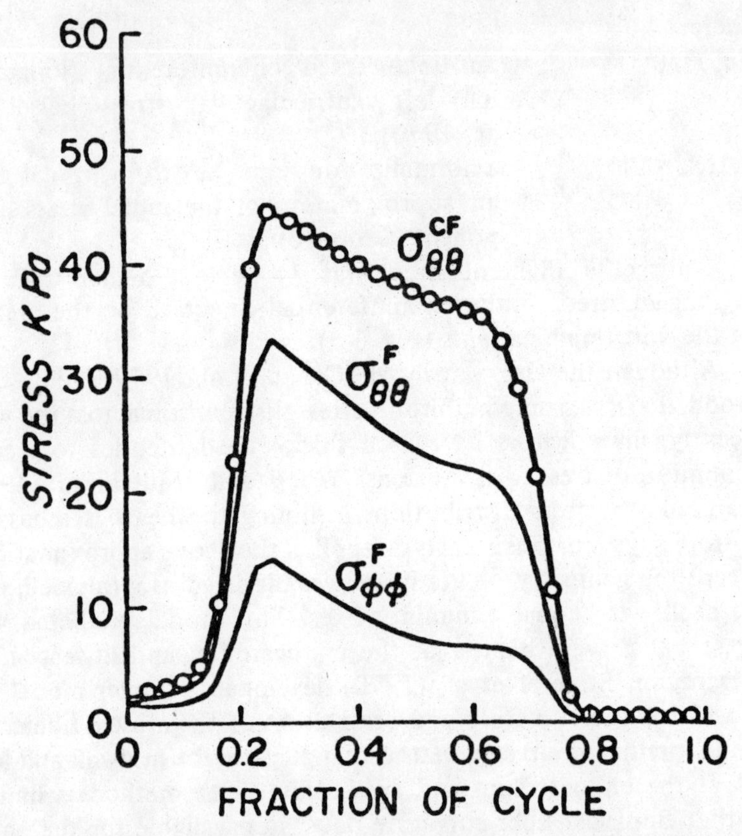

Figure 6-5. Average fiber corrected wall stress and contractile filament stress calculated during a cardiac cycle. The circumferential stresses are approximately twice the longitudinal stresses. The fiber corrected wall stress, $_{\sigma}F$, is less than the corresponding contractile filament stress, $_{\sigma}CF$, due to the negative (compressive) stress in the binder. From E.S. Grood, C.A. Phillips, and R.E. Mates, Contractile Filament Stress in the Left Ventricle and Its Relationship to Wall Stress. *Journal of Biomechanical Engineering, 101*:225-231, 1979.

dium. Mirsky (1976) showed in a very simple analysis that since strain (endocardial) is greater than strain (midwall) which, in turn, is greater than strain (epicardial), and if constant wall stress is assumed, then modulus of elasticity (E) increases from endocardium to epicardium for an incompressible material.

Phillips et al. (1981) investigated the effects of anisotropy and nonhomogeneity on LV circumferential stress distribution as represented by a linear increase in the midwall effective modulus ($\bar{E}_A$) (Fig. 6-6). The proportionality constant defining the $\bar{E}_A$'s rate of change as a function of wall thickness was defined as $\bar{A}$ (Fig. 6-7). Biological studies of the collagen distribution suggest an increase in the effective modulus from endocardium to epicardium. $\bar{A}$ was constrained by the assumption that stress per unit sarcomere length is constant. The rate of change of $\bar{A}$ with strain,

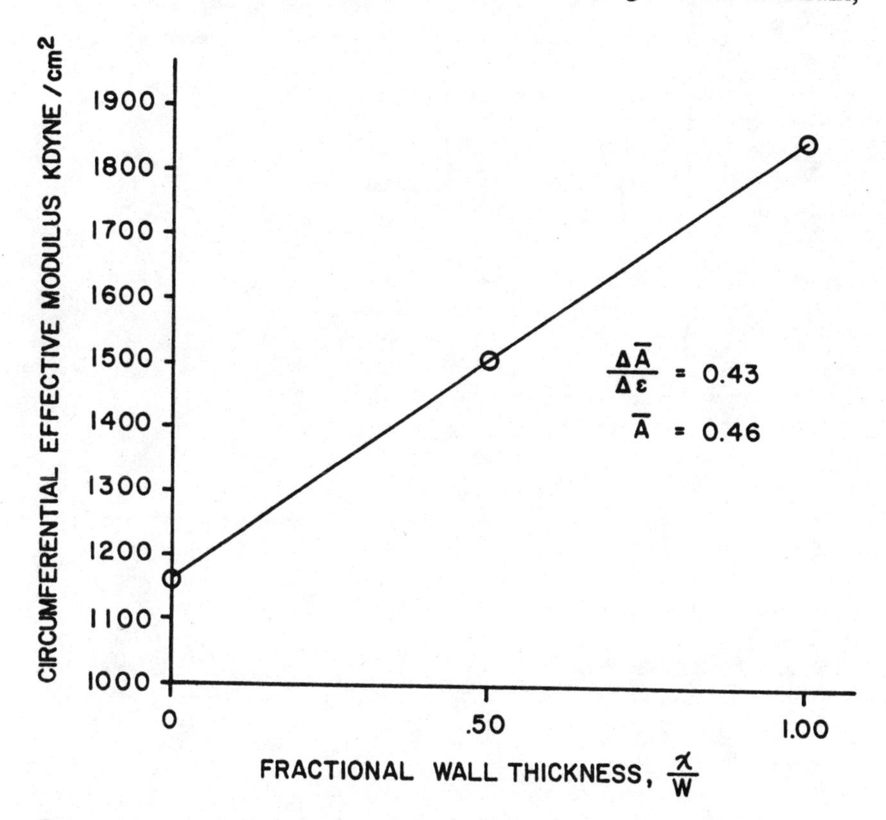

Figure 6-6. A linear increase in mid wall effective modulus ($\bar{E}_A$) as a function of fractional wall thickness (X/W).

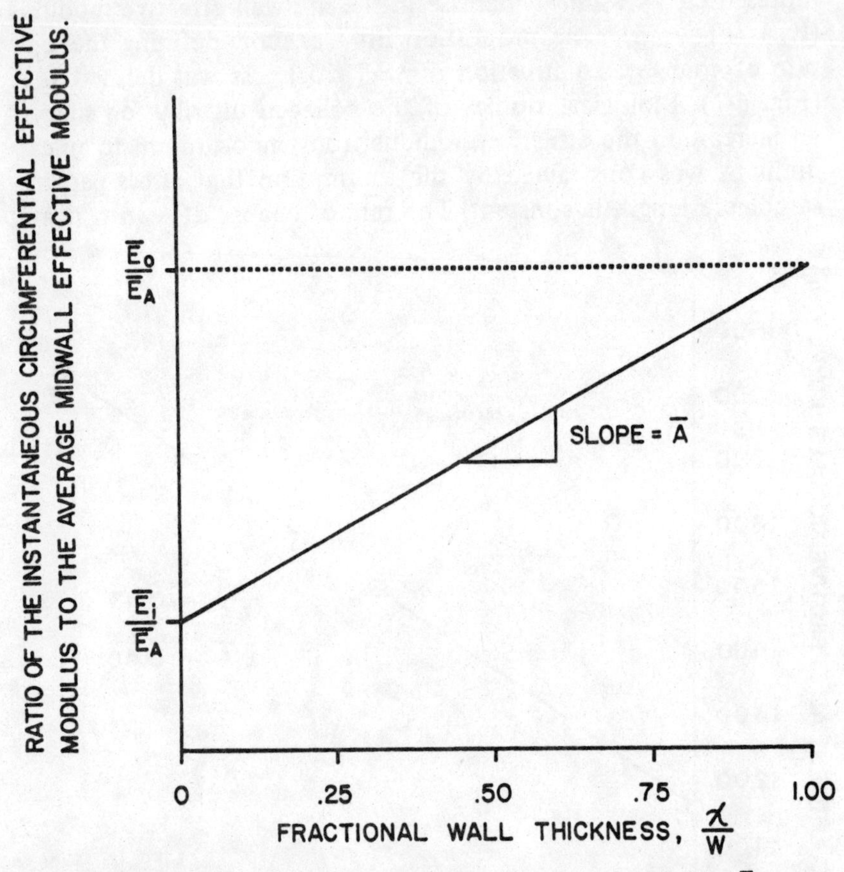

Figure 6-7. Definition of $\bar{A}$ as a proportionality constant defining $\bar{E}_A$'s rate of change as a function of wall thickness. In this figure $\bar{E}_i$ and $\bar{E}_o$ are the instantaneous circumferential effective moduli at the endocardium and epicardium respectively.

defined as $\Delta A / \bar{\Delta}\epsilon$, was 0.43 for twelve functionally normal cases, and 1.09 for nine functionally abnormal cases ($p < 0.0005$). Assuming the myocardium has constant material properties that do not change with functional decompensation, the stress distribution for the nine functionally abnormal cases was calculated with $\Delta A/\bar{\Delta}\epsilon$ equal to 0.43, and then recalculated with $\Delta A/\bar{\Delta}\epsilon$ equal to 1.09. The stress distributions in the first case exhibited higher stresses through the inner half of the myocardium and lower stresses through the outer half than the stress distribution in the second case, indicating that the inner fibers are overloaded and the outer fibers underloaded in left ventricular decompensation.

Obviously, further work in the study of myocardial contractile stress versus shortening strain must await new studies of the actual material properties of the systolic left ventricle. Such "active" material properties will certainly be quite different from the passive diastolic material properties (Mirsky and Parmley, 1974b).

MYOCARDIAL CONTRACTILE POWER

It has been established over the last fifteen years that the left ventricle should be viewed as a muscle as well as a pump in order to obtain a comprehensive evaluation of LV function (Sonnenblick et al., 1969). Work is an important part of the total myocardial oxygen consumption and must be considered in any overall energy balance (Badeer, 1969).

Muscle element work, defined as the time integral of contractile element (CE) power or series elastic (SE) power, allows ventricular work to be defined in terms of the contractile and elastic properties of the muscle itself. Although originally thought to correlate with myocardial oxygen consumption (Britman and Levine, 1964), CE work was later shown to have a variable response to oxygen consumption in both the canine left ventricle (Graham et al., 1967) and the isolated papillary muscle strip (Coleman, 1968).

Phillips et al. (1979a) developed the appropriate mathematical description for (a) the muscle element work and power in terms of the actual contractile filament stress seen by the CE and SE and (b) translating such work and power equations in terms of standard cineangiographic pressure and geometry measurements. The contractile element power (P_{CE}) and series elastic power (P_{SE}) were individually defined:

$$P_{CE} = -(\sigma_\theta + \sigma_r) \frac{V_{CE}}{l_{CF}} \qquad (12)$$

$$P_{SE} = -(\sigma_\theta + \sigma_r) \frac{V_{SE}}{l_{CF}} \qquad (13)$$

where:

σ_θ = circumferential contractile filament stress (*vide supra*),

σ_r = a radial stress component, approximated as one-half the transmural pressure,

V_{CE}, V_{SE} = velocities of the contractile element and series elastic element respectively, and

l_{CF} = length of a hypothetical midwall circumferential fiber.

CE and SE work is simply the time integral of CE and SE power.

Phillips and Grood (1979) appear to be the only study in the literature that have examined peak power of both the CE and the SE in the circumferential direction during both systole and diastole and also used this information to evaluate various functional abnormalities of the human left ventricle. In their study of thirty-nine patients (comprising five clinical groups), they noted that there are two distinct time phases of energy development-dissipation (Fig. 6-8): the first time phase (1) corresponds roughly to ventricular systole, and the second time phase (2) corresponds roughly with ventricular diastole. Each time phase, in turn, is characterized by an energy development element (a) and an energy dissipating element (b). The four resultant "characteristic curves" are then identified as energy being generated by the CE during systole (1a), energy being stored by the SE during systole (1b), energy being released by the SE during diastole (2a), and energy being dissipated by the CE during diastole (2b).

Referring to Table 6-I, reproduced here from Table 3 of Phillips and Grood (1979), only two of the work and power parameters are necessary to distinguish the four pathological groups from the normals: stored SE power and dissipated CE power. Both represent secondary phases of power development-dissipation, and all are shifted in the direction of an increased (less negative) value. One other power parameter provides additional information about the cardiomyopathy group. Generated CE power represents a primary phase of power development-dissipation and is shifted in the direction of a decrease (less positive). Three of the pathological groups (decompensated volume overload, compensated pressure overload, and congestive cardiomyopathy) have significant changes in the distribution between internal and external peak power as compared to normals (Table 6-I). A physical interpretation of these characteristic changes has been offered by the authors.

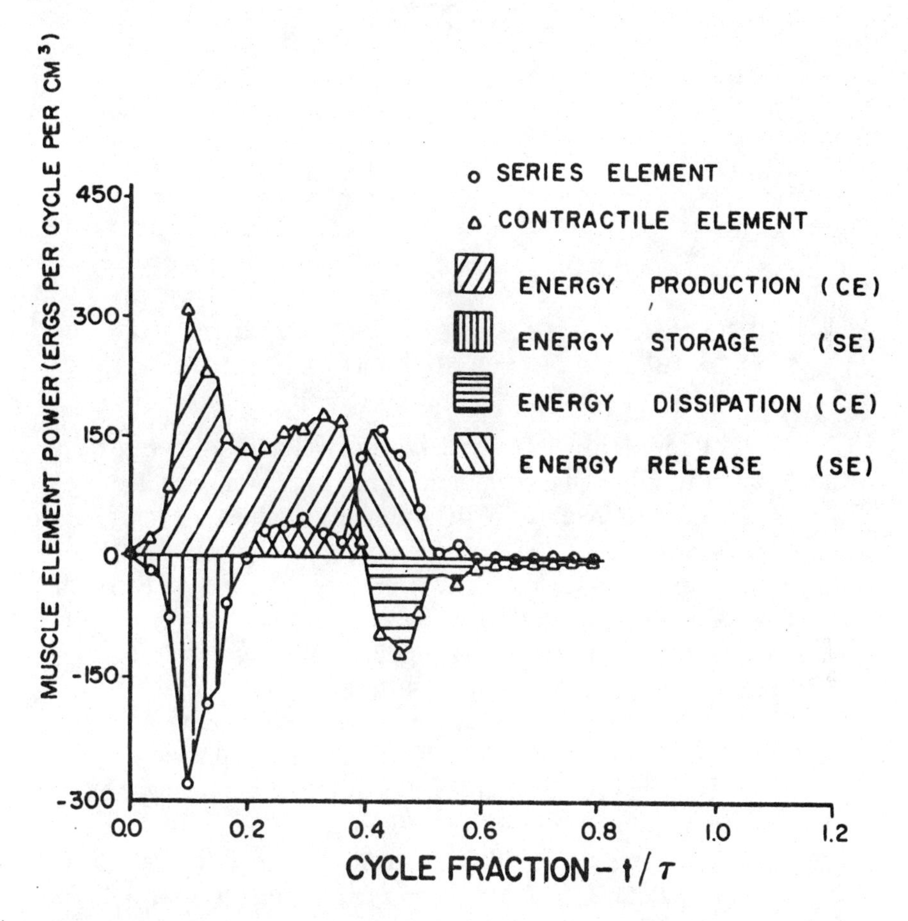

Figure 6-8. The four phases of energy development and dissipation throughout a cardiac cycle. Reprinted with permission from *Journal of Biomechanics, 12*:559-566, C.A. Phillips and E.S. Grood, Contractile Filament and Series Elastic Work and Power–II. Copyright 1979, Pergamon Press, Ltd.

Table 6-I

CHANGES IN POWER, WORK, AND INTERNAL, EXTERNAL POWER AND WORK WHICH DISTINGUISH THE PATHOLOGICAL GROUPS

Parameter	Compensated volume overload	Decompensated volume overload	Compensated pressure overload	Cardiomyopathy
Stored *SE* power		↑	↑	↑
Dissipated *CE* power	↑			↑
Generated *CE* power				↓
Internal power (%)		↓	↓	
External power (%)		↑	↑	
Internal work (%)			↓	↑
External work (%)			↑	↓

Reprinted with permission from *Journal of Biomechanics*, 12:559-566, C. A. Phillips and E. S. Grood, Contractile Filament and Series Elastic Work and Power—II. Copyright 1979, Pergamon Press, Ltd.

RESULTING PRESSURE DISTRIBUTION AND MOTION OF THE FLUID (WALL MOTION) IN THE LEFT VENTRICULAR CHAMBER

The determination of blood motion velocity distribution in the left ventricular chamber merely requires wall velocity data, which can be obtained from the frame-by-frame left ventricular imaging. Likewise, the relative pressure distribution can be obtained from merely the dynamic geometry data, whereas for the absolute pressure distribution we need the value of the catheter reading. The instantaneous wall velocity distribution can be derived from the successive frames' outlines, by adopting the mode of translation of endocardial wall points based on the tracking of implanted wall markers in the walls of dog hearts by Ingels et al. (1975). A rigorous two-dimensional finite element analysis of the motion of a viscous incompressible fluid contained in the irregular chamber has been performed (as a steady flow between successive frames) by Ray et al. (1979), using a velocity-pressure formulation of the nonlinear Navier-Stokes equations.

Based on this analysis, the pressure distribution relative to the pressure at the aortic outflow tract has been computed for patients with and without myocardial disease. In Figure 6-9, which illustrates the results for a patient with coronary lesion before and after coronary bypass surgery, we can see how we can employ this result to demonstrate an improved myocardial contractility, after coronary bypass surgery, in the form of a more

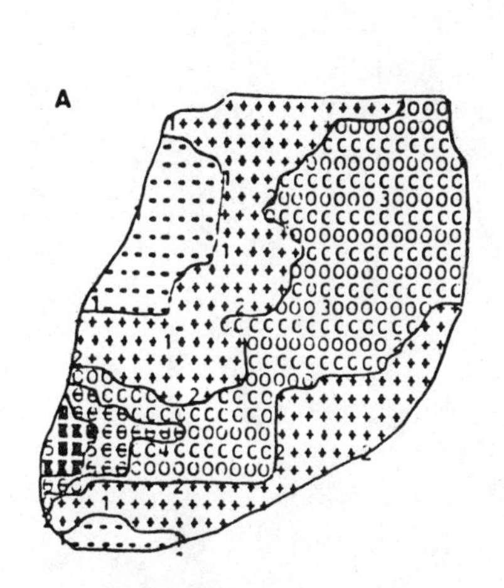

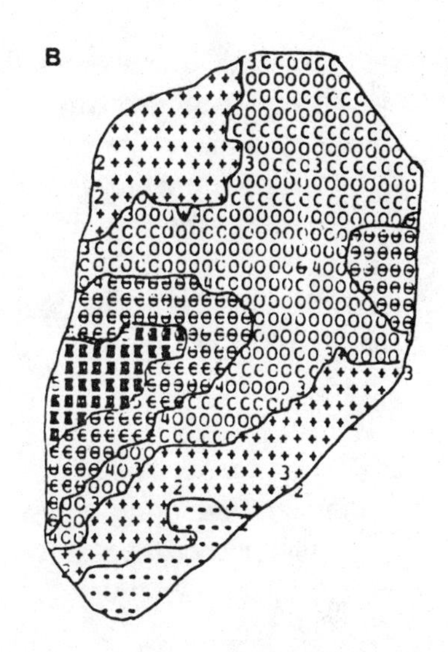

NONDIMENSIONAL PRESSURE		SYMBOL	% LV CHAMBER AREA AT PRESSURE LEVEL		COMMENTS (A NEO INTRINSIC MEASURE OF IMPROVED CONTRACTIVITY AFTER CORONARY BYPASS)
LEVEL	VALUE (P/P_o) (P_o = 1, AT AV BOUNDARY OF LV CHAMBER)		BEFORE BYPASS	AFTER BYPASS	
1	0.80 – 0.90	- - -	12.7 } = 56.3	5.8 } = 32.4	DECREASE OF ZONE OF LOW PRESSURE LEVEL (1 & 2) BY 23.9% AFTER BYPASS
2	0.90 – 0.98	+ + +	43.6 }	26.6 }	
3	0.98 – 1.02	0 0 0	38.0 = 38.0	45.1 = 45.1	INCREASE OF ZONE OF PRESSURE LEVEL (3) BY 7.1% AFTER BYPASS
4	1.02 – 1.08	*0 0 0*	3.8 } = 5.7	17.4 } = 22.5	INCREASE OF ZONE OF HIGH PRESSURE LEVELS (4 & 5) BY 16.8% AFTER BYPASS
5	1.08 – 1.16	■ ■ ■	1.9 }	5.1 }	

Figure 6-9. Regional pressure gradients in the left ventricular chamber of a patient with a coronary artery lesion before (A) and after (B) by-pass surgery (*see* text). From D.N. Ghista, G. Ray, and H. Sandler, Cardiac Assessment Mechanics: 2. *Medical and Biological Engineering and Computing, 18*: 344-352, 1980.

widespread pressure gradient towards the aortic outflow tract, resulting from a return to normal wall kinematics of the originally hypokinematic diseased or infarcted wall segment.

A new concept of contractility has thus been introduced. This is a more mechanically functional, graphically observable and hence diagnostically attractive index. It circumvents the need for modifying the traditional analyses of the "myocardial stress versus velocity-of-shortening" type of index, empirically based on the one-dimensional muscle dynamics analogs (Mirsky and Parmley, 1974a; Mirsky et al. 1974) to incorporate the anatomically realistic concept of the myocardial medium comprising activable fibers embedded in a passive stress-strained matrix.

The experimental counterpart of this analytical method of intraventricular pressure (and velocity) distribution determination is the technique of monitoring intracardiac blood flow with a combined use of the ultrasonic pulsed doppler technique and two-dimensional echocardiography. This technique, which permits imaging of the general features of flow distribution in the left ventricle, has been advanced by Nimura et al. (1977) and Oddon et al. (1978). It can be employed, in conjunction with the above mentioned finite-element method, to provide mutual checking of the results of the two techniques as well as a comprehensive quantitative on-line graphically available diagnostic tool to detect valvular disorders, to provide on-line visualization of the effect of a coronary lesion on the intra-ventricular flow patterns (and the contractility index value) and its alteration following coronary bypass surgery, and to check the *in vivo* performance of an implanted prosthetic heart valve.

Phillips et al. (1978) have defined the ratio of Lagrangian to Eulerian stresses as an area ratio—the ratio of instantaneous area to reference area—and interpreted this as a measure of the amount of local tissue deformation. The area ratio is an important correction factor by which the force per unit area (Eulerian stress) can be converted to the force for a given number of fibers (Lagrangian stress). In their study of thirty-six patients comprising four clinical groups (Phillips et al., 1978), left ventricular area ratios and their time course appear to provide additional information about left ventricular function in different disease states. Specifically, they are an index of the ventricle's ability to compensate for volume overload, as well as a sensitive indicator of the hypocontractile ventricle.

Single indices of cardiac function have included both ventricular volume and velocity changes as well as ventricular flow (ejection rate). These parameters have been studied with respect to their ability to distinguish patient populations with various types of cardiac pathology: (1) the percentage change in the internal circumferential radius from end-diastole to end-systole (Gould et al., 1974); (2) the circumferential fiber shortening velocity computed as the sum of the rate change of the internal radius and wall thickness (Gault et al., 1968); (3) item 2 normalized for the instantaneous circumferential fiber length (Karliner et al., 1971); and (4) item 2 normalized by the end-diastolic fiber length (Hammermeister et al., 1974). However, there is a significant degree of overlap among patient groups that limits the usefulness of such single indices to distinguish any individual patient with cardiac dysfunction from normals. The velocity-strain relationships has been defined as a test of cardiac function in which the circumferential fiber-shortening velocity is plotted against circumferential fiber strain during a single cardiac cycle, including both systole and diastole (Phillips, 1977). The parameters of interest are peak systolic and diastolic circumferential fiber velocity ($+V_{CF}$ and $-V_{CF}$), systolic strain at $+V_{CF}$ (ϵ_S) and diastolic strain at $-V_{CF}$ (ϵ_D) (Figure 6-10), and the systolic and diastolic time constants (τ_S and τ_D) defined as

$$\tau_S = \frac{\epsilon_S}{+V_{CF}} \tag{14}$$

$$\tau_D = \frac{\epsilon_D}{-V_{CF}} \tag{15}$$

Phillips (1977) found in thirty-nine patients studied that two inequalities, $+V_{CF} \leqslant -V_{CF}$ and $\epsilon_D \leqslant \epsilon_S$, distinguish individuals in the depressed cardiac function group from individuals with normal or compensated cardiac function (Table 6-II reproduced from Table 4 of Phillips, 1977).

The velocity-strain relationship can perform this individual discrimination to a high degree because it allows each patient to

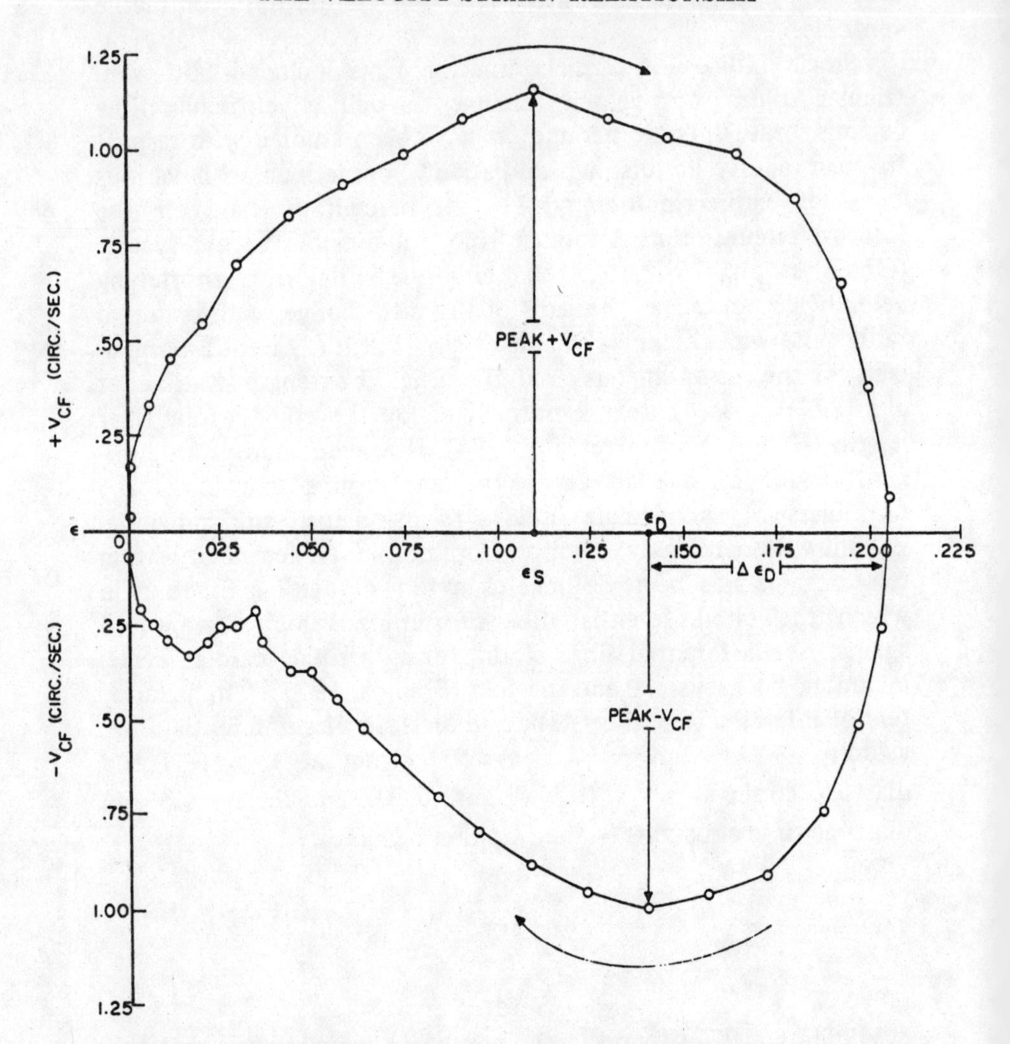

Figure 6-10. The velocity-strain relationship for a normal patient indicating peak $+V_{CF}$, peak $-V_{CF}$, ϵ_S and ϵ_D (*see* text). From C.A. Phillips, The Velocity-strain Relationship: Application to Normal and Abnormal Left Ventricular Function. *Annals of Biomedical Engineering, 5*:329-342, 1977.

be his own control, i.e. systolic events are compared to diastolic events. In patients with normal or compensated cardiac function, the inequality, $+V_{CF} \leqslant -V_{CF}$, is variable, but is very often satisfied in patients with depressed cardiac function. Diastolic events established the baseline peak $-V_{CF}$ and systolic events represent the change from baseline. Thus, in a patient with depressed cardiac function, a reduced peak $+V_{CF}$ often occurs, as compared to its peak $-V_{CF}$, possibly indicating a reduced ability of the ventricle to actively contract. However, this response is variable in normal or compensated ventricles, so an additional inequality is required.

Table 6-II

VELOCITY, STRAIN, AND TIME INEQUALITIES FOR ALL 39 PATIENTS

Case	$+V_{CF} \leq \lvert -V_{CF} \rvert$	$\epsilon_D \leq \epsilon_S$	$\tau_D \leq \tau_S$
Group 1: Normals			
1	−[a]	−	−
2	+[b]	+	−
3	+	−	+
4	−	−	−
5	+	−	−
6	−	−	−
7	−	+	+
8	+	−	+
9	−	−	−
10	+	−	−
11	−	−	−
12	−	−	−
13	−	+	−
14	+	−	−
15	+	+	+
Total[c] ++: 2 out of 15			
Group 2: CVO			
16	+	+	+
17	+	−	+
18	−	−	−
19	−	−	−
20	+	+	+
21	−	−	−
Total ++: 2 out of 6			
Group 3: DVO			
22	+	−	+
23	+	+	+
24	+	+	+
25	+	+	+
26	+	+	+
27	−	−	−
28	−	+	−
29	+	+	+
30	+	+	+
Total ++: 6 out of 9			
Group 4: CPO			
31	+	−	+
32	+	+	+
33	−	−	−
Total ++: 1 out of 3			
Group 5: Cardiomyopathy			
34	+	+	+
35	+	+	+
36	+	+	+
37	−	−	−
38	+	+	+
39	+	+	+
Total ++: 5 out of 6			

[a] Inequality is not satisfied.
[b] Inequality is satisfied.
[c] ++ indicates that both inequalities ($+V_{CF} \leq \lvert -V_{CF} \rvert$ and $\epsilon_D \leq \epsilon_S$) are satisfied.

From C. A. Phillips, The Velocity-strain Relationship: Application to Normal and Abnormal Left Ventricular Function. *Annals of Biomedical Engineering*, 5:329-342, 1972.

In patients with depressed cardiac function, the inequality, $\epsilon_D \leqslant \epsilon_S$, is very often satisfied, but infrequently satisfied with normal or compensated cardiac function. In this case, systolic events establish the baseline ϵ_S and diastolic events represent a change in ϵ_D. Thus, in patients with depressed cardiac function, a reduced ϵ_D often occurs, as compared to its ϵ_S, indicating a prolonged fractional increase in the midwall circumference (prior to reaching peak $-V_{CF}$). This results in a peak $-V_{CF}$ being reached at a lower amount of strain, and τ_D (proportional to ϵ_D) is, therefore, less than τ_S (proportional to ϵ_S).

The system time constants τ_D and τ_S are also inversely proportional to $-V_{CF}$ and $+V_{CF}$, respectively. Since $-V_{CF} \leqslant +V_{CF}$ for the decompensated left ventricle, this further contributes to the tendency for τ_D to be less than τ_S. A smaller system time constant is characteristic of a decreased viscoelastic compliance, i.e. increased stiffness, in a first-order linear system. As might be expected in the decompensated left ventricle, it is the passive diastolic stiffness that is increased compared to its baseline systolic stiffness.

The velocity-strain loop is independent of intraventricular pressure, making it possible to acquire this information noninvasively by using echocardiography or cardiac nuclear scintigraphy (see Chapter 9).

REFERENCES

Badeer, H.S. (1969) Work and energy expenditure of the heart. *Acta Cardiol.*, 24:227-241

Bloomfield, M.E., Katz, A.I., Reddy, R.V., Gold, L.D., and Moreno, A.H. (1971) Myocardial contractile state in terms of power generation. *Clinical Research*, 19:306

Bloomfield, M.E., Gold, L.D., Reddy, R.V., Katz, A.I., and Moreno, A.H. (1972) Thermodynamic characterization of the contractile state of the myocardium. *Circulation Research*, 30: 520-534

Bornhorst, W.J. and Minardi, J.E. (1970a) A phenomenological theory of muscle contraction. I. Rate equations at a given length based on irreversible thermodynamics. *Biophysical*

Journal, 10:137-154

Bornhorst, W.J. and Minardi, J.E. (1970b) A phenomenological theory of muscle contraction. II. Generalized length variations. *Biophysical Journal*, 10:155-171

Britman, N.A. and Levine, H.J. (1964) Contractile element work: A major determinant of myocardial oxygen consumption. *J. Clin Invest.*, 43:1397-1408

Caplan, S.R. (1966) A characteristic of self-regulated linear energy converters. The Hill force-velocity relation for muscle. *J. Theoretical Biology*, 11:63-86

Caplan, S.R. (1968a) Autonomic energy conversion. I. The input relation: Phenomenological and mechanistic considerations. *Biophysical J.*, 8:1146-1166

Caplan, S.R. (1968b) Autonomic energy conversion. II. An approach to the energetics of muscular contraction. *Biophysical J.*, 8:1167-1193

Coleman, H.N. (1968) Effect of alterations in shortening and external work on oxygen consumption of cat papillary muscle. *Am. J. Physiol.*, 214:100-106

Curtin, N.A., Gilbert, C., Kretyschmar, K.M., and Wilkie, D.R. (1974) The effect of the performance of work on the total energy output and metabolism of muscle. *J. Physiology*, 238: 455-472

Falsetti, H.L., Mates, R.E., Grant, C., et al., (1970) Left ventricular wall stress calculated from one-plane cineangiography. *Circulation Res.*, 26:71-83

Fung, Y.C.B. (1971) Comparison of different models of heart muscle. *J. Biomechanics*, 4:289

Fung, Y.C.B. (1970) Mathematical representation of the mechanical properties of the heart muscle. *J. Biomechanics*, 3:381

Gault, J.H., Ross, J., Jr., and Braunwald, E. (1968) Contractile state of the left ventricle in man. *Circulation Research*, 22: 451-463

Ghista, D.N. (1974) Rheological modelling of the intact left ventricle. In *Cardiac Mechanics*, (Ed.) I. Mirsky, D.N. Ghista and H. Sandler. New York, Wiley-Interscience

Ghista, D.N., Ray, G. and Sandler, H. (1980) Cardiac assessment mechanics: 2. *Medical and Biological Engineering and Comput-*

ing, 18:344-352

Gould, K.L., Lipscomb, K., Hamilton, G.W., and Kennedy, J.W. (1974) Relation of left ventricular shape, function and wall stress in man. *Am. J. Cardiology*, 34:627-634

Glantz, S.A., (1977) A three-element description for muscle with viscoelastic passive elements. *J. Biomechanics*, 10:5

Graham, T.P., Jr., Ross, J., Jr., Covell, J.W., Sonnenblick, E.H., and Clancy, R.L. (1967) Myocardial oxygen consumption in acute experimental cardiac depression. *Circulation Res.*, 21: 123-138

Grood, E.S., Phillips, C.A. and Mates, R.E. (1979) Contractile filament stress in the left ventricle and its relationship to wall stress. *J. Biomechanical Engineering*, 101:225-231

Hanna, W.T. (1973) A simulation of human heart function. *Biophys. J.*, 13:603-621

Hammermeister, K.E., Brooks, R.C., and Warbasse, J.R. (1974) The rate of change of left ventricular volume in man. *Circulation*, 49:729-738

Ingels, N.B., Daughters, G.T., Stinson, E.B., and Alderman, E.L. (1975) Measurement of midwall myocardial dynamics in intact man by radiography of surgically implanted markers. *Circulation*, 55:859-867

Janz, R.F. and Waldron, R.J. (1975) Some implications of a constant fiber stress hypothesis in the diastolic left ventricle. *Bulletin of Mathematical Biology*, 38:401-413

Karliner, J.S., Gault, J.H., Eckberg, D., Mullins, C.B. and Ross, J., Jr. (1971) Mean velocity of fiber shortening. *Circulation*, 44:323-333

Mirsky, I. (1969) Left ventricular stresses in the intact human heart. *Biophysical J.*, 9:189-208

Mirsky, I. (1970) Effects of anisotropy and non-homogeneity on left ventricular stresses in the intact heart. *Bulletin of Mathematical Biophysics*, 32:197-213

Mirsky, I. (1976) Assessment of passive elastic stiffness of cardiac muscle: mathematical concepts, physiologic and clinical considerations, directions for future research. *Progress in Cardiovascular Diseases*, 18:227-308

Mirsky, I. and Parmley, W.W. (1974a) Force-velocity studies in isolated and intact heart muscle. In *Cardiac Mechanics* (Ed.) I. Mirsky, D.N. Ghista, and H. Sandler. New York, Wiley-Interscience.

Mirsky, I. and Parmley, W.W. (1974b) Evaluation of passive elastic stiffness for the left ventricle and isolated heart muscle. In *Cardiac Mechanics*. (Ed.) I. Mirsky, D.N. Ghista, and H. Sandler, New York, Wiley-Interscience

Mirsky, I. Pasternac, H., Ellison, R.C., and Hugenholtz, P.G. (1974) Clinical applications of force-velocity parameters and the concept of a normalized velocity. In *Cardiac Mechanics*. (Ed.) I. Mirsky, D.N. Ghista, and H. Sandler, New York, Wiley-Interscience

Moreno, A.H. (1976) Myocardial contractility and isovolumic power and rate of change of power. *Am. J. Cardiology*, 38: 123-124

Nimura, Y., Matsuo, H., Kitabatake, A., Hayaski, T., Asso, M., Terao, Y., Senda, S., Sakakibara, H., and Abe, H. (1977) Studies on the intracardiac blood flow with a combined use of the ultrasonic pulsed doppler technique and two-dimensional echocardiography from a transcutaneous approach. In *Ultrasound in Medicine*, Vol. 3B. (Ed.) D. White and R.E. Brown, New York, Plenum Pub. Corp.

Oddon, C., Brun, P., Dantan, P., and Kulas, A., (1978) Relation entre mechanique du fluide intracardiaque et dynamique de la valve mitral. *J. Fr. Biophysics et Med. Nucl.*, 1:61

Panda, S.C. and Natarajan, R. (1977) Finite element method of stress analysis in the human left ventricular layered wall structure. *Medical and Biological Eng. and Computing*, 15:67

Phillips, C.A. (1977) The velocity-strain relationship: Application to normal and abnormal left ventricular function. *Ann. Biomed. Eng.*, 5:329-342

Phillips, C.A., Cox, T.L., and Petrofsky, J.S. (1981) Active material properties of the myocardium: Correlation with left ventricular function in man. *Ohio J. Sci.*, 81:251-260

Phillips, C.A. and Grood, E.S. (1978) Contractile filament stress: Comparison of different disease states in man. *Ohio J. Sci.*, 5: 259-266

Phillips, C.A. and Grood, E.S. (1979) Contractile filament and series elastic work and power–II. Clinical correlations with left ventricular dysfunction. *J. Biomech.*, 12:559-566

Phillips, C.A., Grood, E.S., Mates, R.E., and Falsetti, H.L. (1978) Left ventricular function: Correlation with deformation of the myocardium. *J. Biomech. Eng.*, 100:99-104

Phillips, C.A., Grood, E.S., and Schuster, B. (1979a) Contractile filament and series elastic work and power–I. Mathematical development and application to cineangiographic measurements. *J. Biomech.*, 12:551-557

Phillips, C.A., Grood, E.S., Scott, W.J., and Petrofsky, J.S. (1979b) Cardiac chemical power: I. Derivation of the chemical power equation and determination of equation constants. *Med. Biol. Engr. Comp.*, 17:503-509

Phillips, C.A., Grood, E.S., Scott, W.J., and Petrofsky, J.S. (1979c) Cardiac chemical power: II. The application of chemical power, work and efficiency equations to characterize left ventricular energetics in man. *Med. Biol. Engr. Comp.*, 17:510-517

Powell, T. (1971) The validity of applying irreversible thermodynamics to muscular contraction. *Physics in Medicine and Biology*, 16:233-242

Ray, G., Ghista, D.N., and Sandler, H. (1979) Mechanocardiography: theory, evaluation and significance of the regional distributions of myocardial constitutive properties and blood pressure in the left ventricular chamber. In *Cardiovascular Engineering* (Ed.) D.N. Ghista, E. VanVollenhoven, W.J. Yang, and H. Reul, Baden-Baden, Germany, Gerhard Witzstrock

Sonnenblick, E.H., Parmley, W.W., and Urschel, C.W. (1969) The contractile state of the heart as expressed by force-velocity relations. *Am. J. Cardiol.*, 23:488-503

Stein, P.D. and Sabbah, H.N. (1975) Ventricular performance in patients based upon rate of change of power during isovolumic contraction. *Am. J. Cardiology*, 35:258-263

Stein, P.D., McBride, G.G., Sabbah, H.N., and Buck, E.F. (1976) Reply. *Am. J. Cardiology*, 38:124

Streeter, D.D., Vaishnav, R.N., Patel, D.J., Spotnitz, H.M., Ross, J., Jr., and Sonnenblick, E.H. (1970) Stress distribution in the canine left ventricle during diastole and systole. *Biophysical J.*,

10:345-363

Wong, A.Y.K. (1971) Mechanics of cardiac muscle based on Huxley's model: Mathematical simulation of isometric contraction. *J. Biomechanics*, 4:529

Wong, A.Y.K. and Rautaharju, P.M. (1968) Stress distribution in the left ventricular wall approximated as a thick ellipsoidal shell. *Am. Heart J.*, 75:649-662

Chapter 7

A PERSPECTIVE OF VENTRICULAR WALL STRESS ANALYSES

D.N. GHISTA, F.C.P. YIN, and G. JAYARAMAN

CONSIDERABLE study of left ventricular myocardial stress has been carried out to indirectly contribute to an understanding of left ventricular function in health and disease. This chapter reviews the shortcomings and applicability of various left ventricular myocardial stress analyses.

BACKGROUND

Cardiologists, physiologists, and engineers have long been intensely interested in quantification of the forces or stresses acting in the wall of the heart. The extent of this interest is evident in the voluminous literature containing references to ventricular wall stress since Wood's publication late in the last century (Wood, 1892). There are many reasons for this interest in wall stress: (1) Myocardial wall stress is one of the primary determinants of myocardial oxygen consumption (Sarnoff et al., 1958); this correlation may be linked to the dependence of myocardial perfusion variation in the ventricular medium to the variation of the hydrostatic stress (equal to one-third the sum of the three normal stresses). (2) In diseases characterized by abnormal loading of the heart, normalization of wall stress is thought to be the feedback signal that governs the rate and extent of ventricular hypertrophy (Alpert, 1971). Cardiac decompensation is thought to result when this feedback loop dysfunctions. (3) Insight into the fundamental principles underlying ventricular mechanics requires knowledge of the relationship between the stresses and deformations acting on the muscle comprising the wall. Our understanding of ventricular muscle mechanics is, to a large extent, based on studies of the stress or force-length-time relationships of isolated cardiac muscle. Assuming that extrapolation of these concepts to the intact heart is valid, these ventricular stress or force-velocity-length relationships form one of the cornerstones in our understanding of overall ventricular mechanics (Sonnenblick et al., 1969; Levine and Britman, 1964).

Because of the complex structure of the ventricular wall, its highly nonlinear material properties, the large deformations involved, the complex geometry of the ventricle, and the inherent difficulty in accurately measuring most of these parameters, a completely satisfactory approach to quantification of wall stress still eludes us. The approaches that have been used to determine wall stress can be lumped into three major categories. In the first category, a simplified geometric shape, which can be described in terms of a few parameters, such as a sphere, spheroid, or ellipsoid, is used to approximate the shape of the ventricle. Assuming that the cavity pressure represents the loading acting on the wall and assuming certain factors about the material properties of the wall, one can then derive an expression for wall stress in terms of the geometric parameters and intracavitary pressure (Sandler and Dodge, 1963; Wong and Rautaharju, 1968; Ghista and Sandler, 1969; Falsetti et al., 1970; Walker et al., 1971; Mirsky, 1969; 1970; 1973; Moriarity, 1980). Obviously, the information that can be obtained from such an approach is limited because of the simplifying assumptions used. Nevertheless, because of the relative ease of performing the calculations, these approaches are useful in providing some important insight into ventricular mechanics.

Secondly, direct measurement of wall force in the intact heart using various types of strain gauge transducers coupled directed to the myocardium has been attempted over the years (Hefner et al., 1962; Feigl et al., 1967; Burns et al., 1971; Lewartowski et al., 1972; McHale and Greenfield, 1973; Robie and Newman, 1974; Huisman et al., 1980a). The measured values were compared to the values predicted by one or more of the models described previously and, if the comparison was reasonable, the directly measured values were assumed to be valid. Recently, some of the limitations and problems associated with direct measurement techniques have been clearly reviewed by Huisman et al. (1980a), who sought to quantify the uncertainty in the measured values of wall stress related to the degree of coupling between the transducer and the muscle wall. An improved transducer they designed was compared to another type (Feigl et al., 1967). This study concluded that (1) the uncertainties produced by the coupling of the instrumentation to the muscle do not allow a reliable direct quantification of wall stresses and (2) that measurement of intramyocardial pressures are unlikely to reveal much insight into wall stresses since these techniques do not account for directionality of the stress components, all methods involve considerable surgical trauma to the wall; and the methods do not measure tensile stresses. However, direct measurement of intramyocardial pressures or hydrostatic stress should not be expected to yield information concerning wall stress. Rather, its variation in the wall may be monitored, and correlated with wall perfusion variation and even with the instantaneous distributions of active and passive myocardial elements.

With the advent of high speed digital computers, numerical approximation schemes such as the finite-element method of structural analysis (Zienkiewicz, 1977) have evolved, resulting in the third category of methods for assessing ventricular wall stress. The finite element method of structural analysis evolved in the late 1950s as the need arose in the aerospace industry for structural analysis of structures with complex geometrics, nonlinear material properties, and large deformations. This method is a numerical approximation scheme in which a continuous structure is conceived as being comprised of a finite number of subunits of elements. Each element interacts with neighboring elements only at certain points called nodes so that forces and displacements are transmitted only at the nodes. The relationship between the forces and displacements within each element is derived by applying a fundamental theorem of mechanics. By enforcing the conditions that a node that is common to several elements must have a single value of force and displacement and that the entire structure must be in physical equilibrium, one arrives at a set of algebraic equations relating the forces and displacement at each node. Solving this set of equations, subject to the loading and boundary conditions acting on the structure, enables calculation of the stresses and strains throughout the structure. Inhomogeneity of the structure is incorporated by allowing each individual element to be composed of different material. Regions of large geometric changes can be accounted for by using very small elements to approximate the geometry as closely as desired. This method of structural analysis has been directly validated in the field of solid mechanics by both comparison with closed-form solutions and experimental data and, to a large degree, has enabled the aerospace and many other industries to perform structural analysis on complex structures that would otherwise have been impossible (Zienkiewicz, 1977).

This powerful method is probably the best approach for a realistic quantitative analysis of regional variations in ventricular wall stress, in that it can take into account variations in structure, geometry, and material properties of the myocardium. Before this detailed analysis can be accomplished, however, we need to be able to accurately define the regional geometry, structure, and material properties of the intact heart. To date, there has been only one study that has demonstrated the method for determining the regional variations in the material properties of the myocardium in the passive state (Ghista et al., 1980); no such study has been made for the contracted state. Attempts are being made to improve our ability to accurately measure the three-dimensional geometry of the ventricle (Sandler and Alderman, 1974; Rankin et al., 1976; Vinson, 1979; Ritman et al., 1980), but realization of this difficult goal in an easily implemented manner with high resolution is still not available. Hence, we are faced with the dilemma of possessing the necessary tools for accurately predicting regional stress variations but are unable to fully utilize these tools because of lack of sufficient data. Once these data become available, it remains to validate the predicted variation of hydrostatic stress with directly measured values, so as to derive confidence in being able to analytically obtain an accurate picture of the regional wall stresses in the intact heart.

The present review will focus primarily on method categories one and three. The specific assumptions, limitations, and usefulness of each of the models in category one will be examined sequentially from the simplest to the more complex models. The models in category three will be examined in roughly chronologic order with particular emphasis on the new insights that can be obtained with the use of these more complex models. Thus this review will summarize the available methods for calculating ventricular wall stress and provide the reader with enough information to allow evaluation of the trade-off between detail of information desired versus computational effort needed to suit a particular case. Predictions of wall stress in association with various clinical states will not be emphasized.

THIN-WALLED VENTRICULAR MODELS

The basic assumption of thin-walled models is that the wall is thin relative to the local ventricular diameter. Typically the thickness-to-radius ratio should be less than 1:10 in order for this approximation to be valid. Assuming a thin wall implicitly assumes that there are stress components acting only in the plane of the surface and that these act in the meridional or circumferential direction. That is, there are no radial or transverse shear components so that there are also no bending stresses. Another result of this assumption is that there is no variation of the in-plane stresses through the wall, i.e. the stress is uniform across the wall thickness.

LaPlace (1806) derived the relationship relating the pressure inside a membrane to the radii of curvature and wall tension (in this case the normalized force is expressed as tension rather than stress since there is no cross-sectional area):

$$P = \frac{T_1}{R_1} + \frac{T_2}{R_2} \tag{1}$$

where R_1 and R_2 are the two principal radii of curvature, and T_1 and T_2 are the principal wall tensions.

Wood (1892) assumed that the heart could be modeled by a sphere and calculated the wall tension as a function of the radius and internal pressure. For a sphere $T_1 = T_2$ and $R_1 = R_2$ thus

$$P = \frac{2T}{R} \tag{2}$$

A spherical model adds the additional assumptions that the ventricular wall is isotropic and homogenous. Obviously the left ventricle does not fulfill any of the assumptions used in deriving eq. 2. Even so, this simplest of all models provides some insight into ventricular mechanics. Based on eq. 2, Wood postulated that one of the functions of the papillary muscles was to help eject blood by pulling along a cord of the sphere, which would be more effective than depending solely on tension developed in the wall to eject blood. This would be most beneficial at the onset of systole since the heart is largest and the developed pressure is lowest at that time. Since tension is directly proportional to radius, he also postulated that the trabeculae carnae reduced the tension in each fiber in the wall, since the local radius of the trabeculae is small relative to the chamber radius. It is remarkable that the insight originally demonstrated by Wood still pervades our thinking in terms of the relationship between size, pressure, and wall force in assessing ventricular wall mechanics.

Sandler and Dodge (1963) assumed that the left ventricle could be approximated by a thin-walled ellipsoid composed of isotropic, homogeneous material. Applying equilibrium conditions at the inner boundary, they obtained an equation for any section of the wall in the form

$$L_1 L_2 P = 2hL_2\, \sigma_{\phi\phi} \sin d\phi/2 + 2\,hL_1\, \sigma_{\theta\theta} \sin d\theta/2 \tag{3}$$

where h is wall thickness, P is uniform internal pressure and the other variables are shown in Figure 7-1. Since $\sin\theta = \theta$ for small angles and $L_1 = Rd\phi$ and $L_2 = rd\theta$, the equation can be simplified to the following form of an infinitesimally small section

$$Rr\, d\theta d\phi P = \sigma_{\phi\phi}\, d\phi d\theta + hR\sigma_{\theta\theta}\, d\phi d\theta. \tag{4}$$

Simplifying this equation, it becomes

$$\frac{P}{h} = \frac{\sigma_{\phi\phi}}{R} + \frac{\sigma_{\theta\theta}}{r} \tag{5}$$

which represents the generalized form of LaPlace's equation that is valid only for thin shells.

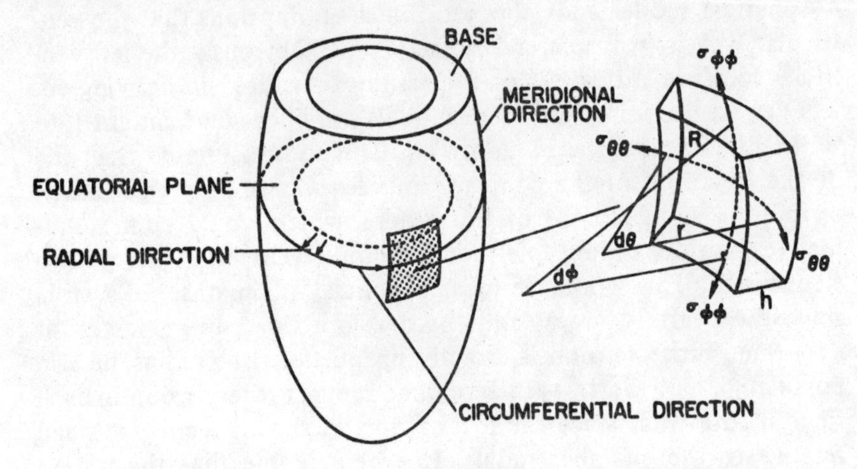

Figure 7-1. Illustration of the principal directions and corresponding normal stress components in a spherical coordinate system. R is the meridional radius of curvature and r is the circumferential radius of curvature.

At the equator of the ellipsoid, force balance requires that

$$\pi r^2 p = \sigma_{\phi\phi}\pi\,[(r+h)^2 - r^2] \tag{6}$$

or, on simplification,

$$\sigma_{\phi\phi} = \frac{Pr^2}{h\,(2r+h)}\,. \tag{7}$$

At the equator of an ellipsoid with semiminor axis b and semimajor axis a, the radii of curvatures are $r = b$ and $R = a^2/b$. Substituting eq. 7 into eq. 5 and expressing the results in terms of the axes of the ellipsoid, we obtain an expression for the circumferential stress in the form

$$\sigma_{\theta\theta} = \frac{Pb}{h}\quad 1 - \left[\frac{b^3}{a^2\,(2b+h)}\right] \tag{8}$$

Equations 7 and 8 are commonly used for calculations of wall stresses in clinical studies where the ventricle is approximated by an ellipsoid. However, eq. 8 is incorrect as is evident from the fact that, in the limiting case of a sphere where $a = b$, $\sigma_{\theta\theta}$ must equal $\sigma_{\phi\phi}$ and such is not the case unless $h \ll b$.

A correct derivation of the circumferential stress formulas was given by Falsetti et al. (1970) and later by Walker et al. (1971). This derivation differed from the original Sandler and Dodge (1963) one in that force equilibrium of the small element was enforced at the midwall so that the equilibrium equation is

$$L_1 L_2 P = 2\,h\,L_2'\,\sigma_{\phi\phi}\sin(d\phi/2) + 2hL_1'\,\sigma_{\theta\theta}\sin(d\theta/2) \qquad (9)$$

where L_1' and L_2' are the midwall arc lengths:

$L_1' = [R + \frac{h}{2}]\,d\phi$ and $L_2' = [r + \frac{h}{2}]\,d\theta$. Using these terms,

the generalized form of the Laplace equation becomes

$$\frac{P}{h} = \frac{\sigma_{\theta\theta}\,(r_\phi + \frac{h}{2})}{r_\theta\, r_\phi} + \frac{\sigma_{\phi\phi}\,(r_\theta + \frac{h}{2})}{r_\theta\, r_\phi} \qquad (10)$$

and the resulting formula for circumferential stress becomes

$$\sigma_{\theta\theta} = \frac{Pb}{h}\,[1 - 1/2\,(b/a)^2]\,[\frac{1}{1 + (h/2b)\,(b/a)^2}] \qquad (11)$$

which is identical to equations of Falsetti et al. (1970) and Walker et al. (1971).

There can be considerable differences between the stresses calculated from eqs. 8 and 11 depending on the geometry and the wall thickness. For example, for a sphere whose wall thickness is one-half the radius, eq. 8 predicts a stress that is 50 per cent greater than that predicted by eq. 11. For an ellipsoid with $a/b = 1.75$ and $h/b = 0.75$, eq. 8 predicts a stress that is 18 per cent greater than eq. 11. Clearly, eq. 11 and not eq. 8 is the one that should be employed in calculations of averaged circumferential wall stress of uniform ellipsoids or spheres of finite thickness.

The implicit assumption of a constant stress across the wall used in these models precludes their use in estimating radial variation of stresses. These models can only calculate stresses that act at the midwall averaged across the wall. Because of uncertainties as to the accuracy of the results of the thick-walled models to be discussed in the next section and because of the tremendous increase in computational effort needed to use these thick-walled models, the constant wall stress models should be employed whenever conditions are suitable. Their use should suffice in most clinical situations in which one is interested in a global picture of an averaged wall stress.

Indeed, use of the constant stress formulae have been applied widely and have demonstrated several important points: (1) wall stress increases markedly with LV dilatation unless wall thickness also increases and is deemed to act as a feedback mechanism for compensatory hypertrophy, (2) averaged wall stresses exceed cavity pressure throughout the cardiac cycle, and (3) peak wall stress may occur at a different time than peak pressure with the difference being dependent on the geometry.

THICK-WALLED VENTRICULAR MODELS

The Lame (1866) derivation for wall stress in a pressurized uniform thick-walled sphere was a landmark in the field of solid mechanics because for the first time it predicted radial variation of stresses. Since the left ventricle is also a thick-walled structure, there must also be a variation of stress across its wall. The desire to gain insight into the nature of this stress variation is the primary motivation behind the many thick-walled ventricular models.

Wong and Rautaharju (1968) derived formulas for the stress distribution in a thick-walled ellipsoid using the following major assumptions: (1) The two minor semiaxes were equal. (2) The myocardium was isotropic, linearly elastic, homogeneous, and in equilibrium. (3) Distortion occurred only in the radial direction, thus bending moments and transverse shear stresses were ignored. (4) The meridional radius of curvature equaled the inner radius of curvature plus the wall thickness. (5) The only load on the ventricle was an internal pressure. They derived closed form solutions for the circumferential, meridional, and radial stress components at any point in the wall. Numerical calculations based on angiographic data revealed several interesting facets of the stress distribution.

At both the apex and equator, all three stress components decreased monotonically from endocardium to epicardium. There was very little difference between radial stresses from apex to equator. There was a threefold difference in circumferential stress that remained relatively constant from apex to equator with the maximum values occurring at the equator. The meridional stress at the equator showed no radial dependence, whereas it decreased by 50 per cent from endo- to epicardium at the apex. The hoop stresses at the equator were more than twice the meridional stress. The pattern of stress did not change with either dilatation or increased pressure load, although both conditions increased the absolute level of stress (pressure more so than dilatation). Increasing wall thickness decreased all of the stress components needed to maintain the same cavity pressure. For example, the maximum stress at a pressure of 110 mmHg was 430×10^3 dyne/cm^2 with a wall thickness of 0.83 cm and was 267×10^3 dyne/cm^2 with a thickness of 1.43.

Ghista and Sandler (1969) used a different approach than Wong and Rautaharju (1968) to derive the stress distribution in an approximately ellipsoid thick-walled model of the ventricle. Their major assumptions were that (1) the only loading was an internal pressure and (2) the material was isotropic, incompressible, homogeneous, and linearly elastic. Their model could account for shear stresses in the wall but this was obtained at the expense of a slight inexactness in geometry such that the model was not mathematically an exact ellipsoid. Their results demonstrated roughly the same pattern of stress distribution as found by Wong and Rautaharju in that there was a monotonic decrease in the magnitude of all stress components from endo- to epicardium. The results differed quantitatively from Wong and Rautaharju's results in that the circumferential stress at the equator only exceeded meridional stress by 30 per cent rather than 200 per cent. In their analysis, the stress distribution is independent of the elastic properties of the wall, so that the model would also apply to a viscoelastic material. The model was also employed to compute the instantaneous myocardial modulus by matching the instantaneous incremental strains of the model with the *in vivo* values in response to the incremental chamber pressures; this will be discussed further in Chapter 8.

Mirsky (1969) used still a different approach to calculate the stress distribution in the left ventricle approximated by a thick-walled shell of prolate spheroid shape. He also assumed isotropy, incompressiblity, homogeneity, and linear elasticity. The effects of transverse stress and shear were included in the derivation. The solution was obtained by using a displacement field that was linear in the meridional and quadratic in the radial direction and substituting this displacement field into the appropriate equilibrium equations. The resulting differential equations were solved by numerical integration for a thick-walled shell of revolution. A simplified set of formulas for the stress distributions at the equator was obtained by retaining terms only up to second order in an asymtotic expansion approach to the solution. Numerical results based on geometric data obtained from ventriculography in a human patient demonstrated the following: Endocardial circumferential stress increased about 10 per cent whereas meridional stress decreased about 50 per cent from apex to base. Midwall circumferential stress increased about 100 per cent and meridional stress increased about 30 per cent from apex to base. These results differed both quantitatively and qualitatively from those of Wong and Rautaharju (1968) and Ghista and Sandler (1969).

Mirsky (1970) investigated the qualitative influence of anisotropy and inhomogeneity on left ventricular stress. The basic analysis was similar to that used by Wong and Rautaharju (1968) and assumed the heart to be an incompressible prolate spheroid whose wall thickness remained constant throughout the cardiac cycle. Axisymmetric deformation was also assumed so that bending moments and shear forces were ignored. The material was assumed to be linearly elastic and orthotropic. A closed formed solution was obtained for the case where the elastic constants were of the form $E = \alpha R^m$ where α is a constant, R is the meridional radius of curvature, and m is a real exponent. If the boundary condition consisted solely of an internal pressure the radial stress was somewhat independent of the elastic constants. However, if the boundary conditions accounted for pericardial pressure as well, the radial stress was then dependent on the elastic constants. Numerical results indicated that maximum and midwall stress increased as anisotropy increased, but midwall stresses were relatively independent of the degree of inhomogeneity. The maximum circumferential stress for an anisotropy ratio of 10 was only 20 per cent greater than with an isotropic material. A special case of inhomogeneity was taken into account by assuming a parabolic distribution for the circumferential elastic modulus. This particular choice of inhomogeneity was chosen to mimic the effect of inhomogeneity based upon a study (Streeter et al., 1969) indicating a roughly parabolic variation of fiber angles across the wall; numerical results indicated that this degree of inhomogeneity had a greater effect than anisotropy upon the stress variation across the ventricular wall. The circumferential stress distribution across the wall was also parabolic, being maximum near the midwall.

Then, Mirsky (1973) investigated the specific effect of large deformations on stresses in the left ventricle which, for this analysis, was approximated by a thick-walled sphere composed of isotropic, homogeneous, and incompressible material that was loaded only by an internal pressure. The stresses at a given pressure level were calculated from a strain energy density function that was a function of the strain invariants. Using data from dog studies, this function was evaluated at the midwall based on strains and cavity and wall volumes at that particular cavity pressure. The nonlinear relation between stress and strain was assumed to be of exponential form. The major finding of this study was that inclusion of nonlinear stress-strain properties predicted a very high stress concentration at the endocardium, which was almost ten times higher than that predicted by the linear theory (Mirsky, 1969). This stress concentration increased markedly as the cavity pressure was increased.

Moriarity (1980) also used a thick-walled spherical model of the ventricle to investigate the effect of nonlinear constitutive relations on wall stress. He assumed that the wall was isotropic and incompressible; two different types of materials with nonlinear constitutive relations were investigated: a Valanis-Landel type and a Rivlin-Saunders type. For calculation of stress distribution across the wall, it was first assumed that the material was homogenous across the wall. A uniaxial constitutive relation was assumed for each material type in the form

$$S = C(\lambda^{a} - 1/\lambda^{a/2})$$

and (12)

$$S = C_1(\lambda^2 - 1/\lambda) + C_2(\lambda - 1/\lambda^2)(2\lambda + 1/\lambda^2 - 3)^2$$

respectively for the Valanis-Landel and Rivlin-Saunders material where S is stress, and λ is stretch or strain. Similarly to Mirsky's results, the predicted wall stress distribution (for a diastolic pressure of 24 gm/cm^2) showed a concentration near the endocardium that was nearly six times as high as that predicted by the exact solution for a thick-walled sphere composed of linear material. At the epicardium, the nonlinear material predicted a stress that was slightly lower than that predicted by the linear material. A Valanis-Landel type material predicted stresses that were about 10 per cent higher than 3 gm/cm^2. The pressure-volume relationships predicted by each of the models differed considerably despite being based on the same geometrical and material property data. One reason for this disparity may relate to the fact that a uniaxial constitutive relation does not uniquely define the material property of tissue.

All of these thick-walled models predicted a variation of stress components across the wall. The exact magnitude of the stresses relative to chamber pressure depended upon the geometric model chosen and the other assumptions used in the derivation. Although we were unable to measure the stress distribution across the wall to validate any of these models, qualitatively they seem to agree, with the exception of the meridional stress predicted by the Mirsky model (1969), that there is an endo- to epicardial stress gradient but there is considerable quantitative variation in the stresses.

Huisman et al. (1980b) recently published an informative review comparing the stresses predicted from several models when the same angiographic data obtained in a number of disease states was used as input to each model. They compared the results obtained from a thin-walled sphere (essentially Wood's model), a thick-walled sphere (Lame model), the Sandler and Dodge (1963) thin ellipsoid model, the Falsetti (1970) modification of the thin ellipsoid model, and the thick-walled models of Wong and Rautaharju (1968) and Ghista and Sandler (1969). In each case, the material was assumed to be isotropic, homogeneous, and incompressible. Their findings can be summarized as follows: (1) The absolute magnitude of the calculated wall stress depended significantly on the model used. For the thick-walled models, mean circumferential stresses differed by up to 20 per cent. The difference in mean stress predicted by a thin compared to a thick-walled sphere was about 12 per cent at end-diastole and 20 per cent at end-systole. For the ellipsoidal models, the stresses between Ghista and Sandler model and Mirsky models differed by about 35 per cent. (2) Although the models differed from one another, among the disease states examined (cardiomyopathy, mitral stenosis, mitral regurgitation, aortic stenosis, and aortic regurgitation) the relative differences among models remained relatively constant. In the descending order of magnitude of the predicted circumferential stresses, the models were Sandler and Dodge, Falsetti, Wong and Rautaharju, Mirsky, Ghista and Sandler, thin-walled sphere, and thick-walled sphere (Fig. 7-2). For longitudinal stresses, this descending order is reversed to thin-walled sphere, thick-walled sphere, Ghista and Sandler, Wong and Rautaharju, Sandler and Dodge, Falsetti, and Mirsky. (3) The thick-walled models yielded considerably different results (Fig. 7-3), with the circumferential stress gradient across the wall decreasing by 43 per cent and 74 per cent and the longitudinal stress gradient increasing by 36 per cent and decreasing 26 per cent for the Mirsky and Ghista and Sandler models, respectively. Without experimental verification, the discrepancy between the thick-walled models makes it difficult to know which one is most accurate. Thus, we cannot rely on these idealized geometry models for providing us representative levels and variations of ventricular wall stress.

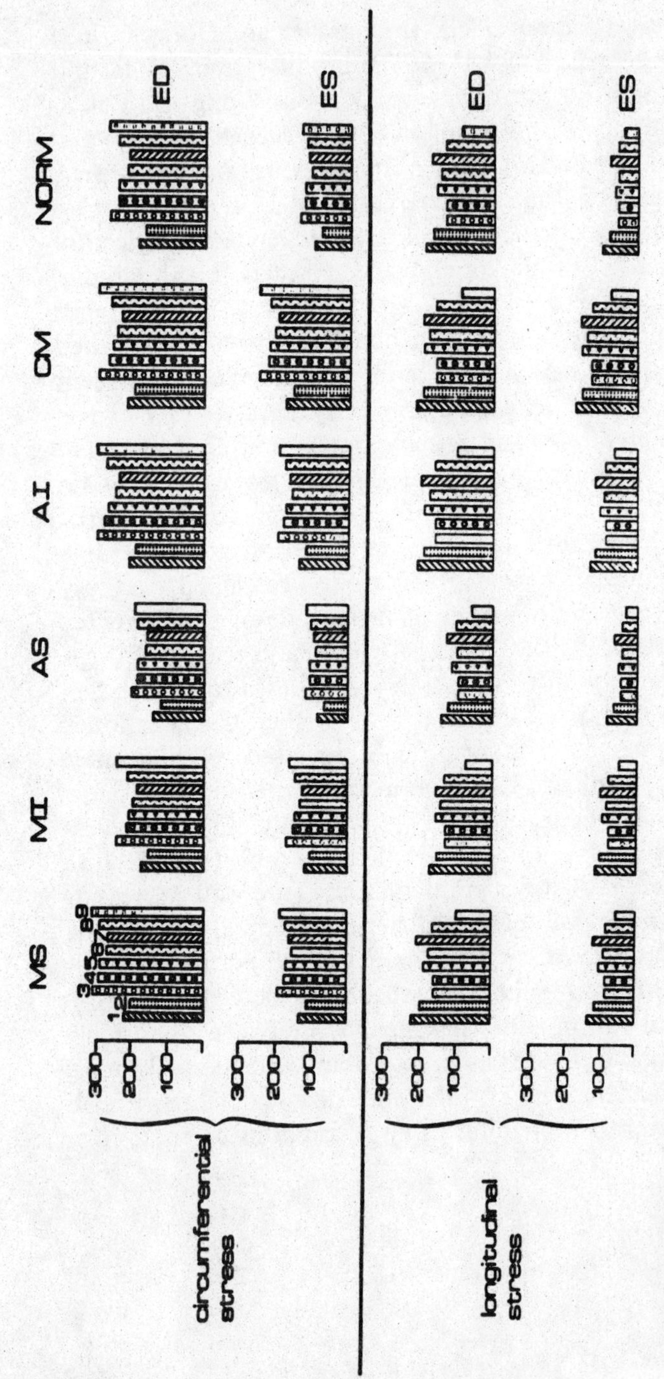

Figure 7-2. Comparison of end-systolic and end-diastolic wall stresses predicted by various analytical models for several disease states. From R.M. Huisman, P. Sipkema, N. Westerhof, and G. Elzinga, Comparison of Models Used to Calculate Left Ventricular Wall Force. *Medical and Biological Engineering and Computing, 18*:133-144, 1980.

Mean stress values (relative to presure) for the studied models under several conditions: MS = mitral stenosis; MI = mitral insufficiency; AS = aortic stenosis; AI = aortic insufficiency; CM = cardiomyopathy; NORM = normal human heart; ED = end-diastolic; ES = end-systolic.

The stress values are represented by columns 1 through 9: 1 = thin-walled sphere; 2 = thick-walled sphere; 3 = Sandler and Dodge model; 4 = Falsetti model; 5 = Wong model; 6 = WONCO model; 7 = Ghista model; 8 = Mirsky model; 9 = Streeter model. (*See* text.)

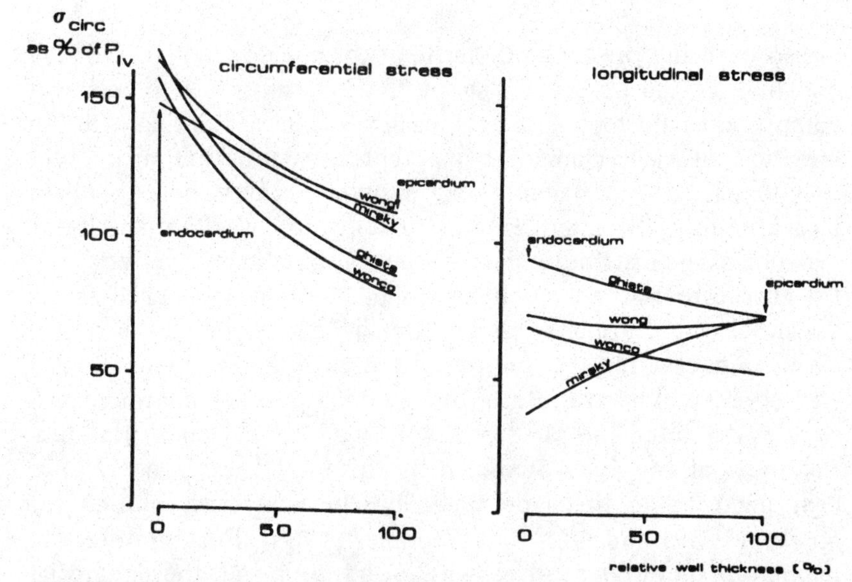

Figure 7-3. Comparison of the stress distribution across the wall for the thick-walled ellipsoid models demonstrating marked quantitative and qualitative differences between the models. The WONCO model is a modification of the model of Wong and Rautaharju (1968) with transverse shear taken into account. From R.M. Huisman, P. Sipkema, N. Westerhof, and G. Elzinga. Comparison of Models Used to Calculate Left Ventricular Wall Force. *Medical and Biological Engineering and Computing, 18*:133-144, 1980.

FINITE ELEMENT MODELS

All of the models discussed thus far approximate the ventricular geometry by a simple axisymmetric geometric figure that is everywhere concave toward the cavity. To take complicated geometry into account, particularly when complex changes in curvature occur as in the case with the actual ventricular geometry, an approximate and more complex method must be used. This, plus the ability to introduce regional inhomogeneity, anisotropy, and nonlinear material properties are the rationale for use of the finite-element models of the left ventricle.

Gould et al. (1972) applied the finite-element method for stress analysis of the internally pressurized left ventricular model of linearly elastic material. Geometric data were obtained from single plane ventriculograms with the ventricular shape approximated by using the long silhouette of the angiogram and revolving one half of the angiogram about the long axis. The model was comprised of thirteen elements, each of which was a ring with the meridion of each element being represented by a fourth order polynomial so that bending and transverse shear could be taken into account. The major finding of the study was that the linear diastolic stress distribution across the wall thickness reversed when the curvature of the wall changed signs. That is, instead of an endocardial to epicardial stress gradient in regions where the ventricle was concave inward, the opposite was obtained where the wall was concave outward. Both the meridional and circumferential stresses exhibited the same behavior. In addition, the circumferential stress at the apex predicted by this model was about 25 per cent higher than those predicted by the analytical solutions of Mirsky (1969) and Ghista and Sandler (1969). Furthermore, the maximum meridional stress was found to be at the equatorial endocardium whereas the two analytical models found the maximum stress to be at the apex, the reversal of the stress gradient predicted by the simple geometric models.

Also beginning in 1972, Janz and his colleagues published a series of studies using the finite element method for structural analysis of the left ventricle with progressive complexities taken into account. The ventricle was assumed to be axisymmetric and to deform axisymmetrically. The initial model (Janz and Grimm, 1972) employed linear elastic theory but allowed for heterogeneity of wall structure. The material at the base was assumed to be isotropic with properties comparable to that of collagen. The remainder of the ventricle was assumed to be composed of two layers: one that was isotropic with the constitutive properties of muscle (Young's modulus = 60 gm/cm^2, Poisson ratio = 0.45, shear modulus = 20.7) and one layer that was transversely isotropic with the transverse modulus one-half that of the passive muscle. The finite elements were quadrilateral cross-section rings of revolution. Use of 198 elements produced a close approximation to the actual state of deformation in rat hearts arrested in diastole at pressures up to 12 cm H_2O. At 24 cm H_2O the predicted deformation significantly exceeded that observed. They postulated that this discrepancy was due to not incorporating nonlinear stress-strain relationships into the model. Further analysis with this model revealed that inhomogeneity and anisotropy of the wall had only a small effect on predicted deformations with an isotropic model predicting deformations of the equator that were 8 per cent less than the heterogenous model. Despite the close agreement in deformations, the stresses predicted by the isotropic model differed by factors of 2 to 3 from those predicted by the heterogenous model.

Janz and Grimm (1973) expanded the analysis to take into account nonlinear stress-strain properties. They proposed a strain energy function in the form

$$W(\epsilon_1, \epsilon_2, \epsilon_3) = \frac{E}{\beta^2 (1+\nu)} \left[\sum_{i=1}^{3} \exp \beta\epsilon_i + \left(\frac{1-2\nu}{\lambda}\right) \exp\left(-\frac{\beta\nu}{1-2\nu} \sum_{i=1}^{3} \epsilon_i\right) - \left(\frac{1+\nu}{\nu}\right) \right] \tag{13}$$

where E is the Young's modulus, ν is Poisson's ratio, and β is a free parameter. This function was chosen because it produced an exponential uniaxial stress-strain curve that closely approximate previously observed rat myocardial data. The model was still axisymmetric and assumed isotropic material but allowed for heterogeneity as in their previous model. An incremental loading approach was used to approximate the nonlinearity in material properties. The major differences obtained by using nonlinear compared to linear material properties are shown in Figures 7-4 and 7-5, in which the diastolic pressure-volume and pressure-strain curves are shown. There is fairly close agreement between the pressure-volume and pressure-strain curves for volumes up to 12 cm H_2O pressure, but the pressure-strain curves differ considerably. Although they did not calculate the regional wall stresses, it is clear that stresses predicted from using the model with nonlinear constitutive properties would differ considerably from those obtained with the model using only linear properties. This example illustrates the insensitivity of the pressure-volume curve to marked variations of wall properties.

Janz et al. (1974) analyzed another aspect of nonlinearity by

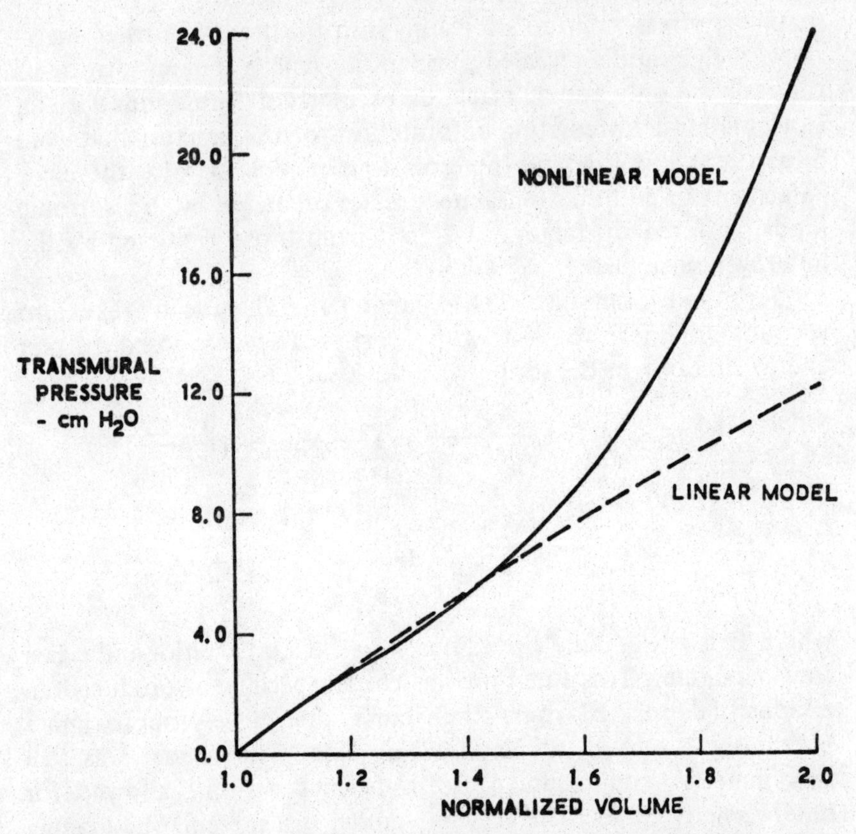

Figure 7-4. Comparison of diastolic left ventricular pressure-volume curves obtained when linear versus nonlinear material constitutive relations were employed in a finite-element model of the rat heart. From R.F. Janz and A.F. Grimm, Deformation of the Diastolic Left Ventricle. 1. Nonlinear Elastic Effects. *Biophysical Journal, 13*:689-704, 1973.

extending their previous analyses to include large deformation theory by retaining the nonlinear terms in the strain-displacement relations. This model was axisymmetric, heterogeneous, isotropic, and nonlinearly elastic. The result of including large deformation effects compared to using small deformation theory are shown in Figures 7-6 and 7-7. Incorporation of nonlinear effects produced a lower chamber stiffness than with linear theory. As with the previous study, there were small differences in the volumes predicted at a given pressure, with greater difference occurring

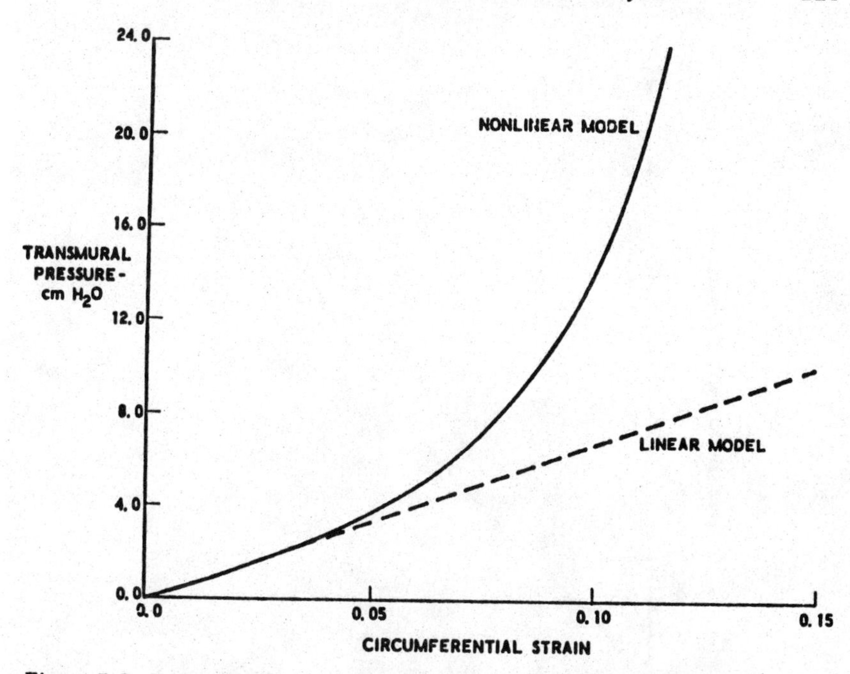

Figure 7-5. Comparison of the diastolic left ventricular pressure-strain curves obtained when linear versus nonlinear material constitutive relations were employed in a finite-element model of the rat heart. From R.F. Janz and A.F. Grimm, Deformation of the Diastolic Left Ventricle. 1. Nonlinear Elastic Effects. *Biophysical Journal, 13*:689-704, 1973.

at higher pressures. Comparison of strains, however, revealed much larger differences with as much as 100 per cent difference in strain at the apex at a pressure of 12 cm H_2O. They attributed this difference to rotational effects near the apex not being accounted for in the linear theory. Once again, this study demonstrated the inability of the pressure-volume relationship to reflect the state of deformation in the ventricle.

Recently Janz and Waldron (1978) used the finite element method to specifically examine the effects of apical aneurysms on ventricular deformations and on the ventricular pressure-volume curve. In this model they used axisymmetric elements and assumed that the wall was regionally homogeneous but did allow for nonlinear stress-strain properties and large deformations. The properties of aneurysmal tissue were assumed to those obtained

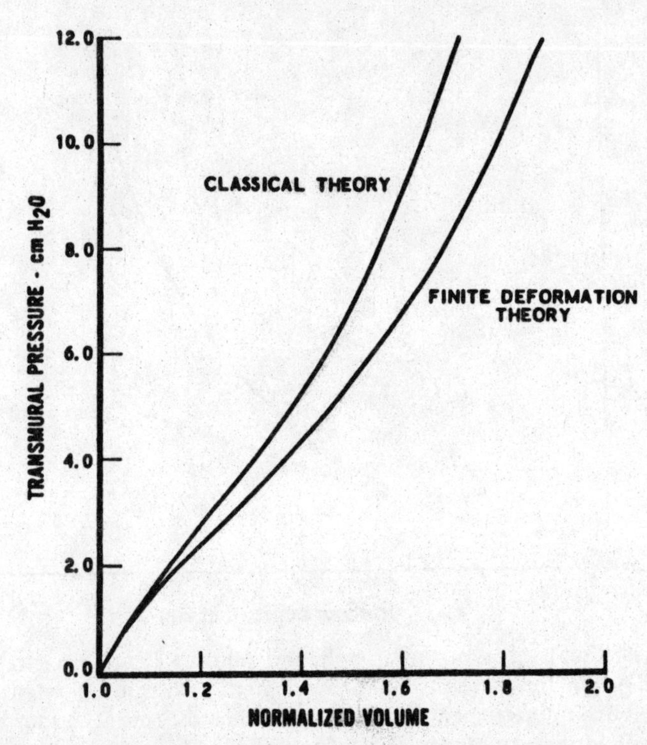

Figure 7-6. Comparison of the diastolic left ventricular pressure-volume curves obtained when large versus small deformation theory was employed in a finite element model of the heart. Reprinted with permission from *Journal of Biomechanics,* 7:509-516, R.F. Janz, B.R. Kubert, and T.F. Moriarity, Deformation of the Diastolic Left Ventricle–II. Nonlinear Geometric Effects. Copyright 1974, Pergamon Press, Ltd.

from an earlier study by Parmley et al. (1973), who performed uniaxial length-tension tests on human aneurysmal tissue of muscular, mixed fibromuscular, and fibrous composition. Their major finding was that, in the region of a fibrous or fibromuscular aneurysm, the outward expansion during diastole was less than in the normal tissue (Fig. 7-8). In addition, the presence of the aneurysm severely tethered neighboring normal tissue causing a reduced expansion in the areas near the aneurysm.

All of these findings occurred at low diastolic pressure of 12 mmHg. At higher diastolic pressures the tethering effect was still present near the aneurysm but in regions far from the aneurysm

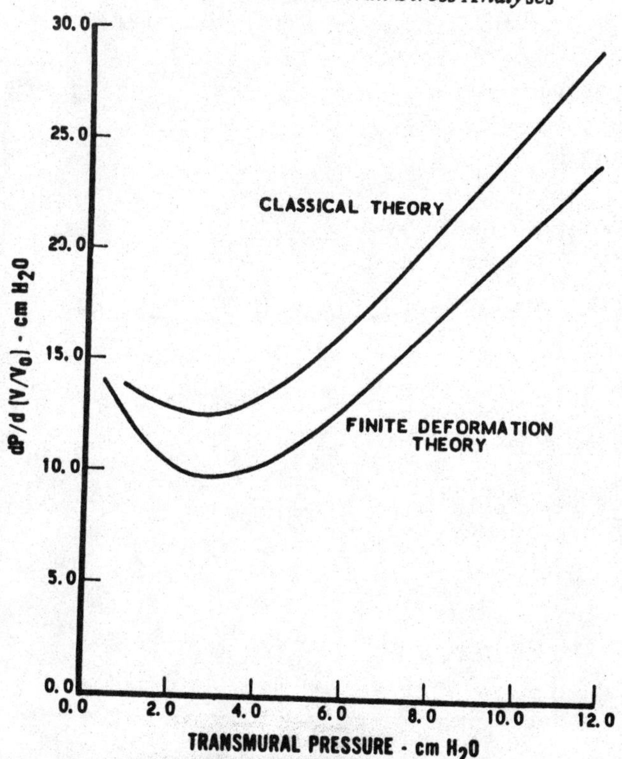

Figure 7-7. Comparison of diastolic stiffness-pressure curves of the left ventricle when large versus small deformation theory was employed in a finite-element model of the heart. Reprinted with permission from *Journal of Biomechanics,* 7:509-516, R.F. Janz, B.R. Kubert, and T.F. Moriarity, Deformation of the Diastolic Left Ventricle—II. Nonlinear Geometric Effects. Copyright 1974, Pergamon Press, Ltd.

there was even more elongation than in the normal ventricle. Regional stresses produced by the presence of the aneurysm would be highly dependent on the constitutive properties of the aneurysm as well as its size but detailed stress analysis was not performed in this study. They examined the pressure-volume relationships due to apical aneurysms and found that, as expected, the P-V curves shifted toward the left with increasing size of the aneurysm. However, when the P-V curves were quantified by an exponential relationship of the form

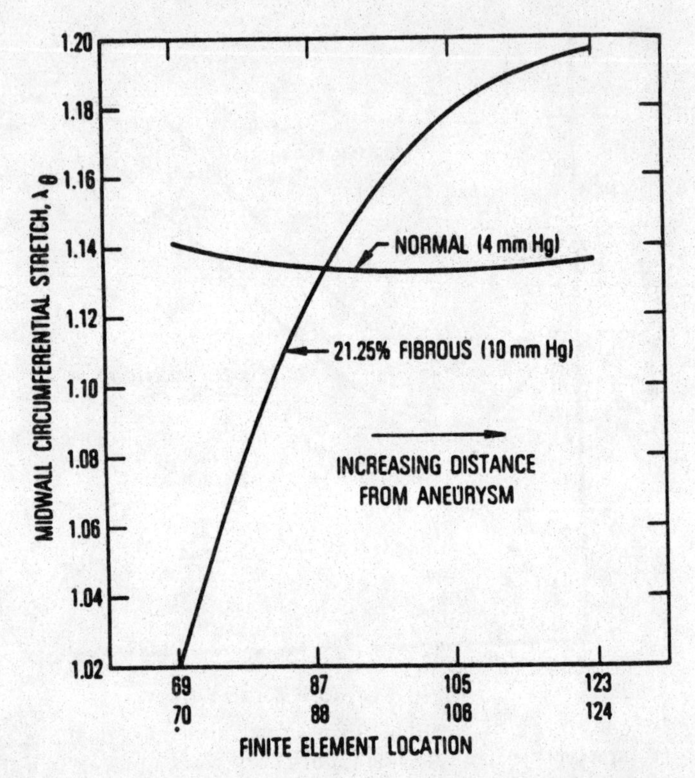

Figure 7-8. Predicted effect of a fibrous apical aneurysm comprising 21 per cent of the left ventricular wall on regional stretch at various distances from the aneurysm compared to the stretch in a ventricle without an aneurysm. The tethering effect near the aneurysm as well as the compensatory effects at portions remote from the aneurysm are demonstrated. From R.F. Janz and R.J. Waldron, Predicted Effect of Chronic Apical Aneurysms on the Passive Stiffness of the Human Left Ventricle. *Circulation Research, 42*:255-263, 1978. Reprinted by permission of the American Heart Association, Inc.

$$\frac{dP}{\left(\frac{V}{V_o}\right)} = aP + \gamma, \tag{14}$$

the parameter a was found to depend on both aneurysm size and stiffness. Thus, they felt that the utility of a as an index of aneurysm properties was limited.

Pao et al. (1974) employed triangular cross-sectioning elements in their finite-element model of the left ventricle, whose geometry was obtained from orthogonal angiograms in an isolated, supported dog heart. Using the approach of Gould et al. (1972), their model was a solid of revolution, created by rotating one half of the angiogram about the long axis. The wall material was assumed to be homogeneous, isotropic, and linearly elastic; small deformation theory was implicitly utilized. Their analysis revealed that the maximum circumferential stress occurred at the endocardial surface at the base of the heart and was 5.5 times the intracavitary pressure. Results (Fig. 7-9) qualitatively similar to the previously discussed nonlinear, thick-walled geometric models were found with both meridional and hoop stresses varying nonlinearly across the wall with highest values occurring at the endocardium.

Pao et al. (1976) used a plane-strain approximation in order to examine in more detail the stress distribution at a cross-section of a left ventricle not connected to a right ventricle. The plane-strain assumption implies that the left ventricle is a long irregular cylinder whose configuration at any cross-section is the same as that used in the analysis. Again multiplanar x-rays were used to obtain the geometric data from an isolated dog heart. Three hundred and eight triangular elements were employed with the material assumed to be isotropic, homogeneous, and elastic. The results demonstrated that maximum stresses occurred at the endocardial surface only at the anterior and posterior regions of the cross-section. In the septum and free wall, the maximum circumferential stress occurred at the epicardial surface. In addition, some compressive stresses were found to occur in these regions where the curvatures become convex inward. Thus, like the study of Gould et al. (1972), this study demonstrated that regions where curvatures change rapidly or change sign are associated with dramatic alterations in the wall stress patterns.

Panda and Natarajan (1977) modelled the human left ventricle as a thick-walled layered shell of revolution in a study designed to assess the effects of fiber orientation on wall stress. They assumed that the fiber directions in the inner, middle, and outer wall were 40, −5, and −50 degrees respectively from the horizontal. They assumed that the material properties along the fiber direction were orthotropic, but that the transverse properties were a certain

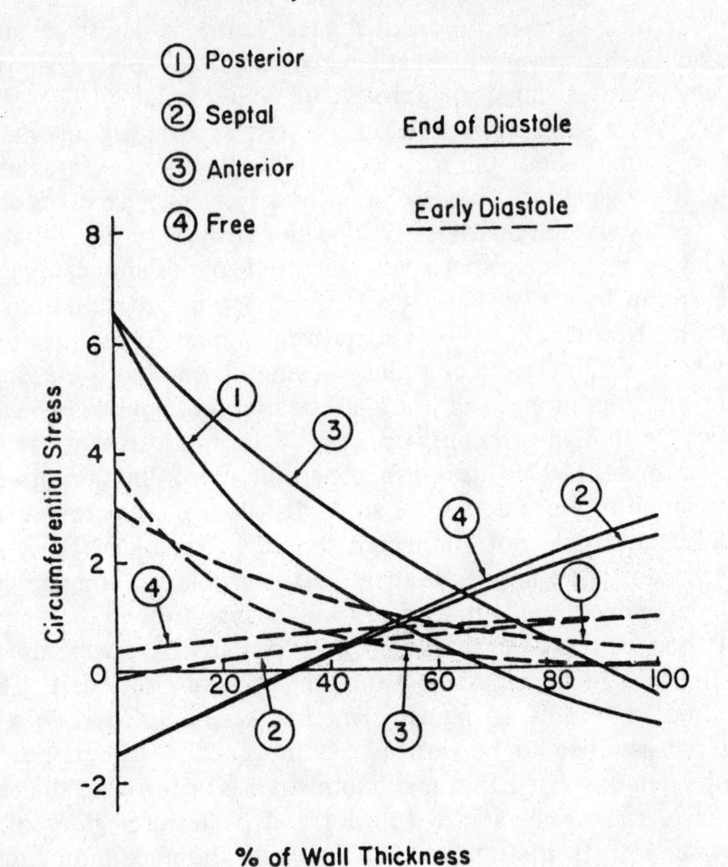

Figure 7-9. Distribution of circumferential stress across the wall in a plane-strain finite-element model of the left ventricle demonstrating the directionally different gradients from endocardium to epicardium in the posterior and anterior walls compared to the septum and free wall. Reprinted with permission from *Annals of Biomedical Engineering, 4*:232-249, Y.C. Pao, R.A. Robb, and E.L. Ritman, Plane-Strain Finite-Element Analysis of Reconstructed Diastolic Left Ventricular Cross Section, Copyright 1976, Pergamon Press, Ltd.

fraction of these in the fiber direction. For the isotropic case, their results demonstrated higher circumferential stress values than those predicted by the solutions of Ghista and Sandler (1969). There was a meridional stress concentration at the apex that was three- to fourfold greater than in the previous studies.

For the layered model, they found that the maximum circumferential endocardial and epicardial stresses diminished and midwall stress increased as the ratio of the longitudinal to transverse stiffness increased. However, the meridional stress increased in both the inner and outer layers and diminshed in the middle layer with increasing stiffness.

Heethaar et al. (1976) employed the finite-element method of analysis to study the stress distribution in both a single left ventricle and in a combined right and left ventricle model. The restricting axisymmetric model constraint was removed in the models of Hamid and Ghista (1974), Ghista and Hamid (1977), and Ghista et al. (1980) wherein isoparametric finite elements were employed. These models could be employed to obtain realistic wall stresses, provided the nonlinear elastic properties of the finite elements are first determined by matching the instantaneous volumes of the left ventricular model and the actual filling chamber as described in Ghista et al. (1980). This model was much more realistic than all previous models, since each chamber was not restricted to being axisymmetric and since the effect of the right ventricle was included. They employed up to 7000 tetrahedral elements to model the two ventricles. The geometric data for left ventricular studies were obtained from multiple x-ray views of an isolated working dog heart in which the heart pumped fluid containing a radiopaque compound. Geometric data for the combined right and left ventricle model was obtained from a dead heart that was filled with the contrast fluid and placed in the x-ray field. The myocardium was assumed to be isotropic and homogeneous and linear elasticity was assumed. The diastolic single left ventricle was shown to have highest stresses at the endocardial surface which decreased transversely. The combined ventricle model also demonstrated largest stresses at the endocardium but, unlike previous models, this one predicted marked regional variations in stresses with a large stress concentration at the septal-right ventricular junction as well as compressive stresses in the septum.

Vinson et al. (1979) used a left ventricular model composed of thirty-six brick type elements composed of six layers each. The geometry of the ventricle was obtained by reconstruction from biplane ventriculograms. The effects of the pericardium and right ventricle were not considered. Anisotropy was allowed by assuming the wall to be twice as stiff in both the circumferential and radial directions as in the meridional direction in one case and by assuming a transverse stiffness to be one-half of the tangential stiffness in another case. Inhomogeneity was accounted for by allowing the elements near the base and apex to have variable material properties. Numerical results demonstrated maximum stress concentrations at the right anterior and left posterior regions of the endocardium which were of the same order as those found by Ghista and Hamid (1977). In addition, they found that the most important determinants of the diastolic deformation were the ratio of circumferential to meridional stiffness, the variation of fiber angle, the variation of shear modulus, and the relative stiffness of the basal portions.

These finite element models demonstrate a powerful method of examining regional details of the stress distribution in a ventricle which is not feasible with any of the analytical models discussed previously. The studies thus far provide some information as to the effects of nonlinear material properties, regional heterogeneity of the wall, and large deformations. In addition to the need for some validation by direct measurement of hydrostatic stress, our ability to utilize this powerful method of analysis is currently limited by our inability to very accurately define the instant-to-instant three-dimensional geometry of the ventricle tissue.

MYOCARDIAL CONSTITUTIVE RELATIONS

A necessary requisite for stress prediction from analytical models discussed earlier is a knowledge of the constitutive relations of myocardial tissue. In actuality, very little is known about myocardial constitutive relations. The need for more data has been emphasized in many recent publications (Bergel and Hunter, 1979; Huisman et al., 1980a; Phillips et al., 1981). Two basic approaches have been used to quantify myocardial constitutive relations: (1) direct measurement from excised tissue and (2) prediction of global material properties from measuring loading (pressure) and deformations (volume or strains). Thus far, the direct measurement of material properties has been limited to uniaxial tests conducted on papillary muscles or strips of trabeculae carnae. In these tests, the specimen is pulled longitudinally while either its overall length or a segmental length is measured while the force developed at the ends are recorded. Many such studies (Hill, 1950; Sonnenblick et al., 1964; Pinto and Fung, 1973; Mirsky and Pasipoularides, 1980) have been performed and the results are in general agreement that the passive uniaxial stress-strain relationship can be accurately described by an exponential function of the form

$$\sigma = Ae^{B\lambda} + C \tag{15}$$

where λ is a strain quantity and A, B, and C are numerical coefficients. The numerical values of the coefficients vary somewhat from study to study partly due to different methods of normalizing the strain parameter. The values of the coefficients are altered by certain interventions including hypertrophy (Mirsky, 1976; Mirsky and Pasipoularides, 1980), infarction (Parmley et al., 1973), and aging (Janz et al., 1976; Spurgeon et al., 1977). A recent summary (Mirsky, 1976) presents an excellent review of the myocardial properties in the passive state, and the reader is referred to this work.

Recently, several studies have examined the constitutive relations during activation of these muscle strips (Templeton, 1973; Loeffler and Sagawa, 1975; Spurgeon et al., 1977) and have found that the relationship between stress and strain remains exponential during muscle activation. The values of the coefficients change from the resting to the active state so that the muscle becomes stiffer during activation and viscous effects become more prominent. Yin et al. (1980) recently demonstrated, for example, that, with aging, there was no change in the resting constitutive relation, but that the active aged muscle was stiffer than the young muscle. A portion of this increase in active stiffness was attributable to age alone, but a portion was accounted for by underlying cardiac hypertrophy with aging.

While these uniaxial studies have provided much insight into the pathogenesis of various disease processes, the quantitative extrapolation of these data to the entire ventricle is subject to question because the tissue in the ventricular wall is not subject to uniaxial stresses. A strain energy function is a requisite for the general formulation of the constitutive relations of tissue undergoing large deformations such as occur in the heart. From a theoretical standpoint, uniaxial stress-strain data cannot be extrapolated to yield a strain energy function that is valid for multiaxial states (Fung, 1973; Abe et al., 1978; Moriarity, 1980). Even when the ventricle is assumed to be a simple structure such as a thick-walled sphere, extrapolation from uniaxial stress-strain data to the entire ventricle produces spurious results.

Moriarity (1980) demonstrated that a single uniaxial stress-strain curve produced multiple pressure-volume relationships. Conversely, Abe et al. (1978) utilized a single pressure-volume relationship and demonstrated that nonunique uniaxial stress-strain curves were obtained depending on one's choice of the form of the strain energy function whereas a unique biaxial stress-strain curve was obtained that was independent of the choice of the form of the strain-energy function.

When one considers the properties of the heart wall during contraction, even less data are available, and more complications arise. Streeter's (1969) demonstration of the variation of fiber orientation across the wall has led to numerous attempts to calculate stresses along the fiber directions at various times in the cardiac cycle (Voukydis, 1972; Janz, 1980; Streeter, 1970). However, these theoretical studies are all predicted on the assumption that stresses both at rest and during contraction occur only along the fiber direction and not in the transverse direction. This assumption seems attractive because the forces produced by contraction in a single muscle fiber should be directed predominantly along the fiber axis. However, interfiber connections could produce significant transverse or shear stresses both at rest and during contraction and the magnitude of these is as yet unknown. Thus, such experimental data on the fiber orientation and its changes during the cardiac cycle (at every location in the heart) are needed for incorporation in the model.

There have recently been some studies published (Nikravesh, 1976; Ghista et al., 1980) in which the authors employ the finite element method, not for stress analysis, but for predicting the constitutive relations of the tissue. The idea is based on the principle that the finite element solution will converge to the real solution if the proper loading conditions, boundary conditions, deformations, and constitutive relations are employed and equilibrium and compatibility are enforced. The approach is to measure the loading, boundary conditions, and deformations and use an iterative approach to successively approximate the constitutive relation of the tissue. When the predicted and measured deformations are close enough the constitutive relations giving the predicted deformation is considered to represent that of the tissue.

Based thereupon, the ranges of the constitutive parameters, for the constitutive relations, for normal and infarcted myocardium were obtained as follows:

$$E(N/m^2) = (1.09 \times 10^3 - 7.85 \times 10^3) + (2.116 - 7.996)\,\sigma\ (\text{Pascal}),$$

$$\text{for the normal myocardium} \tag{16}$$

$$E(N/m^2) = (13.56 \times 10^3 - 35.36 \times 10^3) + (8.806 - 14.306)\,\sigma\ (\text{Pascal}),$$

$$\text{for infarcted myocardium}$$

It would be useful to apply this method to a pressurized structure with known material properties, so as to validate the predicted material properties with directly measured values in order to make a convincing argument that the iterative scheme works. Computerized tomography (Ritman et al., 1980) or direct visualization of the heart borders with multiple markers (Shoukas et al., 1980) would in any case enable more validation of this method, so that practical application of these methods may soon be initiated.

CONCLUSIONS

Quantification of ventricular wall stress is necessary for an understanding of both normal and pathological ventricular mechanics. At present there are no reliable means to directly measure wall stresses in the intact ventricle. Thus, we must rely on predicting these stresses based on various models. The models used in this chapter to calculate wall stresses were reviewed with the following major conclusions: (1) For most applications, detailed knowledge of the stress distribution across the wall is not essential. The mean stress averaged across the wall is adequate. The Falsetti (1970) thin-walled ellipsoidal formulae are the best to use in terms of correctness, ease of computation, and incorporation of a realistic albeit greatly oversimplified geometry. The thin-walled or thick-walled spherical formulae underestimate circumferential stresses by about 20 to 40 per cent and overestimate longitudinal stresses by about 10 per cent to 20 per cent respectively. (2) When details of the stress distribution across the wall are desired, the wide qualitative and quantitative differences predicted by the various analytical thick-walled formulae make it difficult to judge which gives the most realistic representation of actual stresses and are thus of little practical utility. (3) The finite element method offers an extremely powerful method for analyzing regional variations in stress. This method can account for variations in material properties, complex geometry, anisotropy, and variations in fiber angle from region to region. However, until faster and better methods of accurate three-dimensional reconstruction of the heart are available and until more data on the multiaxial constitutive properties of myocardium are obtained, this method cannot be utilized to its full capabilities. In addition, the effort and expense involved in using the finite-element method are considerable and may limit its widespread application. (4) Accurate and reliable method to measure wall stress in an intact heart needs to be developed to validate the predictions of the models.

REFERENCES

Abe, H., Nakamura, T., Motomiya, M., Konno, K., and Arai, S. (1978) Stresses in left ventricular wall and biaxial stress-strain relation of the cardiac muscle fiber for the potassium arrested heart. *Tran. ASME, J. Biomech. Eng.*, 100:116-121

Alpert, N.R. (1971) *Cardiac Hypertrophy*. New York, Academic Press

Bergel, D.A. and Hunter, P.J. (1979) The mechanics of the heart in quantitative cardiovascular studies. In *Clinical and Research*

Applications of Engineering Principles, (ed.) N.I.C. Hwang, D.R. Gross, and D.J. Patel, Baltimore, MD, University Park Press, 151-213

Burns, J.W., Covell, J.W., Myers, R., and Ross, J., Jr. (1971) Comparison of directly measured left ventricular wall stress calculated from geometric reference figures. *Circ. Res.*, 28:611-621

Falsetti, H.L., Mates, R.E., Grant, C., Greene, D.G., and Bunnell, I.L. (1970) Left ventricular wall stress calculated from one-plane cineangiography. *Circ. Res.*, 26:71-83

Feigl, E.O., Simon, G.A., Fry, D.L. (1967) Auxotonic and isometric cardiac force transducers. *J. Appl. Physiol.*, 23:597-600

Fung, Y.C.B. (1973) Biorheology of soft tissue. *Biorheology*, 10: 139-155

Ghista, D.N. and Sandler, H. (1969) An analytic elastic-viscoelastic model for the shape and the forces in the left ventricle. *J. Biomech.* 2:35-47.

Ghista, D.N. and Hamid, S. (1977) Finite element analysis of the human left ventricle. *Computer Programs in Medicine*, vol. 7, No. 3

Ghista, D.N., Ray, G., and Sandler, H. (1980) Cardiac assessment mechanics: 1. Left Ventricular mechanomyocardiography. A new approach to the detection of diseased myocardial elements and states. *Med. Biol. Eng. and Comput.*, 18:271-280

Gould, P., Ghista, D.N., and Brombolich, L. (1972) In vivo stresses in the human left ventricular wall: analysis accounting for the irregular 3-dimensional geometry and comparison with idealized geometry analyses. *J. Biomech.*, 5:521-539

Hamid, M.S. and Ghista, D.N. (1974) Finite element analysis of human cardiac structures. *Finite Element Methods in Engineering*, Univ. of New So. Wales, pp. 1-12

Heethaar, R.M., Robb, R.A., Pao, Y.C., and Ritman, E.L. (1976) Three-dimensional stress and strain in the intact heart. *Proc. San Diego Biomed. Symp.*, 337-342

Hefner, L.L., Sheffield, L.T., Cobbs, G.C., and Klip, W. (1962) Relation between mural force and pressure in the left ventricle of the dog. *Circ. Res.*, II:654-663

Hill, A.V. (1950) Discussion on muscular contraction and relation: their physical and chemical basis. *Proc. R. Soc. Lond (Biol),*

137:40-87

Huisman, R.M., Elzinga, G., Westerhof, N., and Sipkema, P. (1980a) Measurement of left ventricular wall stress. *CV Res.*, 14:142-153

Huisman, R.M., Sipkema, P., Westerhof, N., and Elzinga, G. (1980b) Comparison of models used to calculate left ventricular wall force. *Med. Biol. Eng. & Compt.*, 18:133-144

Janz, R.F. and Grimm, A.F. (1972) Finite Element Model for the Mechanical Behavior of the Left Ventricle. *Circ. Res.*, 30:244-252

Janz, R.F. and Grimm, A.F. (1973) Deformation of the diastolic left ventricle. 1. Nonlinear elastic effects. *Biophys. J.*, 13:689-704

Janz, R.F., Kubert, B.R., and Moriarity, T.F. (1974) Deformation of the diastolic left ventricle–II. Nonlinear geometric effects. *J. Biomech.*, 7:509-516

Janz, R.F., Kubert, B.R., Mirsky, I., Korecky, B., and Taichman, G.C. (1976) Effect of age on passive elastic stiffness of rat heart muscle. *Biophy. J.*, 16:281-290

Janz, R.F. and Waldron, R.J. (1978) Predicted effect of chronic apical aneurysms on the passive stiffness of the human left ventricle. *Circ. Res.*, 42:255-263

Janz, R.F. (1980) Effect of chamber eccentricity on equatorial fiber stress during systole. *Fed. Proc.*, 39:183-187

Lame, G. (1866) *Leçons sur la théorie mathématique de l'élasticité des corps solides*, ed. 2. Paris, Gauthier-Villars, 211-213

LaPlace, P.S. (1806) *Théorie de l'action caprillarie en traité de mecanique Celéste*, suppl. au livre X, Paris, Coarcien

Levine, H.J. and Britman, N.A. (1964) Force-velocity relations in the intact dog heart. *J. Clin Invest.*, 43:1383-1396

Lewartowski, B., Sedek, G., and Okolska, A. (1972) Direct measurement of tension within left ventricular wall of the dog heart. *CV Res.*, 6:28-35

Loeffler, L. and Sagawa, K. (1975) A one-dimensional viscoelastic model of cat heart muscle studied by small length perturbations during isometric contraction. *Circ. Res.*, 498-512

McHale, P.A. and Greenfield, J.C., Jr. (1973) Evaluation of several geometric models for estimation of left ventricular circumferential wall stress. *Circ. Res.*, 33:303-312

Mirsky, I. (1969) Left ventricular stresses in the intact human heart. *Biophys. J.*, 9:189-208

Mirsky, I. (1970) Ventricular and arterial wall stresses based on large deformation analyses. *Biophys. J.*, 10:1141-1159

Mirsky, I. (1973) Effects of anisotropy and nonhomogeneity on left ventricular stresses in the intact heart. *Bull Math Biophys.*, 32:197-213

Mirsky, I. (1976) Assessment of passive elastic stiffness of cardiac muscle: Mathematical concepts, physiologic and clinical considerations, directions of future research. *Prog. CV Dis.*, 18: 277-308

Mirsky, I. and Pasipoularides, A. (1980) Elastic properties of normal and hypertrophied cardiac muscle. *Fed. Proc.*, 39:156-161

Moriarity, T.F. (1980) The law of LaPlace, its limitations as a relation for diastolic pressure, volume, or wall stress of the left ventricle. *Circ.*, 46:321-331

Nikravesh, E. (1976) *Optimization in Finite Element Analysis with Special Reference to Three-Dimensional Left Ventricular Dynamics.* Ph.D. Thesis, Tulane Univ.

Panda, S.C. and Natarajan, R. (1977) Finite-element method of stress analysis in human left ventricular layered wall structure. *Med. and Biol. Eng. and Comput.*, 15:67-71

Pao, Y.C., Robb, R.A., and Ritman, E.L. (1976) Plane-strain finite element analysis of reconstructed diastolic left ventricular cross section. *Ann. Biomed. Eng.*, 4:232-249

Parmley, W.W., Chuck, L., Kivowitz, C., Matloff, J.M., and Swann, H.J.C. (1973) In vitro length-tension relations of human ventricular aneurysms. *Am. J. Cardiol.*, 32:889-894

Phillips, C.A., Cox, T.L., and Petrofsky, J.S. (1981) Active material properties of the myocardium: Correlation with left ventricular function in man. *Ohio J. Sci.*, 81:153-160

Pinto, J.G. and Fung, Y.C. (1973) Mechanical properties of the heart muscle in the passive state. *J. Biomech.*, 6:597-616

Rankin, J.S., McHale, P.A., Arentzen, C.E., Ling, D., Greenfield, J.C., and Anderson, R.W. (1976) The three-dimensional dynamic geometry of the left ventricle in the conscious dog. *Circ. Res.*, 39:304-313

Ritman, E.L., Kinsey, J.H., Robb, R.A., Gilbert, B.K., Harris, L.D., and Wood, W.H. (1980) Three-dimensional imaging of heart, lungs and circulation. *Science*, 210:273-280

Robie, N.W. and Newman, W.H. (1974) Effect of altered ventricular load on the Walton-Brodie strain gauge arch. *J. Appl. Physiol.*, 36:20-27

Sandler, H. and Alderman, E. (1974) Determination of left ventricular size and shape. *Circ. Res.*, 34:1-8

Sandler, H. and Dodge, H.T. (1963) Left ventricular tension and stress in man. *Circ. Res.*, 13:91-104

Sarnoff, S.J., Braunwald, E., Welch, G.H. Jr., Case, R.B., Stainsby, W.N., and Macruz, R. (1958) Hemodynamic determinants of oxygen consumption of the heart with special reference to the tension-time index. *Am. J. Physiol.*, 192:148-156

Shoukas, A.A., Sagawa, K., Maughan, W.L., Ebert, W., and Garrison, J.B. (1980) Multiple marker implantation for biplane cineventriculography (abstr) *Cir.*, 62:111

Sonnenblick, E.H. (1964) Series elastic and contractile elements in heart muscle: Change in muscle length. *Am. J. Physiol.*, 207:1330-1338

Sonnenblick, E.H., Parmley, W.W., and Urschel, C.W. (1969) The contractile state of the heart as expressed by force-velocity relations. *Am. J. Card.*, 23:488-503

Spurgeon, H.A., Thorne, P.R., Yin, F.C.P., Shock, N.W., and Weisfeldt, M.L. (1977) Increased dynamic stiffness of trabeculae carneae from senescent rats. *Am. J. Physiol.*, 232:H373-H380

Streeter, D.D., Spotnitz, H.M., Patel, D.J., Ross, J., Jr., and Sonnenblick, E.H. (1969) Fiber orientation in the canine left ventricle during diastole and systole. *Circ. Res.*, 24:339-347

Streeter, D.D., Vaishnav, R.N., Patel, D.J., Spotnitz, H.M., and Sonnenblick, E.H. (1970) Stress distribution in the canine ventricle during diastole and systole. *Biophys. J.*, 10:345-363

Templeton, G.H. (1973) Dynamic stiffness of papillary muscle during contraction and relaxation. *Am. J. Physiol.*, 224:692-698

Vinson, C.A., Gibson, D.G., Yettram, A.L. (1979) Analysis of left ventricula behaviour in diastole by means of finite element method. *Br. Heart J.*, 41:60-67

Voukydis, P.C. (1972) Fiber stress profiles in the left ventricle of the heart during diastole: effects of distension and hypertrophy. *Bull Math Biophys.*, 34: 379–392

Wong, A.Y.K. and Rautaharju, P.M. (1968) Stress distribution within the left ventricular wall approximated as a thick ellipsoidal shell. *Am. Heart J.*, 75:649-662

Wood, R.H. (1892) A few applications of physical theorem to membranes in the human body in a state of tension. *J. Anat. Physiol.*, 26:362-370

Yin, F.C.P., Spurgeon, H.A., Weisfeldt, M.L., and Lakatta, E.G. (1980) Mechanical properties of myocardium from hypertrophied rat hearts. A comparison between hypertrophy induced by senescence and by aortic banding. *Circ. Res.*, 46: 292-300

Zienkiewicz, O.C. (1977) *The Finite Element Method*, 3rd ed. London, McGraw-Hill Ltd.

Chapter 8

LEFT VENTRICULAR PROPERTIES AND PERFORMANCE ASSESSMENT ANALYSES

D.N. GHISTA, F.C.P. YIN, and G. JAYARAMAN

ANALYSES for evaluation of cardiac mechanical performance are provided in terms of (1) intrinsic indices that can characterize either the material properties of the myocardial or valvular tissue, (2) the performance of the left ventricular chamber as a pump or a valvular disorder (such as valvular stenosis or insufficiency), and (3) expected sizes of aneurysm that can develop from infarcts of various sizes.

In the framework of this chapter, we will characterize cardiac mechanical pumping performance in terms of the myocardial capability (in turn dependent on the myocardial constitutive property) to contract effectively and to set up appropriate pressure distributions in the chamber to facilitate proper emptying of the chamber. We will thus concern ourselves with the chamber pressure distributions developed during systole (computed by finite element analysis), influenced by the myocardial excitation-contraction process (and its alteration or loss in the case of a coronary lesion and resulting myocardial ischemia and infarction). However, differentiation of ischemic-infarcted segments will be shown to be possible in terms of (and by determining) the passive regional properties during diastole. During diastole, the left ventricle behaves like a pressurized inflated thick-walled structure (akin to a mechanical structure) and lends itself to a conventional mechanics formulation; whereas detection of infarcted segments requires determination of the distributed myocardial properties, characterization of ventricular compensation to pressure-volume overload (resulting from valvular disorders) can be done reasonably sensitively in terms of the gross or effective modulus of the left ventricle.

The data that can be employed for these analyses, on myocardial (regional or gross) property determination, are the dynamic geometry of the left ventricular structures, the left ventricular chamber pressure, and the heart sounds' frequency spectra. As regards the determination of the left ventricular geometry, cineangiography (entailing pefusing the chamber with radiopaque contrast medium) delineates the endocardial boundary but does not distinctly demarcate the epicardial as well as the valvular boundaries. Sector scan ultrasonoechocardiography can edge out the irregularities of both the inner and outer boundaries of only a portion of the left ventricular chamber; although its field of view and resolution are currently limited by the state-of-the-art of acoustic element array technology, it has the best potential to enable 3-dimensional (computerized) reconstruction of the complete chamber from a number of 2-dimensional planar scans.

The phonocardiographically monitorable heart sound frequency spectra are related to the vibrational frequency spectra of the left ventricular myocardial wall and valve structures (particularly the mitral valve, which constitute a major portion of the chamber boundary). Thus, it should occur to us to perform vibrational analyses of the myocardial wall and mitral valve segments of the left ventricular chamber, determine their primary frequency ranges and obtain the basis of selecting the heart sound spectral frequencies that are associated with their primary vibrational modes. Hence, in the case of the mitral valve, we can (by employing its vibrational and stress-deformation, at the instant of occurrence of the heart sound) determine the leaflet's Young's modulus and stress, and thereby its constitutive parameters that characterize it as normal and pathologic. The foundations of phonocardiography, to acquire the pertinent heart sound data requisite for the above mitral valve analysis, are provided in the chapter by Van Vollenhoven et al. (1979).

THE LEFT VENTRICULAR MYOCARDIAL PRESSURE DEPENDENT INSTANTANEOUS OVERALL ELASTIC MODULUS

A number of analyses exist that enable determination of the instantaneous value of the so-called overall elastic modulus of the left ventricle. The earliest of these analyses were published by Ghista and Vayo (1969) and later again by Ghista and Sandler (1975), and Mirsky and Parmley (1974). The instantaneous left ventricular elasticity (static) model of Ghista and Vayo (1969) and Ghista and Sandler (1975) employed (in keeping with almost ellipsoidal shape of the chamber) superposition of a Line dilatation stress system (made up of uniform dilations of strengths Ad ξ distributed over $-a<y<a$) given by the displacement

$$U = \int_{\xi=-a}^{\xi=a} A \frac{Xi_x + (y-\xi)i_y + Zi_z}{(x^2 + (y-\xi)^2 + z^2)^{3/2}} d\xi \qquad (1)$$

and a Hydrostatic stress system, whose intensity parameters are obtained by having the monitored endo- and epicardial boundaries (of the single plane left ventricular cineangiographic image) represent two sets of stress trajectories of the superimposed stress systems. The left ventricular wall modulus value, at each instant, can be computed by equating the model derived instantaneous wall displacements expressions (derived as functions of the known instantaneous catheterized chamber pressure increment and the to-be-determined instantaneous elastic modulus) with the cineangiographically obtained values of the chamber wall's instantaneous displacements.

For clinical applications, this (infinitesimal elasticity) model needs to be employed with frame-by-frame left ventricular geometry and associated incremental pressure. When applied during diastole, it yields at each instant a value of the Young's modulus of the passive left ventricular myocardium that is representative of the overall or a quasi-average property of the medium; it is recognized that the model is a closed chamber whereas the actual left ventricle is not closed during diastolic filling. The Young's modulus value increases as the left ventricular chamber is filled and its pressure increases.

Strictly speaking, the model should only be applied to diastolic data. Whereas during diastole the myocardial instantaneous or pressure dependent elastic modulus is the result of left ventricular filling or pressurization, during systole an increase in the value of the modulus occurs by myocardial activation. The associated myocardial contraction (resulting from its excitation) occurring against the contained incompressible blood then increases the myocardial stiffness and raises the chamber pressure during systole. During systole, until the maximum chamber pressure is attained, we have the phenomenon of chamber pressure increase associated with either no change in volume (during the isometric contraction phase) or a decrease in volume (during the initial ejection phase). This precludes the applicability of our model. However, if we assume that the instant of attainment of maximum chamber pressure during ejection phase corresponds to the development of maximum contraction or myocardial activation, then, in the remaining period of the ejection phase (when a decrease in chamber pressure corresponds to a decrease in chamber volume), our model can again be employed to now determine the instantaneous (stress dependent) effective modulus of the activated left ventricle.

An increase in the diastolic modulus value can portray increased resistance to filling. During systole, a valvular disorder (such as a regurgitant aortic valve) can result in a volume-pressure overload, necessitating a high effort on the part of the myocardium to contract adequately, i.e. to increase the value of the systolic modulus, so as to appropriately raise the chamber pressure value above the aortic pressure value (in order to empty the left ventricular chamber). In this situation, the development of myocardial hypertrophy or wall thickening can obviate the necessity of an undue increase in systolic modulus. Thus a normal value of systolic modulus, in the face of left ventricular chamber enlargement (due to a chronic valvular disorder), can denote a compensated left ventricular myocardium. On the other hand, an unduly high value of the systolic modulus denotes an uncompensated overly exerted left ventricle; this situation, in the long run, can result in loss of contractility (or dysfunction) in the most severely stressed segments. This analysis employs an instantaneous value of the chamber pressure.

It should be evident from hemodynamic considerations that, due to the intraventricular blood motion, the chamber pressure will not be constant, but have an average variation of about 10 to 15 per cent from the base to the apex. Hence, it is important to account for the location of the catheter tip and determine the chamber pressure field (and hence the variation of pressure along the endocardial wall), in terms of the pressure at the catheter tip, by solving the governing blood flow equation in the left ventricular chamber from its known wall motion data, derived from either cineangiography or two-dimensional echocardiography. Based on such an analysis, Wang and Sonnenblick (1979) have advocated that, in order to minimize dynamic effects close to the root of the aortic valve and to average the gravity effect, the pressure sensor should be located in a horizontal plane at midventricle.

TOWARDS A DIFFERENTIAL INDEX FOR LV COMPENSATED-UNCOMPENSATED MYOCARDIAL HYPERTROPHY AND MYOCARDIAL PATHOLOGY

The intent here is to (1) employ noninvasively determinable primary first sound frequency (ω, taken equal to the left ventricular primary vibrational frequency) and model chamber geometry (size r_i and wall thickness h) and (2) develop diagnostic domains in the (ω, r_i, h) coordinate space. To this end, the values of myocardial pathology reflecting stiffness (or Young's modulus, E) are computed, from a formula involving cineangiographically obtained left ventricular (LV) pressure and dimensions, by means of a dynamic LV model to be presented in this section. These values of E are designated on the (ω, r_i, h) coordinate space, in which (LV vibrational model derived) constant E contours are plotted. Thus, diagnostic domains are developed in the (ω, r_i, h) coordinate space, characterize normal and athlete's LV(s), hypertensive and enlarged compensated and noncompensated hypertrophic LV(s), and infarcted LV(s).

For modeling purposes, the left ventricle is simulated (at an instant) as a fluid-filled thick-walled small displacement sustaining elastic sphere, whose (1) instantaneous inner volume equals that of the LV chamber; (2) instantaneous wall thickness is such that the wall mass or volume equals that of the left ventricle; (3) the myocardial wall medium's instantaneous elastic modulus is E(t) at time t; (4) instantaneous inner and outer radii are r_i and r_e, respectively; (5) the pressure response to volume change ΔV is ΔP during filling, with the instantaneous point-symmetric displacement being U_r (r, t); and (6) sinusoidal chamber volumetric changes during the period of occurrence of the first heart sound are given by $\Delta V \sin(\omega\tau)$.

By solving the LV model medium's incompressibility condition

$$\epsilon_{kk} = 0 = \nabla^2 U = \frac{1}{r^2}\frac{\delta}{\delta r}(r^2 U_r) = 0 \qquad (2)$$

we obtain

$$U_r = \frac{F(t)}{r^2} \qquad (3)$$

Now for the Instantaneous Pressure-Deformation Analysis, the function F(t) in equation 3 is defined by the boundary condition for $U_r(r_i, t)$ to accommodate the incremental volume change ΔV during filling

$$\frac{4}{3}\pi(r_i^2 + U_{ri})^3 - \frac{4}{3}\pi r_i^3 = \Delta V(t) \qquad (4)$$

so that, from equations 3 and 4, we obtain

$$U_r = \frac{\Delta V(t)}{3V(t)} \cdot \frac{r_i^3}{r^2} = \frac{F(t)}{r^2} \qquad (5)$$

Now for the above U_r expression, the dynamic equilibrium equation is as follows:

$$-\frac{\delta H}{\delta r} = \rho_\omega \ddot{u}_r = \rho_\omega \frac{\ddot{F}(t)}{r^2}; \tag{6}$$

being the wall medium density. This equation (upon integration with respect to r) yields an expression for the hydrostatic stress H(r, t), which, when substituted in the constitutive equation, gives

$$\sigma_{rr} = \rho_\omega \frac{\ddot{F}(t)}{r} - 4\,\frac{F(t)}{r^3} - h(t) \tag{7}$$

By having equation 7 satisfy the boundary conditions: $\sigma_{rr}(r_e, t) = 0$; $\sigma_{rr}(r_i, t) = -\Delta p(t)$, the following expression for myocardial modulus is obtained

$$E(t) = \frac{\Delta p(t) + \rho_\omega \ddot{F}(t)\left(\frac{1}{r_i} - \frac{1}{r_e}\right)}{4F(t)\left(\frac{1}{r_i^{\,3}} - \frac{1}{r_e^{\,3}}\right)} \tag{8}$$

where F(t) is given by equation 5. This equation is employed to compute the values of E of subjects (from available cineangiographically obtained pressure-geometry data) at instants corresponding to the first heart sound occurrence. In order to record these values of E in the (ω, r_i, h). Hence, continuing the above analysis to accommodate the first heart sound induced vibrational volumetric changes $\Delta V \sin \omega \Upsilon$ about the instantaneous volume V(t), we replace the variable t by Υ in the previous analysis. The function $F(\Upsilon)$ and U_r are now obtained as

$$F(\tau) = r_i^{\,3}(t)\,\frac{\Delta V}{3V(t)}\sin(\omega\tau); \quad U_r(r, \tau) = \frac{r_i^{\,3}}{r^2}\,\frac{\Delta V \sin(\omega\tau)}{3V(t)} \tag{9}$$

The radial stress equation, analogous to equation 7 is now

$$\sigma_{rr}(r, \tau) = \rho_\omega \frac{r_i^{\,3}\,\omega^2\,\Delta V \sin(\omega\tau)}{3V(t)} - h(t) - \frac{4}{3}\mu\frac{r_i^{\,3}}{r^3}\,\frac{\Delta V \sin(\omega\tau)}{V(t)} \tag{10}$$

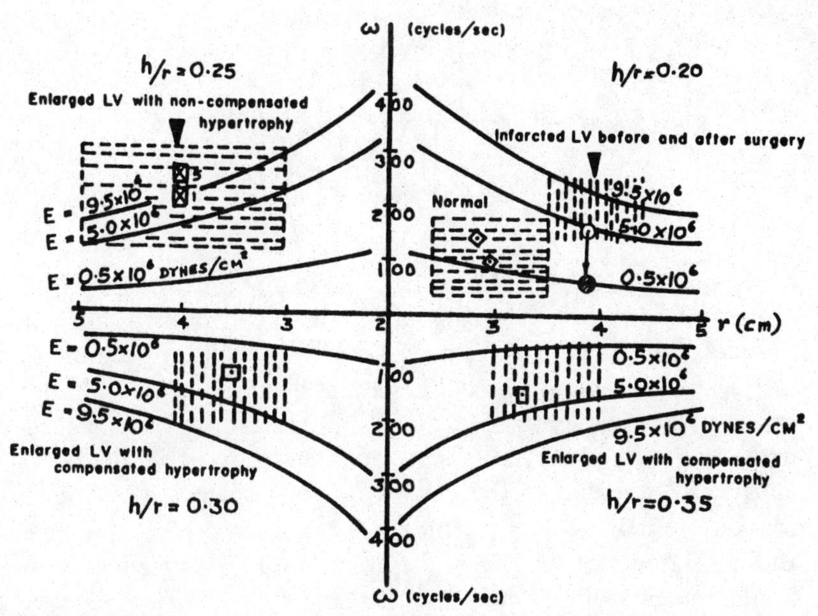

Figure 8-1. Contours of constant E on the 'ω vs r' coordinate space, depicting domains of normal, infarcted and compensated LV (s).

so that satisfaction of the boundary conditions

$$\sigma_{rr}(r_e, \tau) = 0; \quad \sigma_{rr}(r_i, \tau) = -\Delta p \sin(\omega \tau) \tag{11}$$

yields (for $\nu = 0.5$ and $E = 3\mu$)

$$\Delta p = \frac{\Delta V}{3V(t)} \left\{ -\rho_\omega \omega^2 r_i^{\,2} \left[1 - \frac{r_i}{r_e}\right] + \frac{4E(t)}{3} \left[1 - \left(\frac{r_i}{r_e}\right)^3\right] \right\} \tag{12}$$

Since, for free point-symmetric vibrations of the thick-walled spherical ventricular shell filled with fluid (of zero fluid density), Δp is zero, equation 12 yields, for the vibrational frequency,

$$\omega^2 = \frac{4E\left[1 - \left(\frac{1}{1+\frac{h}{r}}\right)^3\right]}{3\rho_\omega r^2\left[1 - \left(\frac{1}{1+\frac{h}{r}}\right)\right]} \tag{13}$$

Now equation 13 is employed to develop and plot constant E contours, in Figure 8-1. With the aid of these contours, points corresponding to values of r_i, h, and E (computed with the aid of equation 8 and cineangiographic data) are designated on the coordinate space, for subjects with myocardial infarcts, and valvular disorders with and without hypertrophy compensation. The preliminary diagnostic domains in the (ω, r_i, h) coordinate space, are illustrated in Figure 8-1. In order to establish more comprehensive and validated domains, considerably more patient simulation is in the process of being carried out.

THE FINITE ELASTICITY LEFT VENTRICULAR SPHERICAL MODEL FOR DETERMINING THE NONLINEAR MYOCARDIAL CONSTITUTIVE PROPERTIES

Instead of employing chamber incremental pressure and volume for determining the myocardial wall modulus at each instant (with the aid of the above two types of infinitesimal static and dynamic models), we can determine the finite elasticity type of constitutive parameters that would hold true for the entire diastole as well as ejection phases.

Under the diastolic pressure loading P, let the spherical chamber undergo point symmetric deformation from radius R to radius r, which for material incompressibility yields

$$\frac{dr}{dR} = \frac{R^2}{r^2} = x^2 \quad \text{and}\ r = (R^3 + A)^{1/3} \tag{14}$$

where A is a constant.

Let the left ventricular chamber volume change from its end-systolic volume V_s (inner and external radii = R_i, R_e) to V (inner and external radii = r_i, r_e) at the diastolic pressure value P, so that

$$V = V_s + \Delta V \quad or \quad x_i^3 = \left(\frac{r_i}{R_i}\right)^3 = \left(1 + \frac{\Delta V}{V_s}\right) \tag{15}$$

and the invariant left ventricular medium volume V_m is given by

$$\frac{V_m}{V_s} = \frac{1}{R^*} - 1; \; R^* = \frac{R_i^3}{R_e^3}. \tag{16}$$

The constitutive properties of the left ventricular medium can be represented by the parameters of its strain energy function W selected as (Demiray, 1976; Mirsky, 1973; Janz et al., 1976)

$$W = \frac{\beta}{2a} e^{a(I-3)}; \; I = 2x^2 + \frac{1}{x^4}; \; x = \frac{R}{r}. \tag{17}$$

When the stress-strain relations,

$$\sigma_{rr} = H + 2\beta x^2 \, e^{a(I-3)}$$

$$\sigma_{\theta\theta} = \sigma_{\phi\phi} = H + \beta(x^2 + \frac{1}{x^4}) \, e^{a(I-3)}, \tag{18}$$

are introducted in the equilibrium equation, we obtain

$$\sigma_{rr} = 2\beta \int_{x_e}^{x} e^{a(I-3)} \, (x + \frac{1}{x^2}) \, dx \tag{19}$$

$$\sigma_{\theta\theta} = \sigma_{\phi\phi} = 2\beta \int_{x_e}^{x} e^{a(I-3)} \left(x + \frac{1}{x^2}\right) dx$$

$$+ \beta\left(\frac{1}{x^4} - x^2\right) e^{a(I-3)}. \tag{20}$$

Now equations 19 and 20, with the usual boundary conditions

$$\sigma_{rr} \text{ (at } x = x_i) = -P, \quad \sigma_{rr} \text{ (at } x = x_e) = 0, \tag{21}$$

yield

$$P = 2\beta \int_{x_i}^{x_e} e^{a(I-3)} \left(x + \frac{1}{x^2}\right) dx \tag{22}$$

wherein (1) x_i is obtained from equation 15 and (2) x_e is obtained from the following global incompressibility condition

$$R_e^3 - R_i^3 = r_e^3 - r_i^3$$

or

$$x_e = \left[1 - R^* + \frac{R^*}{x_i^3}\right]^{-1/3}, \text{ with } R^* = (R_i/R_e)^3 \tag{23}$$

Equation 22 then enables us to determine the values of a and β from knowledge of the data on (1) the left ventricular chamber geometry (and hence chamber and myocardial medium volumes, V_s and V_m) at the start of diastole and (2) the chamber volume increment ratio $\Delta V/V_s$ for two independent readings of the catheterized chamber pressure loading P, by first determining the value of a that satisfies the equation

$$\frac{\displaystyle\int_{x_i^I}^{x_e^I} e^{a(I-3)} \left(x + \frac{1}{x^2}\right) dx}{\displaystyle\int_{x_i^{II}}^{x_e^{II}} e^{a(I-3)} \left(x + \frac{1}{x^2}\right) dx} = \frac{P^{II}}{P^I} \tag{24}$$

wherein x_i is expressed in terms of V_s and ΔV by virtue of equation 15 and x_e is expressed in terms of V_s, ΔV, and V_m by virtue of equations 22 and 23. The value of β is subsequently determined by substituting the value of a in equation 22 for either data set $[P^I, x_i^I, x_e^I)$ or $(P^{II}, x_i^{II}, x_e^{II})]$. Having determined the values of a and β, the values of chamber pressure can be obtained at subsequent instants, with the aid of equation 22, from knowledge of $x_i(t)$ and $x_e(t)$.

Also, the model could conceivably be employed during the phase of systolic ejection, from the instant of attainment of maximum chamber pressure (or systolic blood pressure) to the end of the ejection phase. In that case, a and β would be obtained in terms of the left ventricular chamber and medium volumes at the end of ejection (which are equal to V_s and V_m) and the volume changes ΔV during this late ejection phase. The chamber pressure values during the ejection phase, between the instant of maximum chamber pressure (or systolic blood pressure) and the end-ejection chamber pressure (or diastolic blood pressure), can then be determined from equation 22 in terms of the values of $x_i(t)$ and $x_e(t)$. The values of a and β of course can have diagnostic implications. However, to this end, considerably more patient simulation is required, by determining a and β (during both left ventricular filling and latter phase of ejection) for healthy left ventricles and hypertrophically compensated and uncompensated chronically distended and hypertensive left ventricles. The left ventricular stiffness (dV/dP) can also be employed as an index for at least passive ventricular distensibility. By noting that $dV/dp = (dV/dx_i)/(dP/dx_i)$, the left ventricular chamber stiffness is obtained (in terms of V_m, V_s, and ΔV), by differentiating equation 22 with respect to x_i, to obtain

$$\frac{dV}{dP} = 3\,[e^{a(I_i-3)}(x_i^5 + x_e^2) - R^* e^{a(I_e-3)}(x_e^5 + x_e^2)\,]^{-1}. \quad (25)$$

It would be interesting to do an analysis of pressure measurement errors on the values of the constitutive parameters a, β, and dV/dP. It can be seen from equation 25, that the values of

α, β, and dV/dP are independent of identical orders of percentage errors in the measurement of pressure at the two instants I and II (that could be caused by catheter calibration errors).

The above formulation can also have a significant noninvasive (bedside) continuous monitoring implication as indicated above. If once the left ventricle is catheterized and the values of α and β are determined, then, by monitoring V_s, V_m, and the chamber volume (V) continuously by echocardiography, the beat-to-beat systolic and diastolic blood pressure values can be monitored at the patient's bedside with the aid of a portable echocardioscope (Schmidt and Miller, 1975).

AN INSTANTANEOUS FINITE ELEMENT MODEL FOR THE SMALL PRESSURE INCREMENT DEFORMATION RESPONSE OF THE LEFT VENTRICLE

The models presented thus far help determine (1) the overall left ventricular myocardial elasticity (E) and stiffness (dV/dP) as functions of chamber pressure, volume, or size. While these properties can help effectively characterize ventricular compensation to volume-pressure overload caused by valvular defects, they cannot help detect the locations and extents of myocardial segments. For this purpose, we need to determine regional or segment-by-segment (nonlinear) elastic constitutive property of the passive myocardium during diastole, so that, on the basis of the values of the property parameters, we can detect infarcted segments.

As indicated before, the data can consist of single or biplane left ventricular chamber cineangiographic frames and associated (catheter-derived) chamber pressure during diastole. At each instant, the left ventricular finite element model is developed (corresponding to a single or biplane left ventricular image) to simulate the incremental myocardial stress and deformation due to the monitored incremental pressure. The details of the finite element model construction and formulation are provided in the publication of Ghista and Hamid (1977) as well as in the chapter by Ghista et al. (1981). Tetrahedral linear elastic, small, strain displacement type elements are employed in the finite element discretization.

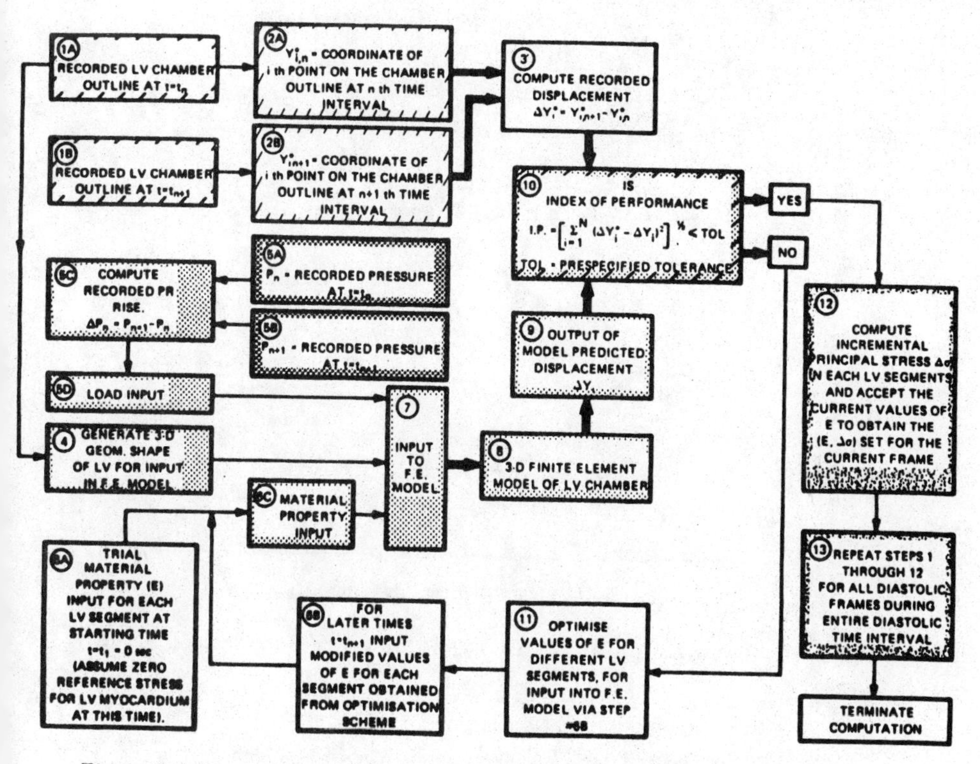

Figure 8-2. The optimization scheme to converge to the instantaneous values of the left ventricular model's finite elements.

For the finite element model corresponding to an instant, the values of the Young's modulus of the finite elements are determined by making the instantaneous model's internal volume, due to the incremental pressure, match the internal volume of the reconstructed left ventricular chamber at the next instant or frame; the methodology and the optimization scheme (to converge to the values of the moduli), are summarized in Figure 8-2.

For each element, the instantaneous sets of values of the modulus (E) and principal stress (σ, obtained by summing the incremental stress, $\Delta\sigma$, over the diastolic time intervals) are plotted in the E-σ domain of normal myocardial elements (Figure 8-3). Thereafter, by analyzing symptomatic subjects, we find that the (E, σ) coordinates of some of their left ventricular myocardial

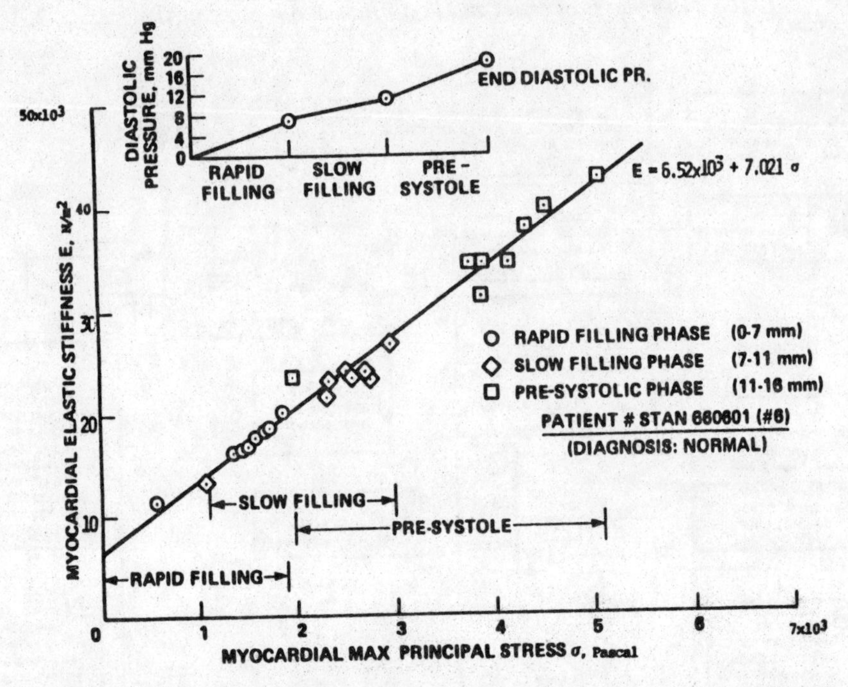

Figure 8-3. 'E vs σ' constitutive relationships for a normal myocardial element.

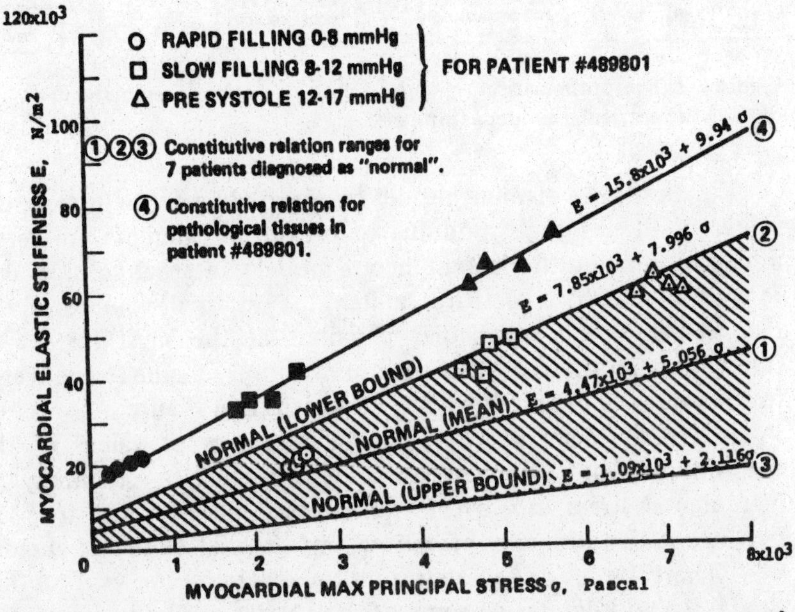

Figure 8-4. (E, σ) coordinate points for the myocardial finite elements of a symptomatic patient, showing that some elements fall in the 'normal' category, whereas the ones outside the normal range could hence be deemed to be infarcted.

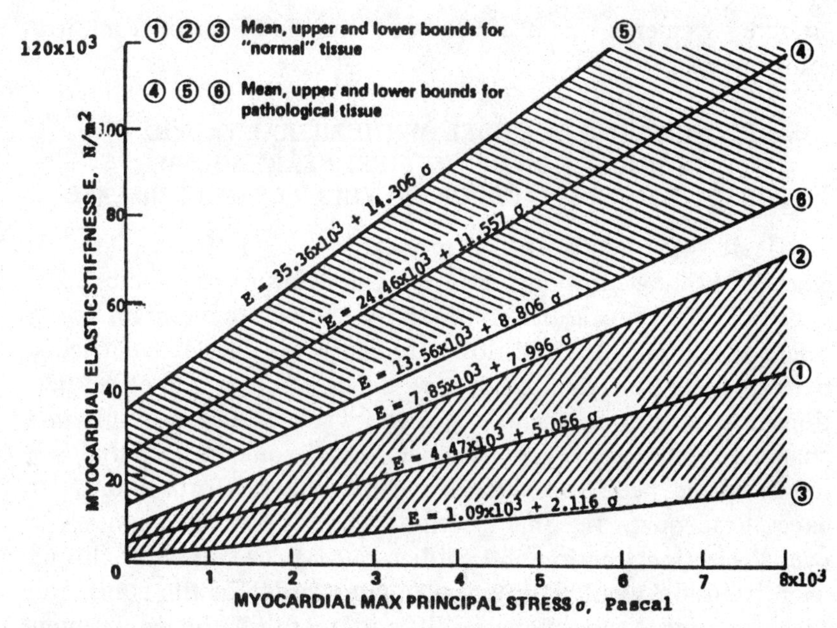

Figure 8-5. Constitutive properties of normal and infarcted myocardial elements.

elements fall outside the normal E-σ domain (Fig. 8-4), the domain for infarcted myocardial elements is thereby determined. Figure 8-5 shows the E-σ domains of normal and pathological myocardial segment. Now, by determining the (E, σ) coordinates of the left ventricular myocardial segments of a subject to be evaluated and ascertaining whether the (E, σ) coordinates fall within the normal or pathological domain, we can detect the location and spread of the infarcted zone. This method indicates how the regional myocardial properties can be determined to designate diseased myocardial segments by employing chamber pressure and dynamic geometry data. It is conceptually possible to introduce minute air bubbles into the left ventricular chamber, invoke the principle of their resonant frequency proportional to the pressure of the surrounding fluid, and thus determine the blood pressure time-variation by acoustically monitoring the variations in the amplitude of the transmitted ultrasonic pulses (Scully, 1979). By combining the two techniques, it should be possible to detect infarcted segments noninvasively with the aid of sector scan ultrasonic imaging.

VIBRATIONAL MODEL OF THE MITRAL VALVE, IN SIMULATION OF THE FIRST HEART SOUND, FOR CHARACTERIZATION OF VALVULAR LEAFLET DISEASE

Heart valve constitutive properties can influence (1) the chamber filling-emptying and pressure-volume dynamics, and hence (2) the myocardial contractility invoked to maintain appropriate aortic pressure and flow rate characterizing left ventricular performance. For example, if a sequence of cause and effect relationships, stemming from a diseased valvular tissue, result in a chamber volume-pressure overload, then some myocardial segments would need to exert high contractility efforts in order to execute adequate regional ejection fractions so as to maintain an overall ejection fraction and cardiac output. The effect of chronic overdistension would prevent adequate overlap of the contractile filaments and hence make it difficult to fulfill the requirement for exertion of adequate contractile force. Yet, the demand for it would eventually cause a breakdown in the contractile machinery and of the exerted myocardial segment and associated loss of its contractility. This would impose additional burden on the surrounding myocardium, which would thereby also lose its contractility. Thus an eventual aneurysm would develop over the incapacitated myocardial zone due to its inability to sustain the high systolic pressure.

Heart sounds provide us with a readily available noninvasive data concerning the vibration of cardiac structures (such as heart walls and valves). Information on the etiology of these heart sounds and murmurs can be derived from Rushmer (1961) and McKusnick (1958). An excellent treatise on the physical basis of cardiovascular sound has been provided by Wiskind and Talbot (1958).

At the end of the filling phase, the contracting left ventricle accelerates the contained blood and sets the closing mitral valve leaflet as well as the left ventricular chamber into vibration. Hence, the first heart sound spectra contains the vibrational frequency of both the mitral valve leaflet and the left ventricular myocardial wall. We thus need an analytical basis for distinguishing spectral peaks due to primary modes of vibration of the left ventricular chamber and the mitral valve leaflet. For this purpose, we reason that since the mitral valve constitutes a portion of the left ventricular chamber, its primary vibrational mode frequency must be higher than that of the myocardial chamber. Thus the lower bound of the (normal-pathological) mitral valve leaflet vibrational frequency needs to be obtained, so that first heart sound spectral peaks greater than and less than this lower bound frequency (of the mitral valve) may be adopted as the primary vibrational mode frequencies of the mitral valve leaflet and the left ventricular myocardial chamber, respectively. We will present how the lower bound frequency can be obtained, although for a detailed analysis the reader is advised to refer to the chapter by Mazumder et al. (1979), which also contains the detailed formulation of the mitral valve vibrational and stress-deformation analyses.

The leaflet constitutive equations,

$$E = 17\,(\sigma + 4{,}083.6)\ \text{dynes/cm}^2,\ \text{for healthy leaflets}$$

and

$$E = 17\,(\sigma + 9{,}714.6)\ \text{dynes/cm}^2,\ \text{for diseased leaflets,} \qquad (26)$$

are derived by stress-strain evaluation of excised mitral valve strips (Ghista and Rao, 1973); the corresponding E, σ constitutive coordinate space is shown in Figure 8-6. In order to determine the leaflet's E-σ characteristics, we need expressions for E and σ in terms of its monitored major leaflet's geometry, deformation, and vibrational (heard sound derived) frequency characteristics. The mitral valve can be idealized as a limacon (Figure 8-7) whose ring boundary is given by

$$R(\theta) = a(1 + 0.5 \cos \theta) \qquad (27)$$

where the size parameter a is related to the longest valve diameter, as indicated by Mazumder et al. (1979).

If u* represents the maximum value of the leaflet's isodisplacement amplitude contours, u(x, y) = constant, and T denotes the

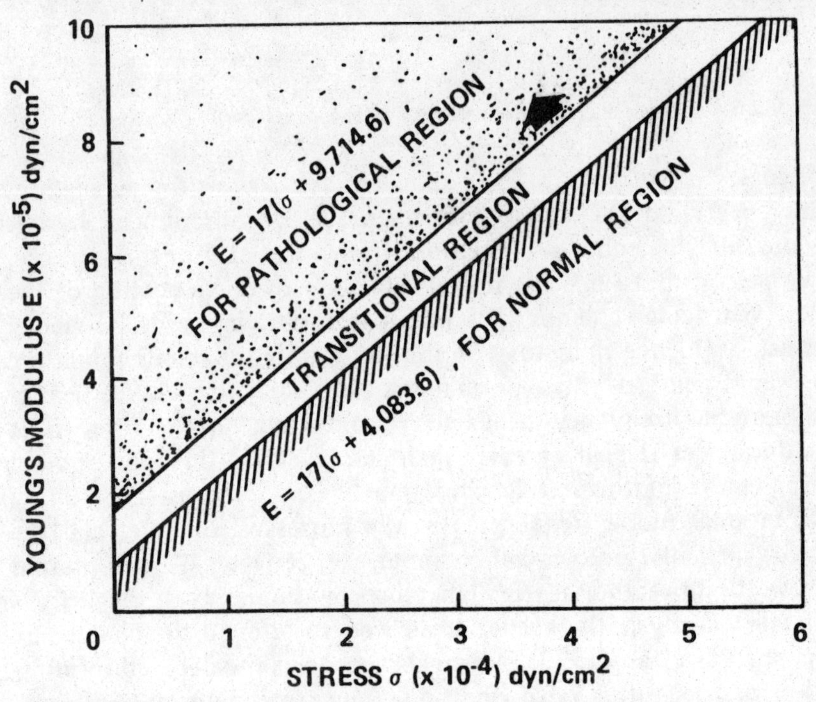

Figure 8-6. The E-σ constitutive property coordinate space for normal and pathological mitral valve tissue.

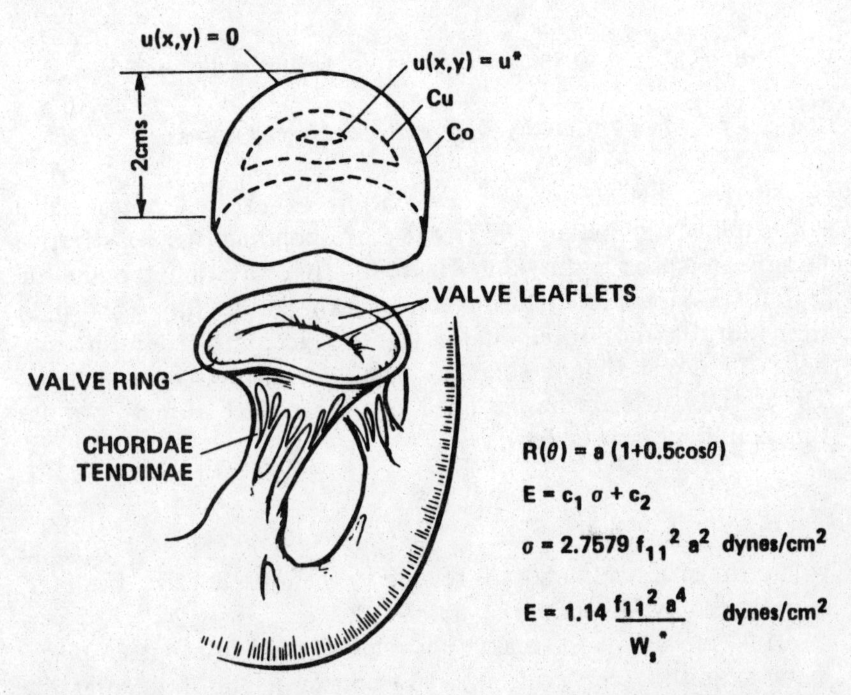

Figure 8-7. The mitral valve configuration and model geometry characterization.

tension in the membrane (whose density per unit area is ρ), then the primary mode vibrational frequency equation of the membrane is given by

$$\omega\left(\frac{\rho}{T}\right)^{1/2} = \frac{2.405}{(2u^*)^{1/2}} \tag{28}$$

Also, the expression of the second vibrational mode frequency of the limacon, which equals the primary vibrational frequency of its major leaflet, is given by (Torvik and Eastep, 1972)

$$a\omega\left(\frac{\rho}{T}\right)^{1/2} = 3.783 \tag{29}$$

so that, from equations 28 and 29 we get

$$u^* = 0.202\ a^2$$

Now by equating the changes in the leaflet membrane surface area due to its deformation under a maximal transmural pressure q, and the membrane tension T, we obtain (Mazumder and Hearn, 1978) the load-stress-deformation relationship

$$\sigma = \frac{Eq^2 u^*}{16h^2(1-\nu)} = \frac{T}{h}$$

where h and ν are respectively the membrane thickness and Poisson's ratio and the transmural pressure q is given by (Polya and Szego, 1951).

$$q = -2T\ \frac{W_s^*}{u^*} \tag{30}$$

On substituting for q and T from equations 30 and 28 into equation 29, the load-stress-deformation relationship becomes

$$E = \frac{23\pi^2 D(1-\nu)f_m{}^2 u^{*2}}{(2.4049)^2 W_s{}^{*2}} \tag{31}$$

where the primary frequency $f_m = \omega/2\pi$ and $D = \rho/h$.

Then by employing the value of u* from equation 29 into the vibrational and load-stress-deformation analyses derived equations 28 and 31 we obtain (for D = 1 gm/cm³)

$$\sigma = \frac{T}{h} = 2.758\ f_m{}^2 a^2\ \text{dyn/cm}^2 \tag{32}$$

$$E = 1.114\ \left(\frac{a}{W_s{}^*}\right)^2 f_m{}^2 a^2\ \text{dyn/cm}^2 \tag{33}$$

which can enable us to determine the leaflet stress and modulus in terms of the monitored leaflet size parameters a and deformation $W_s{}^*$, provided its vibrational frequency f_m can be related to the appropriate first heart sound peak. To that end, by combining equations 26, 32, and 33 we obtain

$$f_m{}^2\ a^2\ [1{,}114\ (\frac{a}{W_s{}^*}) - 46.88] = 69{,}421.2. \tag{34}$$

Then, in order to obtain a lower bound on f_m, we take the anatomical maximal value of a = 1.8167 cm and the physiological minimal value of $W_s{}^* = 0.2$ cm, so that lower bound on $f_m = 66$ Hz. Thus in equation 32 and 33, we take for f_m the value of the first spectral frequency peak that is greater than 66 Hz. The equations 32 and 33 are shown as nomograms in Figure 8-8(a & b). From these nomograms, the values of σ and E are determined (by monitoring f_m, a, $W_s{}^*$) and plotted in the (σ, E) coordinate point in Figure 8-6; depending on whether the coordinate point falls in the healthy or diseased leaflet E-σ domains, the leaflet is termed as healthy or diseased.

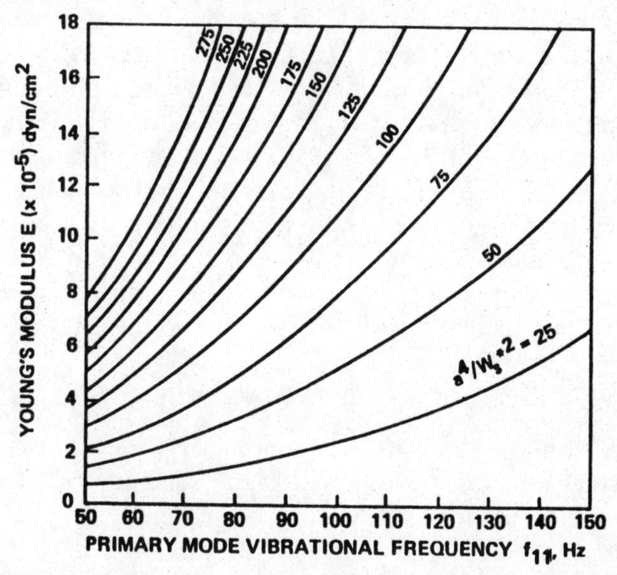

Figure 8-8a. Nomograms of E vs f_{11}, to permit determination of the E value from knowledge of the heart sound frequency, valve size and deformation.

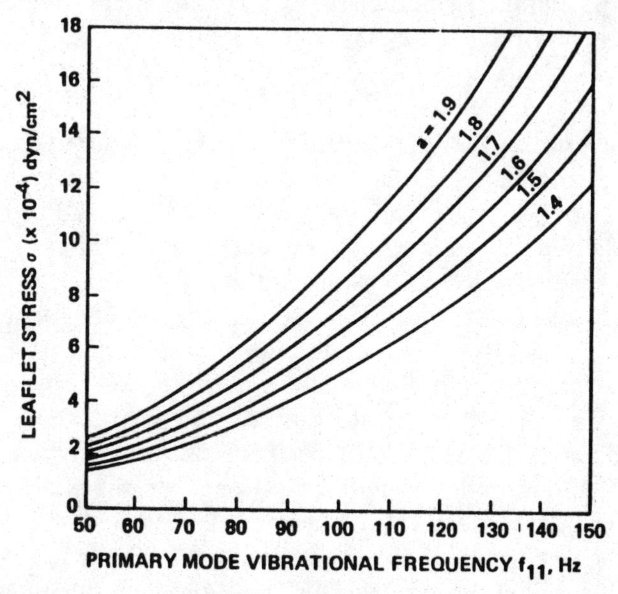

Figure 8-8b. Nomograms of σ vs f_{11}, to permit determination of the leaflet stress from knowledge of the heart sound frequency (f_{11}) and valve size parameter(a).

Considerable improvement in the diagnostically relevant accuracy of the above type of analysis (to determine the elastic properties of the mitral and aortic leaflets) can be obtained by developing more sophisticated formulations of (1) the vibrations of the mitral (and aortic) valve leaflets (incorporating their precise geometry and the effect of the surrounding fluid) and (2) the left ventricular myocardial wall (incorporating its geometry), in order to provide a sounder basis for discrimination of the heart sound's spectra peak frequencies as those associated with the vibrational frequencies of the heart valves and the myocardial wall.

DEVELOPMENT OF THE LEFT VENTRICULAR ANEURYSM

In an earlier section of this chapter, we presented the method of detecting an ischemic-infarcted myocardial zone. The decision to salvage the ischemic segments by coronary bypass surgery would depend on (1) how seriously the infarcted zone (by virtue of its location and size) impairs the generation of suitable chamber pressure distributions during systole so as to effect adequate cardiac output and (2) whether the size of the infarct is big enough to cause an eventual aneurysm.

Background

Aneurysm of the left ventricle (LV) is recognized as a consequence of myocardial infarction, leading to loss of contractility of the involved tissue and eventually resulting in the infarcted myocardium developing a fluidlike consistency. Infarction of myocardium results in loss of its contractility, and progressive destruction with loss of stiffness. In order to compensate for its loss of stiffness and still equilibrate the LV chamber pressure, the infarcted segment of the myocardium deforms into a bulge known as aneurysm. Thus, the distortion of the ventricular wall after infarction may be deemed to occur so as to satisfy the requirements of the mechanical equilibrium between the wall tension in the (thin) infarcted myocardial segment and the chamber pressure, under the condition of incompressibility of the muscle wall. It is this phase of infarction that is considered in the (ventricular chamber causing) wall deformation analysis.

Although the process of aneurysm development is a long-term process of a continual modification of fluid pressure forces, chamber and aneurysm geometry, and alteration in the infarcted tissue properties, we will analyze the aneurysm in its deformed state under the action of an equivalent systolic level pressure (at the instant of ejection, when the chamber is still closed) and determine the relations between the deformation parameters and the infarct and left ventricular chamber geometries. Specifically, we will determine the influence of the size of the aneurysmic bulge and its stress level on the amount of myocardial wall damaged and the percentage thickness of the infarcted layer and its angle of damage for the aneurysmic stress to reach an assumed rupture level.

The LV is assumed to be of ellipsoidal shape with semimajor

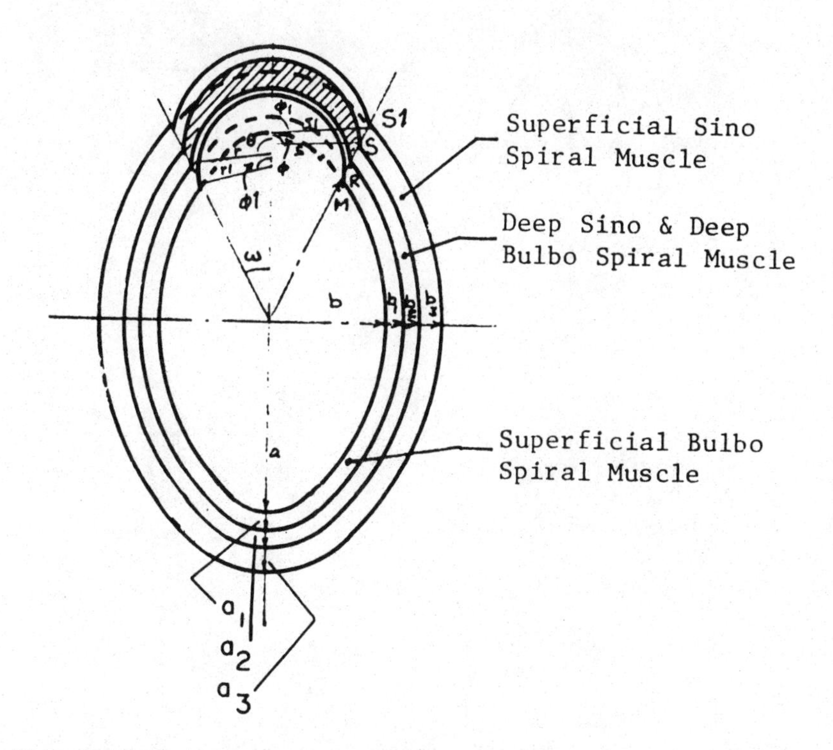

Figure 8-9. Ellipsoidal model of left ventricle (the infarcted region is shown hatched). 2ω is the angle of damage.

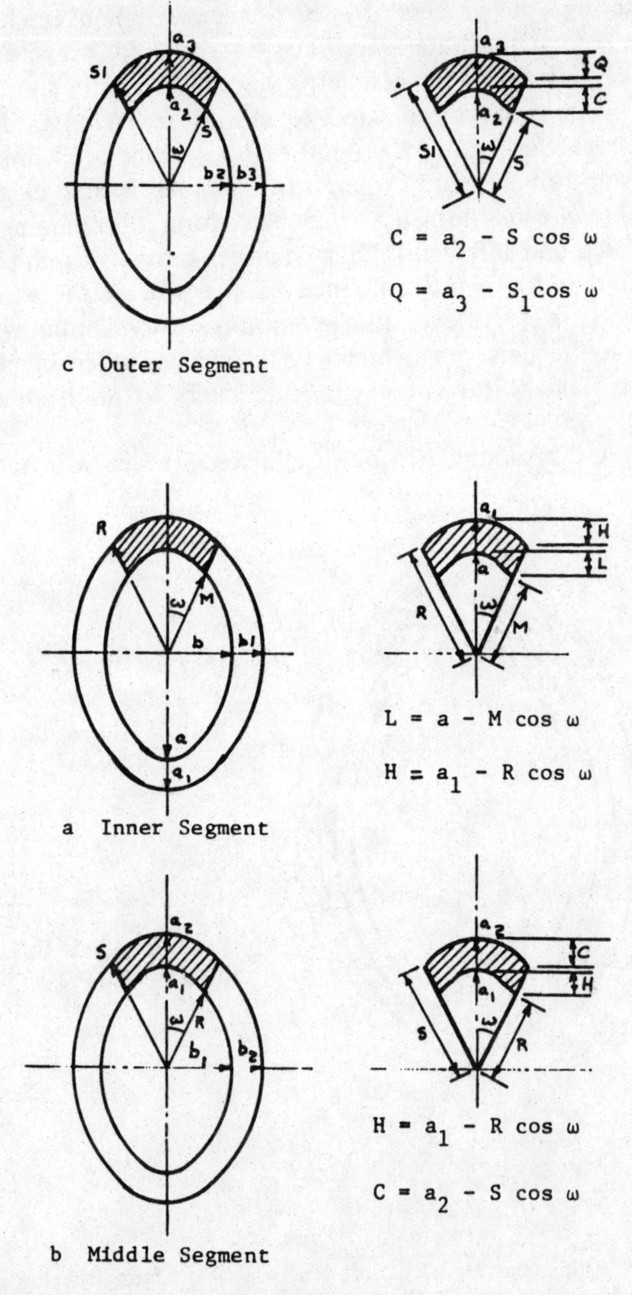

Figure 8-10. Geometry of the infarcted segment.

and minor axes of dimensions a, a_1, a_2, and a_3, and b, b_1, b_2, and b_3, respectively, as shown in Figure 8-9. This ellipsoidal LV model consists of three concentric layers. The geometry (size and location) of an assumed infarcted segment is defined in Figure 8-10 by the ellipsoidal coordinates S and R, (the inner and outer radii of the infarct whose middle layer has a fluid consistency), the angle of damage 2ω, and the ellipsoidal shell of semimajor axes a_1, a_2, and the semiminor axes b_1 and b_2.

The deformation of this infarcted segment into a spherical aneurysm, under the action of the intra-ventricular pressure (p, assumed to be 80 mm Hg), is investigated for varying values of the LVar shell and the infarcted segment's geometrical parameters. The geometrical parameters of the aneurysm, are r_1, r, s, and s_1 (the radii to the four spherical shaped bulged layers) and $2\theta_1$, 2θ, 2ϕ, and $2\phi_1$ (the angles subtended by the bulged infarcted wall segment at the center).

The superficial myocardial layers on the epicardial and endocardial sides of the damaged segment are assumed to be intact, and the volume of the central liquid phase of the infarcted wall segment within the aneurysm is taken equal to the volume of the originally undeformed destroyed muscle tissue. Likewise the volumes of the superficial myocardial layers within the developed aneurysm are (as a result of incompressibility) also assumed equal to the volumes of the corresponding superficial layers in the undeformed stage.

Analysis

The meridional stress, in the healthy left ventricular shell *prior to the development of the infarct*, at the section ψ (the angle made by the shell normal with the axis of the revolution), is given by

$$\sigma_\psi = \frac{Pr_\psi}{2h} \tag{35}$$

wherein P = left ventricular pressure, 'h' = the wall thickness, and r_ψ = the local radius of curvature (for the middle surface of the shell).

The average stress (T) in the portion of the myocardium, from $\psi = 0$ to ψ_o (along which the infarct extends), is given by

$$T = \frac{1}{\psi_o} \int_0^{\psi_o} \sigma_\psi \, d\psi = \frac{Pb^2}{2ah\psi_o} F_1(\psi_o, e) \qquad (36)$$

where

$$F_1(\psi_o, e) = \int_0^{\psi_o} \frac{d\psi}{\sqrt{1 - e^2 \sin^2 \psi}}; \; e^2 = 1 - \frac{b^2}{a^2} \qquad (37)$$

and

$$\psi_o = \tan^{-1}\left(\frac{a^2}{b^2} \tan \omega\right). \qquad (38)$$

We now assume that the middle layer of the myocardial shell wall segment ($-\psi_o < \psi < \psi_o$) gets infarcted and acquires a liquid state with a certain hydrostatic pressure. The thin superficial myocardial layers enclosing this liquid layer are now unable to sustain the intraventricular pressure p (originally sustained by the thick-walled myocardium in its preinfarcted state) and, consequently, blow out into an aneurysm of size permitting them to equilibrate the intraventricular pressure.

Since the muscle medium is assumed to be incompressible, the volume of the three myocardial shell layers (the outer superficial muscle layer, the inner superficial muscle layer, and the middle deep infarcted liquid-phase layer) before and after the development of aneurysm is the same. By equating the volumes of the superficial and deep muscle wall layers of the infarcted wall segment, prior to and following the development of the aneurysm, we obtain the following incompressibility equation:

$$\frac{a_1{}^3 \sin^2\theta_1}{[(a_1/b_1)^2 + \cot^2\omega]^{3/2}} (8 - 9\cos\theta + \cos 3\theta)$$

$$- \frac{a^3 \sin^3\theta\,(8 - 9\cos\theta_1 + \cos 3\theta_1)}{[(a/b)^2 + \cot^2\omega]^{3/2}}$$

$$-4[(b_1/a_1)^2 H^2(3a_1 - H) - (b/a)^2 L^2(3a - L)] \sin^3\theta_1 \sin^3\theta = 0 \quad (39)$$

and two other equations wherein θ, θ_1, b_1, and a_1 are respectively replaced by θ, ϕ, b_2, and a_2 and ϕ, ϕ_1, b_3, and a_3.

The inner muscle layer is acted upon by the LV pressure 'p' on its inner surface and by the liquid pressure 'p_1' (in the infarcted middle segmental layer) on its outer surface; the outer muscle layer is acted upon by the infarcted layer's liquid phase pressure p_1 on its inner surface and is stress free on its outer surface (Fig. 8-11). Let T be the uniform average stress in all layers in the undamaged ventricle, and T_θ and T_ϕ be the average uniform tensile stresses in the inner and outer superficial bulged spherical layers surrounding the middle infarcted layer. Combination of the three equilibrium equations yields

$$\frac{a_3\, P_2 \sin\theta}{a_2\, P_3 \sin\theta_1} \left[1 + \frac{1}{\lambda^2}\left(\frac{a_1 \sin\theta_1 P - \sin\theta\, P_1}{a \sin\theta\, P_1}\right)\left(\frac{T_\theta}{T} - 1\right)\right]$$

$$+ \frac{a}{a_1}\,\frac{P_1}{P}\,\frac{\sin\theta}{\sin\theta_1}\,\lambda\left[\left(\frac{a_3 \sin\phi\, P_2 - a_2 \sin\phi_1\, P_3}{a_3 \sin\phi\, P_2}\right)\left(\frac{T_\phi}{T} - 1\right)\right] - 1 = 0 \quad (40)$$

where

$$P = [(a/b)^2 + \cot^2\omega]^{1/2};\; P_1 = [(a_1/b_1)^2 + \cot^2\omega]^{1/2};$$

$$P_2 = [(a_2/b_2)^2 + \cot^2\omega]^{1/2};\; P_3 = [(a_3/b_3)^2 + \cot^2\omega]^{1/2};$$

$$\lambda^2 = 1 + \frac{2\psi_0}{F_1}\left(\frac{a}{b}\right)\left(\frac{h}{b}\right).$$

T is given by equation 19;

$$T_\theta = (r_1{}^2 p - r^2 p_1)/(r^2 - r_1{}^2);\ T_\phi = s^2 p_1/(s_1{}^2 - s^2) \quad (41)$$

and the stress-strain relations are

$$\frac{T_\theta}{T} = (l_1{}^F - l_1{}^0)/(l_1 - l_1{}^0);\ \frac{T_\phi}{T} = (l_2 F - l_2{}^0)/(l_2 - l_2{}^0) \quad (42)$$

where T, T_θ, T_ϕ, = tensions in the undamaged, inner and outer bulged layer of the infarcted portion; $l_1{}^F$, $l_2{}^F$ = lengths of the inner and outer layers of the infarcted spherical segment assumed spherical in nature; l_1, l_2, = lengths of the inner and outer elliptic middle segment layer before the onset of damage, l_{10}, l_{20}, = lengths of the inner and outer elliptic undamaged middle segment layers at zero stress. The expressions for $l_1{}^F$, $l_2{}^F$, l_{10}, l_{20} are obtained from expression of type

$$l = 2a[E(\pi/2 - a), e)]$$

$$e = 1 - (b^2/a^2)$$

$$E(a, e) = \int_0^a (1 - e^2 \sin^2 \phi)^{1/2}\, d\phi \quad (43)$$

The Bulge Factor is defined as the ratio of height of bulge above a common datum to the preinfarcted height. Thus, for the inner bulge,

$$H_\theta = r(1 - \cos\theta)/\left(\frac{a_1}{R} - \cos\omega\right). \quad (44)$$

For the outer bulge,

$$H_\phi = s(1 - \cos\phi)/\left(\frac{a_1}{S} - \cos\omega\right). \quad (45)$$

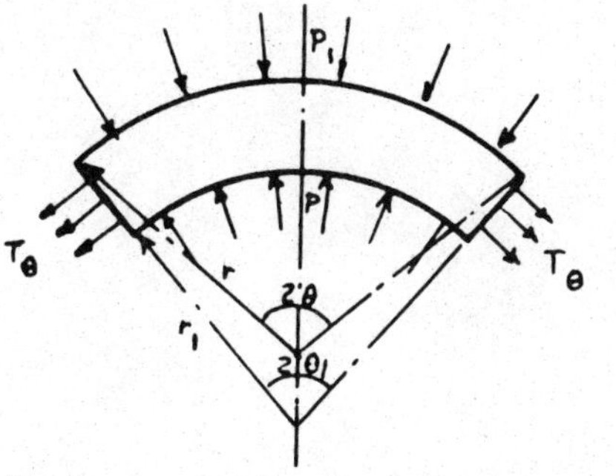

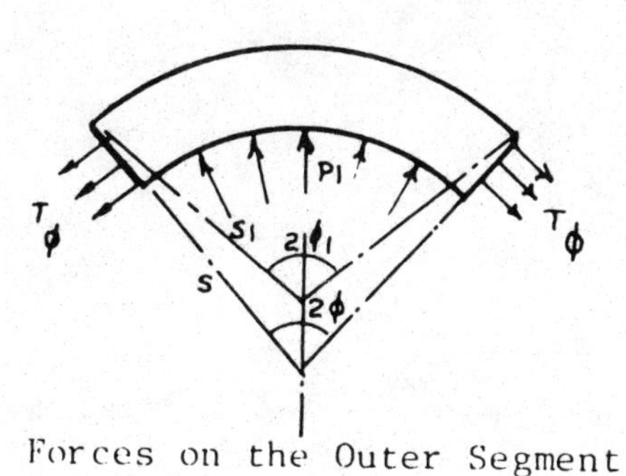

Figure 8-11. Loading of the inner and outer segments surrounding an infarct.

The equations 40, 41, and 43 have four unknowns—θ_1, θ, ϕ, and ϕ_1—that are the angles subtended by the different bulged layers with their centers. This system of nonlinear simultaneous equations is solved (in terms of the preinfarcted geometrical dimensions, the angle of damage and the unknown zero tension parameters) by successive iteration of Newton-Raphson method. From the computed values of the parameters θ and ϕ thus obtained, we determine (1) the tension factors T_θ/T and T_ϕ/T corresponding to the inner and outer bulged layers are calculated and (2) the values of bulge factors H_θ and H_ϕ corresponding to

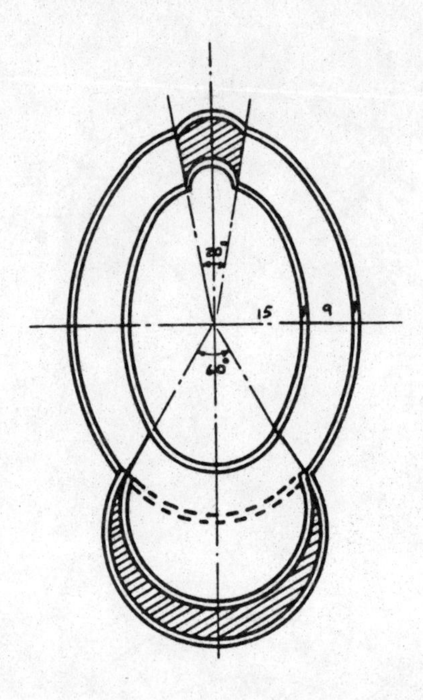

2ω angle of damage	height factor		tension factor	
	inner bulge	outer bulge	inner bulge	outer bulge
	H_θ	H_ϕ	T_θ/T	T_ϕ/T
20°	3.9320	3.2150	2.6912	1.6072
60°	4.4085	3.7714	9.6691	6.0142

Figure 8-12. Damage to DSS (3 mm layer) and DBS (6 mm layer) muscles. Spherical bulge over an ellipsoidal model of thickness constant b/a = 0.6.

the inner and outer layers of the infarcted segment.

Figure 8-12 shows infarction of both D.S.S. and D.B.S. muscles wherein about 82 per cent of the total wall thickness is destroyed. The innermost layer is most severely stressed. The tension factor reaches a very high value even at 60 degree angle of damage, and rupture of ventricular wall may be expected at about 35 degree angle of damage.

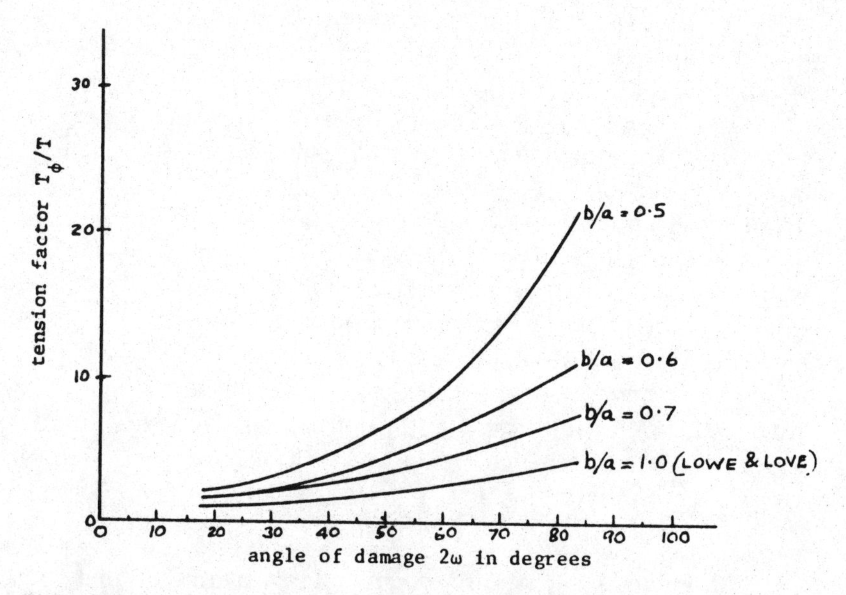

Figure 8-13. Tension factor T_ϕ/T versus Angle of damage, for the case of transmural infarct (82% damage), for various b/a ratios.

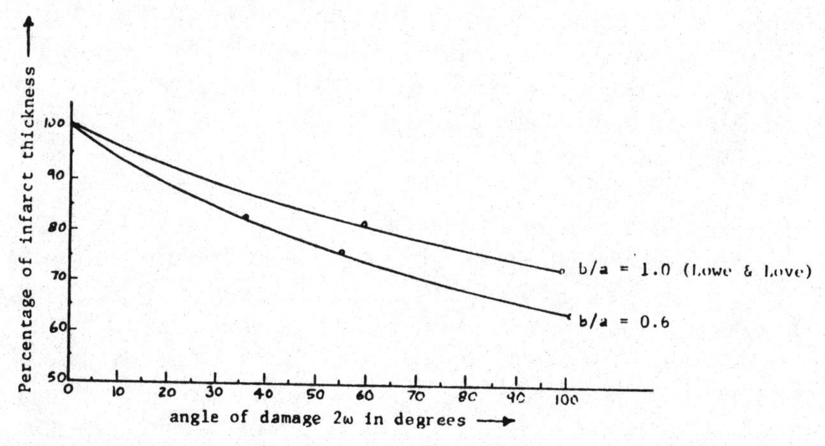

Figure 8-14. Variation of percentage of infarct thickness versus angle of damage for tension factor of five leading to ventricular rupture.

Figure 8-13 shows the variation of T_θ/T extent of damage (2W), for different thickness involvement for b/a = 0.6. It is sure that the tension factor (of 5) leading to ventricular rupture would occur only for infarcts of above 70 per cent thickness. Figure 8-14 shows the variation of percentage of infarct thickness vs. the angle of damage 2ω (for a tension factor value leading to ventricular rupture) for $b/a = 0.6$. One can predict therefrom the percentage of thickness infarct at which ventriuclar rupture would occur.

In summing up, there is no danger of rupture of the ventricular wall for infarcts that are less than about 70 per cent of total wall thickness, for $b/a = 0.6$. The analysis predicts rupture of ventricular walls for transmural infarcts (82 per cent thickness damage) for $b/a = 0.6$ at about 35 degree angle of damage. Finally, the greater the ratio of b/a is for the left ventricular shape, the less danger there is of ventricular rupture at the same angle of damage.

CONCLUSION

Radionuclide angiocardiography and two dimensional ultrasonography have primarily been utilized for qualitative assessment of myocardial function and/or structural abnormality. In short, they are descriptive images. The use of these dynamic images to quantify the constitutive properties of the myocardium has not been fully explored. Fundamental hemodynamic variables such as cardiac output, blood pressure, and aortic flow rate are but the holistic products of left ventricular instantaneous stress generation and wall motion.

On the other hand, there is knowledge of regional wall motion and velocity, left ventricular blood flow patterns and *in vivo* myocardial stress-strain characterization of ventricular function. This would help solve one of the more perplexing problems in clinical cardiology, which is the evaluation of patients with borderline left ventricular dysfunction. The analyses herein described would, it is hoped, result in clinical methods to help distinguish between global left ventricular dysfunction and reduced performance caused by regional wall motion disorders or structural abnormalities.

Current attempts at defining cardiac mechanics employ holistic indices such as pressure-volume, loops, dP/dt, force-velocity curves and passive stress-strain properties of the intact ventricles. However, the most common clinical entity affecting left ventricular performance is coronary artery disease, which is a segmental disorder. Differentiation of contiguous ischemic and nonischemic segments and the effect of ischemic segments on global performance cannot be adequately determined by the above global indices.

Therefore, there is a need to characterize the mechanical properties of the left ventricle in terms of regional contractility. In addition, there is a need to identify and quantify the size of ischemics segments and determine the impact of these segments on overall function.

Despite the many advances made in valvular surgery, one of the most frustrating problems facing the cardiologist and surgeon is the proper timing for surgical intervention. For example, the unloading effect of mitral regurgitation can mask an underlying defect in myocardial contractility. Only when the volume load is corrected by valve replacement does the myocardial component advertise itself in the form of depressed muscle function. Knowledge of the constitutive properties of valvular function (as provided in this chapter) and its effect on both the pumping and myocardial components of ventricular function would be a critical advance. This chapter attempts to develop the analytical foundations for these methods. Then, by using the signals from standard diagnostic noninvasive techniques, one could employ these methods to (1) quantify the regional properties of the left ventricle, determine the size of the infarcted segments during passive filling of the ventricle, determine to what extent the segment is interfering with optimal global performance, and follow these processes in time or in response to specific therapy and (2) determine the *in vivo* constitutive property of the valve leaflet and provide a more educated basis for medical or surgical intervention.

REFERENCES

Demiray, H. (1976) Stresses in ventricular wall. *J. Applied Mechanics. Transactions of ASME*, 194

Ghista, D.N. and Hamid, S. (1977) Finite element analysis of the human left ventricle. *Computer Programs in Medicine*, Vol. 7, No. 3

Ghista, D.N. and Rao, A.P. (1973) Mitral valve mechanics: Stress/strain characteristics. *Medical and Biological Engineering*, 11: 691-702

Ghista, D.N. and Sandler, H. (1975) Elastic modulus of the human intact left ventricle-determination of and physiological interpretations. *Med. and Biol. Eng.*, Vol. 13, No. 2

Ghista, D.N. and Vayo, W. (1969) The time-ranging elastic properties of the left ventricle. *Bulletin of Math. Biophysics*, Vol. 31

Ghista, D.N., Ray, G., and Sandler, H. (1981) Mechanocardiography: theory, evaluation and significance of the regional distributions of myocardial constitutive properties and blood pressure in the left ventricular chamber. In *Cardiovascular Engineering* (ed) D.N. Ghista, E. Van Vollenhoven, W.J. Yang, and E. Reul, Baden-Baden, Germany, Gerhard Witzstrock

Janz, R.F., Kubert, B.R., Mirsky, I., Korecky, B., and Taichman, G.C. (1976) Effect of age on passive elastic stiffness of rat heart muscle. *Biophy. J.*, 16:281

McKusick, V.A. (1958) *Cardiovascular Sound in Health and Disease*. Baltimore, Williams and Wilkins

Mazumdar, J., Hearn, T. and Ghista, D.N. (1979) Determination of *in vivo* constitutive properties and normal-pathogenic states of mitral valve leaflets and left ventrical myocardium from phonocardiographic data. *Applied Physiological Mechanics* (ed.) D.N. Ghista, Harwood Academic Press, 1980

Mazumdar, J. and Hearn, T.C. (1978) Mathematical analysis of mitral valve lealflets. *Biomechanics*, 11, No. 6/7, 291

Mirsky, I. and Parmley, W.W. (1974) Evaluation of passive elastic stiffness for the left ventricle and isolated heart muscle. *Cardiac Mechanics*. (ed) I. Mirsky, D.N. Ghista and H. Sandler, New York, Wiley-Interscience

Mirsky, I. (1973) Effects of anisotropy and nonhomogeneity on left ventricular stresses in the intact heart. *Bull Math Biophys.*, 32:197-213

Polya, G. and Szego, G. (1951) *Isoperimetric Inequalities in Mathematical Physics*. Princeton, UP Princeton

Rushmer, R.F. (1961) *Cardiovascular Dynamics*. Philadelphia, W.B. Sanders

Schmidt, G. and Miller, H. (1975). The Stanford-Ames portable echocardioscope: A case study in technology transfer. In *Cardiovascular Imaging and Image Processing*. (ed.) D. Harrison et al.. Society of Photo-Optical Instrument Engineers

Scully, M.O. (1979) Towards a new ultrasonic technique for cardiac pressure measurements. In *Non-Invasive Cardiovascular Measurements*. (ed.) D.C. Harrison, H. Sandler, and H. Miller, Society of Photo-Optical Instrumentation Engineers

Torvik, P.J. and Eastep, F.E. (1972) A method for improving the estimation of membrane frequencies. *J. Sound and Vibration*, 21(3), pp. 285-294

Van Vollenhoven, E., Suzumura, N., Ghista, D.N., Mazumdan, J., and Hearn, T. (1979) Phonocardiography, analyses of instrumentation, and vibration of heart structures to determine their constitutive properties. In *Cardiogram: Theory and Application*. (ed.) D.N. Ghista, E. Van Vollenhoven, W.J. Yang and H. Reul, S. Karger (Basel)

Wang, C.Y. and Sonnenblick, E.H. (1979) Dynamic pressure distribution inside a spherical ventricle. *J. Biomechanics*, 12, 9

Wiskind, H.K. and Talbot, S.A. (1958) Physical basis of cardiovascular sound–An analytical survey. *AFOSR Tech. Report No. TR58-160*, ASTIA Document AD207 459

Chapter 9

CARDIAC MATERIAL MECHANICS

The Noninvasive Analysis of Myocardial Material Properties

C.A. PHILLIPS and J.S. PETROFSKY

THE subject of mechanics of materials is one expression of the more general subject of mechanics and includes the study of the effects of the elastic properties of structural materials. The central tenet of material mechanics is that all materials when subjected to a force (stress) will react by deforming (strain). The quantitative relationship between these two factors represents a major focus of the mechanics of materials.

CARDIAC MATERIAL MECHANICS

The traditional, i.e. invasive, approach to cardiac material mechanics has been to apply the basic mathematical formulations of the mechanics of materials, e.g. Young's modulus, to the myocardium. This approach has required knowledge of stress as well as strain and has generally been performed with isolated cardiac muscle or the *in vivo* left ventricle. An excellent review of the state of the art in this area, which considers mathematical and physiological as well as clinical concepts, has been reported by Mirsky (1976).

This chapter proposes a nontraditional, i.e. noninvasive, approach to cardiac material mechanics. The approach is based upon two premises. First, deformation, strain, time-rate of strain (velocity) can be measured noninvasively. Second, force, stress, load, etc., which are generally measured invasively, can be neglected if (and this is a critical assumption) the following condition is satisfied: a given clinical population responds to a characteristic set of forces (stress) with a characteristic set of fluxes (strain). For example, a normal set of forces may result in a normal amount of deformation, and a larger (increased) set of forces may result in an increased amount of deformation. In this case, the material property of the myocardial wall remains constant, but reacts to altered forces with altered deformation. The mathematical development and clinical application of this example is presented in the following section where a normal clinical population and a ventricular pressure overload population are compared and contrasted.

A second example is when a normal set of forces (stress) results in an abnormal amount of deformation (strain or time-rate of strain). This situation would indicate that a fundamental change has occurred in the material properties of the system. The mathematical development and clinical application of this example is presented in the final section where a clinically normal population is compared and contrasted with a coronary artery disease population.

With respect to noninvasive cardiac material mechanics, there are two important patient factors that must be satisfied. First, the pathological patient population to be analyzed must have one and only one cardiac disease condition (either extrinsic or intrinsic). Complicating, superimposed disease conditions would potentially alter the analysis. A hypertensive patient with superimposed coronary artery disease could present a very confused picture indeed. Second, the nature of the forces must satisfy a characteristic set. For example, one would expect an elevated intraventricular pressure in patients with hypertension or moderate aortic stenosis; one would also anticipate a normal systolic intraventricular pressure with normals and patients with coronary artery disease.

There are two important bioengineering factors that must also be satisfied. First, the established mechanical equations (incorporating force and flux) must adequately describe the system under study. When dealing with the left ventricle, this requires approximations as to the constitutive properties of the myocardium, geometry of the left ventricular chamber and consideration of the quasistatic nature of the system. Second, the simplified mechanical equations (incorporating only flux terms and without force terms) must also adequately describe the system. This bioengineering factor is approached in an empirical manner by directly applying such mechanical equations to patient populations in the following two sections. However, a word of caution is appropriate at this point. The studies reported in the next two sections are, by nature, preliminary. They represent pilot projects to evaluate the feasibility of a noninvasive approach to cardiac material mechanics. Expanded studies on larger clinical populations are currently being conducted in our laboratory and will be reported in the near future (Phillips and Petrofsky, 1981; 1982). Such studies are essential if the noninvasive analysis of myocardial material properties is to be fully realized.

MYOCARDIAL WALL DEFORMATION

The importance of ventricular geometry has been recognized since the 1890s by Wood (1892) and, more recently, Burton (1957) and Badeer (1964) when they applied the law of Laplace to the intact ventricle. These investigators noted that the size and shape of the left ventricle significantly affect the relation between stress developed by the heart muscle and intraventricular pressures. Falsetti et al. (1970) pointed out that the calculated wall stresses do not directly reflect the forces developed by the muscle fibers because of the cross sectional area changes during the cardiac cycle while the number of muscle fibers remain constant. They defined a "fiber-corrected stress" for the equatorial direction:

$$(\sigma\theta\theta)_F = (\sigma\theta\theta)\frac{\text{instantaneous wall cross-sectional area}}{\text{wall cross-sectional area at end-diastole}} \qquad (1)$$

which is proportional to the average force for a given number of fibers. In this section, we shall define the latter term

$$\frac{\text{instantaneous wall cross-sectional area}}{\text{wall cross-sectional area at end-diastole}}$$

as the left ventricular area ratio (AR). Utilizing a thick-walled ellipsoidal model previously developed by Sandler and Dodge (1963), we define two left ventricular area ratios: the longitudinal area ratio, AR_ϕ, and the circumferential area ratio, AR_θ. The subscript ϕ or θ denotes the face of element upon which the stress acts. For example, AR_ϕ is the cross-sectional area ratio at the equator of the left ventricle.

Area ratios are important for two reasons. First, they are correction factors by which the instantaneous force per unit area (an Eulerian stress) can be converted to a stress that is proportional to force for a given number of fibers (a Lagrangian stress). Falsetti et al. (1970) illustrated that the fiber-corrected stress is significantly higher than the calculated Eulerian stress. Second, and most important, area ratios are an index of the relative change in shape (deformation) of muscle in the ventricular wall, a material property. This local deformation is due to both local fiber contraction and regional myocardial forces. The extent of local tissue deformation will have some complex relationship to overall ventricular contraction and, thus, to ventricular function.

Previously, Phillips et al. (1978) examined the circumferential and longitudinal area ratios in thirty-six patients using a cine-angiographic technique. The patients comprised four clinical groups composed of normal, compensated volume overload, decompensated volume overload, and congestive cardiomyopathy. In a discussion of this study by Walburn and Phillips (1978), it was indicated that the longitudinal area ratio (AR_ϕ) could have been calculated from noninvasive measurements such as routinely obtained by echocardiography.

The purpose of this section is twofold. First, AR_ϕ will be determined for nine normal individuals using the noninvasive procedure of echocardiography. Secondly, an additional clinical population, consisting of five compensated pressure overload patients, will be evaluated using the same echocardiographic methods.

Mathematical Approach

A mathematical expression for the longitudinal area ratio

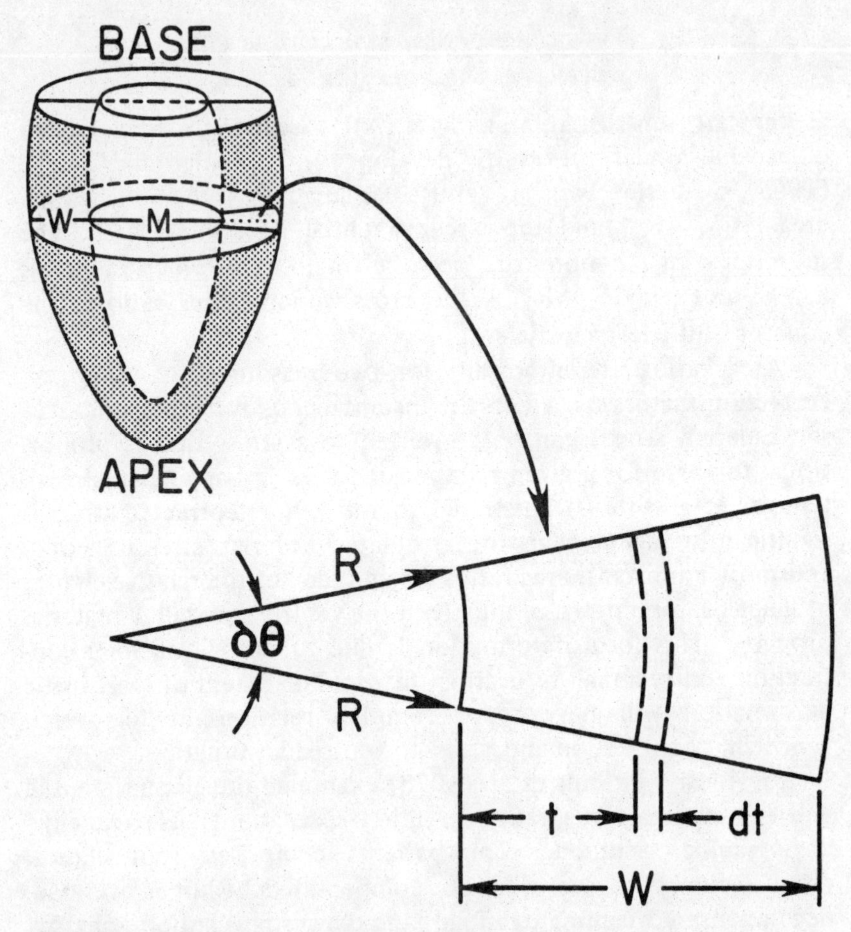

Figure 9-1. Unit element of tissue removed from the myocardial wall.

AR_ϕ is derived by considering a unit element of tissue removed from the myocardial wall (Fig. 9-1). This is a full-thickness element such that the inner surface is endocardium and the outer surface is epicardium. The wall thickness at the equator is denoted by W, while the radius of curvature of the endocardial surface in the circumferential direction is denoted by R. The angle inscribed by the element in the circumferential direction is $\delta\theta$, and the area of the longitudinal face is denoted by A_ϕ. A thick slice of tissue located at a depth, t, from the endocardial surface, has a thickness, dt. The area of the element is the width times the thickness, given by

$$A_\phi = (R + t)\,\delta\theta\,dt \tag{2}$$

The term $(R + t)\,\delta\theta$ is the length of the edge along the radial face of the element. The complete area of the tissue sample is obtained by integrating equation 2 between the endocardial surface, $t = 0$, and the epicardial surface, $t = W$:

$$A_\phi = \int_0^W (R + t)\,\delta\theta\,dt. \tag{3}$$

From any Table of Integrals,

$$A_\phi = \delta\theta\,\left[\frac{(R + t)^2}{2}\right] \tag{4}$$

and solving across the limits

$$A_\phi = \delta\theta\,\left[\frac{(R + W)^2}{2} - \frac{R^2}{2}\right] \tag{5}$$

which after expanding, subtracting, and factoring yields

$$A_\phi = (2R + W)(W/2)\,\delta\theta. \tag{6}$$

This equation gives an area for a small angular portion of a circular annulus. Designating the end-diastolic condition by subscript D, an equation for the ratio of the area at any time during the cardiac cycle to the area at end-diastole is obtained.

$$AR_\phi = \frac{(2R + W)W}{(2R_D + W_D)W_D}. \tag{7}$$

This equation requires the approximation that the contraction is symmetrical in the circumferential direction so that $\delta\theta = \delta\theta_D$.

The derivation can be completed by assuming that the ventricular cavity is a sphere with diameter, M. At the equator, the radius of curvature is then

$$R = M/2. \tag{8}$$

By substituting equation 8 into 7, the longitudinal cross-sectional area ratio for an element of tissue located at the equator of the ventricle is

$$AR_{\phi} = \frac{W(M + W)}{W_D (M_D + W_D)}. \tag{9}$$

Clinical Approach

A total of nine patients were studied and divided into two groups:

Normal–Composed of five patients who were asymptomatic and were studied as part of a routine cardiac screening process. These five patients had no hemodynamic or valvular abnormality of the left ventricle detected by echocardiography.

Compensated Pressure Overload (CPO)–Composed of four patients who had either systemic hypertension (2)–treated–or aortic stenosis (2)–but were clinically compensated–i.e. had no evidence of clinical congestive heart failure.

The echocardiographic transducer was positioned at the fourth intercostal space, left sternal border, and directed posteriorly, slightly inferiorly, and laterally. M-mode scanning began at the apex and proceeded toward the base. The left ventricular cavity was identified when the left side of the intraventricular septum (LS) and the posterior left ventricular wall endocardium (EN) were clearly observed. Recording was at a rate of 10 cm/sec.

A difference of 2 to 3 mm in the left ventricular internal diameter (LVID) between inspiration and expiration has been noted to exist. Therefore, the examination was made during held respiration, such that greater accuracy was obtained (Feigenbaum, 1976).

Upon completion of the recordings, the posterior free wall (FW), posterior endocardium (EN), and left ventricular septum (LS) were outlined over a complete cardiac cycle (R-R interval) (Fig. 9-2).

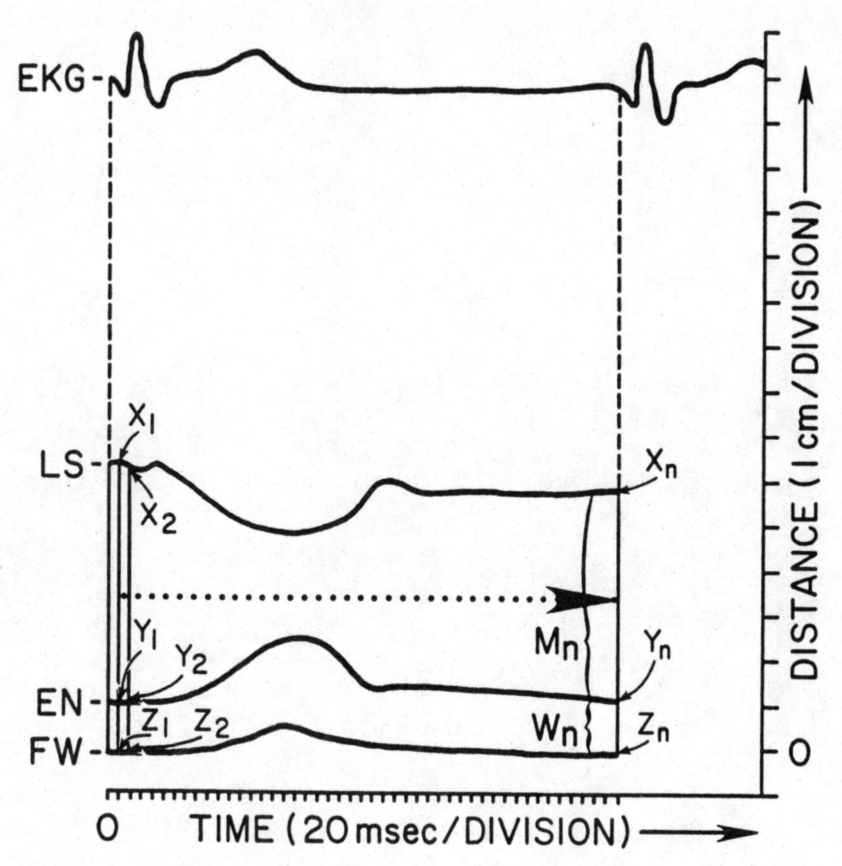

Figure 9-2. Tracing of posterior free wall (FW), posterior endocardium (EN) and left ventricular septum (LS) throughout a cardiac cycle and computed set of lengths (*see* text).

Timing marks occur vertically throughout the tracing at 20 msec intervals (approximately every 2 mm). The oscillographic recordings were then placed on a digitizing tablet connected to a PDP 11/5 computer.

Beginning with the R-waves, an electrostatic pen was used to define the LS point ($x_1, x_2, \ldots, x_n$), the EN point ($y_1, y_2, \ldots, y_n$) and the FW point ($z_1, z_2, \ldots, z_n$) for each 20 msec vertical interval. The procedure was terminated at the next R-wave. The data was automatically entered into the PDP minicomputer where the remaining calculations were performed.

A set of lengths are computed as $M_1 = x_1 - y_1$, $M_2 = x_2 - y_2$, $\ldots$, $M_n = x_n - y_n$ and also as $W_1 = y_1 - z_1$, $W_2 = y_2 - z_2, \ldots,$ $W_n = y_n - z_n$ (Fig. 9-2).

The longitudinal cross-sectional area ratio has been derived in the previous section:

$$AR_\phi = \frac{W(M + W)}{W_D (M_D + W_D)} \tag{9}$$

where subscript D designates the end-diastolic condition (occurring at the R-wave).

The left ventricular midwall circumferential fiber strain, previously derived by Phillips (1977), was computed for each 20 msec interval:

$$\epsilon_{CF} = 1 - \frac{(M + W)}{(M_D + W_D)} \tag{10}$$

Finally, AR_ϕ (ES), the value of AR_ϕ at end-systole, was defined to occur at $|\epsilon_{CF}|_{MX}$, the point at which ECF was a maximum.

Three additional parameters were also calculated:

(a) End Diastolic Volume (EDV) in milliliters,

$$EDV = (\pi/6)(M_D)^3, \tag{11}$$

(b) Ejection Fraction (EF),

$$EF = 1 - (M_S/M_D)^3 \tag{12}$$

(c) Left Ventricular Mass (LVM) in cubic centimeters,

$$LVM = (\pi/6)[(M_D + 2W_D)^3 - M_D{}^3] \tag{13}$$

where the subscript, S, denotes the measurement taken at $|\epsilon_{CF}|_{MX}$, i.e. the end-systolic measurement.

In order to test for the level of significance between the two populations, the Welch test was employed since the two population variances can differ (Mack, 1967). The one-tailed test-function W was evaluated to determine if the AR_{ϕ} (ES) for the CPO group was significantly higher than for the normal group.

Results and Conclusions

The descriptive clinical data for the nine patients, subdivided into normal and CPO groups, are presented in Table 9-I. Ages ranged from eighteen to sixty-two and there was a predominance of males (7 to 2). All patients were normotensive and in sinus rhythm.

Table 9-I

STATISTICAL AND HEMODYNAMIC DATA FOR NINE PATIENTS

PATIENT	AGE	SEX	BSA	BP	HR
NORMAL					
01	62	M	2.07	124/90	61
02	61	M	1.73	152/86	69
03	57	M	2.12	150/96	64
04	59	F	1.88	112/84	81
05	49	M	1.86	140/78	62
COMPENSATED PRESSURE OVERLOAD					
06	18	F	1.77	122/80	76
07	60	M	2.15	130/100	62
08	53	M	2.19	134/86	55
09	32	M	2.02	142/90	78

B.S.A.: Body Surface Area (M^2)

B.P. : Blood Pressure (Systolic/Diastolic), mmHg

H.R. : Heart Rate (Beats Per Minute)

Table 9-II

COMPUTED (EXPERIMENTAL) PARAMETERS
FOR NINE PATIENTS

PATIENT	EDV	EF	LVM	AR_{ϕ}(ES)	$\lvert ECF \rvert_{MX}$
NORMAL					
01	125	.560	157	1.098	.187
02	157	.730	160	1.011	.199
03	125	.736	229	1.549	.181
04	111	.542	144	1.694	.165
05	111	.800	214	1.147	.271
MEAN	126	.674	181	1.300	.201
S.D.	±19	±.115	±38	±.302	±.041
COMPENSATED PRESSURE OVERLOAD					
06	91	.673	243	1.927	.227
07	141	.695	183	2.599	.108
08	141	.668	313	1.502	.174
09	104	.587	236	2.342	.138
MEAN	119	.656	244	2.092	.162
S.D.	±26	±.047	±53	±.481	±.051
SIGNIFICANCE	N.S.	N.S.	p<.05	p<.025	N.S.

The computed parameters for the nine patients (two groupings) is shown in Table 9-II. The CPO group had a significantly elevated AR_{ϕ}(ES) and left ventricular mass ($p < .025$ and $p < .05$ respectively) when compared to normals. There were not any significant differences between the two groups with respect to end-diastolic volume, ejection fraction or $|\epsilon_{CF}|_{MX}$.

Echocardiographically determined estimates of AR_{ϕ} for normal individuals appear to correlate well with previously obtained cineangiographic estimates of AR_{ϕ}. In the original report, Phillips et al. (1978) obtained a mean AR_{ϕ} of 1.27 ± 0.14 (range

1.10 to 1.58) for fifteen normals. In this study, the echocardiographically obtained AR_ϕ had a mean of 1.30 ± 0.30 for five normals. The larger standard deviation in this study may be due to data processing techniques or simply subject variability. Our previous study (Phillips et al. 1978) utilized curve fitting of all the data by the Fourier technique (Figure 2 of Phillips et al. 1978) and calculating peak AR_ϕ from the curve fitted values. In the present study, however, one value of AR_ϕ (at end-systole) has been reported for each subject from an empirical ("eyeball") fit of the data. Mean AR_ϕ by echocardiography in this study is still very similar to mean AR_ϕ by cineangiography (Phillips et al. 1978). Echocardiographically developed estimates of AR_ϕ of compensated ventricular dysfunction appear to correlate well with previously obtained estimates of AR_ϕ. In their original report, Phillips et al. (1978) demonstrated a significant increase in AR_ϕ (1.68 ±0.14) associated with six compensated volume overload patients. This present study indicates that compensated pressure overload is also associated with a significant increase in AR_ϕ (2.09 ± 0.48). The difference between echocardiographically derived AR_ϕ and cineangiographic AR_ϕ in compensated ventricular dysfunction can be reconciled on the basis of differences in curve fitting techniques (*vide supra*), and subject variability. Moreover, a different process of ventricular compensation is involved with respect to the two studies. The volume overload population compensates by way of eccentric hypertrophy, while the pressure overload population compensates by way of concentric hypertrophy (Dodge and Baxley, 1969).

As stated earlier, an area ratio is an index of the amount of deformation of the left ventricle. As part of the compensatory process both in left ventricular volume overload and (from this preliminary study) in pressure overload, an increased amount of myocardial wall deformation apparently occurs. Phillips et al. (1978) demonstrated that when an increased amount of wall deformation does not occur (for example, in decompensated volume overload where AR_ϕ remains normal) or is actually reduced below normal (for example, in congestive cardiomyopathy), then clinical evidence of cardiac decompensation (audible S_3, basilar rales, and dyspnea on exertion) occurs.

VENTRICULAR VISCOELASTIC COMPLIANCE

Various parameters have been studied to evaluate different clinical patient populations including ventricular volume and velocity changes as well as ventricular flow (ejection rate). Specific parameters have been the percentage change in internal circumferential radius from end-diastole to end-systole (%ΔM) (Gould et al., 1974) and the circumferential fiber shortening velocity computed as the sum of rate change of the internal radius and wall thickness (V'_{CF}) (Gault et al., 1968) and normalized for instantaneous circumferential fiber length (V''_{CF}) (Karliner et al., 1971) or normalized for by end-diastolic fiber length (V_{CF}) (Hammermeister et al., 1974).

There is some difficulty, however, in distinguishing any individual patient with cardiac dysfunction due to the significant degree of overlap between patient groups. In order to improve this situation, Phillips (1977) defined the velocity-strain relationship, a test of cardiac function, in which circumferential fiber shortening velocity of the left ventricle was related to circumferential fiber strain during a complete cardiac cycle. That investigator then proceeded to define inequalities between systolic and diastolic time constants and systolic and diastolic strain, which apparently characterized the decompensated left ventricle on an individual basis. Variations in the inequalities were related to changes in ventricular viscoelastic compliance (Phillips, 1977).

The section will focus upon two primary objectives. First, the parameters necessary to compute the velocity-strain relationship will be obtained noninvasively using echocardiography. The velocity-strain relationship will then be computed from noninvasive measurements and compared to previous results obtained invasively by cineangiography. Second, an additional patient population (not previously studied) will be examined noninvasively using echocardiography. Coronary artery disease patients specifically will be selected since ventricular compliance has been shown to be altered in this disease state (Mann et al., 1976).

Mathematical Approach

Derivation of the systolic and diastolic time constants commences with considering a first order, linear system for which the following characteristic equation may be written:

$$F = k \cdot x + c \cdot (dx/dt) \tag{14}$$

Consequently, there are two force components: one required in overcoming viscous forces (proportional to the rate of change of x with time) and one required to overcome elastic forces (proportional to x).

When the left ventricle is considered, F can be related to the intraventricular pressure (P) and x can be a length, i.e. the length of a midwall circumferential fiber at the equator (l_{CF}). This length can be normalized by an initial (end-diastolic) length, $(l_{CF})_D$, in order to directly compare ventricles of various sizes.

$$l_{CF} = \frac{l_{CF}}{(l_{CF})_D} \tag{15}$$

Incorporating the above condition, the coefficient of viscous compliance is

$$c = \frac{dP}{dl_{CF}/dt} = \frac{dP}{V_{CF}}. \tag{16}$$

The coefficient for elastic compliance is

$$k = \frac{dP}{\Delta l_{CF}} = \frac{dP}{\epsilon}. \tag{17}$$

When the ratio of c with respect to k is taken, a pressure independent term is derived, called the system time constant (which has the dimensions of time):

$$\tau = c/k = \frac{\epsilon}{V_{CF}}. \tag{18}$$

Doeblin (1962) asserts that all first order systems have their dynamic characteristics entirely described by this system time constant. The shorter the system time constant, the faster the transient response to a step-input function dies out and the lower is viscoelastic compliance, i.e. the stiffer the system. In physical

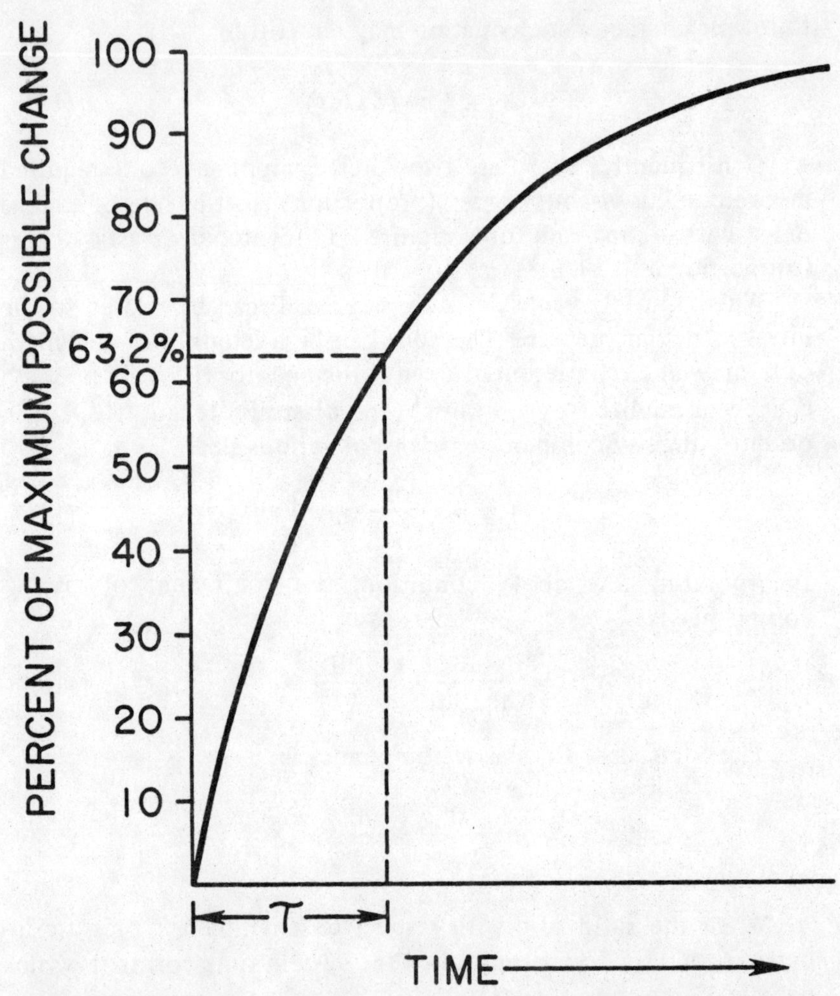

Figure 9-3. Definition of system time constant (τ) as time required for the system to reach 63.2 per cent of the total change which occurs in response to a step input function.

terms, the system time constant corresponds to the duration of time required for a system to reach 63.2 per cent of the total change that occurs in response to a step-input function (Fig. 9-3).

Even though we are dealing with the left ventricle over a complete cardiac cycle, one convenient definition of a systolic time constant (τ_S) could be obtained by substituting eqs. 21 and 27 into eq. 18:

$$\tau_S = \frac{\epsilon_S}{V_{CF}} \tag{19}$$

Similarly, the diastolic time constant (τ_D) could be derived by substituting eqs. 21 and 29 into eq. 18:

$$\tau_D = \frac{\epsilon_D}{|-V_{CF}|} \tag{20}$$

Clinical Approach

M-mode echocardiographic data was acquired on eighteen subjects, fourteen male and four female, ages eighteen to sixty-seven. The patients (Table 9-III) were placed in two groups:

Normal–Composed of eight patients who were asymptomatic and were studied as part of a routine cardiac screening process. The patients had no demonstrable abnormality of their left ventricle by echocardiography.

Coronary Artery Disease (CAD)–Composed of six patients who satisfied one of the following criteria: (1) A history of myocardial infarction documented by characteristic ECG changes and/or serum isoenzyme changes or (2) A history associated with angina pectoris *and* a positive treadmill stress test.

Echocardiographic data of the left ventricle has been acquired and reduced to a series of lengths ($M_1, M_2, \ldots, M_n$ and $W_1, W_2, \ldots, W_n$) by the same method reported in the previous section under *Clinical Approach* (Fig. 9-2).

The various parameters of the velocity-strain relationship have been defined in a previous study (Phillips, 1977). The circumferential fiber velocity (V_{CF}) is the rate of change of a hypothetical fiber located at the midwall of the circumferential axis of the left ventricle (l_{CF}) and normalized by its end-diastolic value $(l_{CF})_D$ reported in circumferences per second:

$$V_{CF} = \frac{d(l_{CF})/dt}{(l_{CF})_D} = \frac{d[\pi(M+W)]/dt}{\pi(M_D + W_D)}$$

$$= (\frac{dM}{dt} + \frac{dW}{dt})/(M_D + W_D) \quad (21)$$

where M_D = minor axis at end-diastole and W_D = wall thickness at end-diastole.

Conventionally, the fiber-shortening rate is a positive velocity, and fiber-lengthening rate is a negative velocity. When a peak systolic value is taken, this term is roughly similar, although not identical, to the peak S(dv/dt)/EDV term of Hammermeister et al. (1974).

Circumferential fiber strain (ϵ') is the amount of change in the minor ventricular circumference, $\pi(\Delta M + \Delta W)$, normalized by end-diastolic circumference, $\pi(M_D + W_D)$ and is dimensionless:

$$\epsilon' = \frac{\Delta_M + \Delta_W}{M_D + W_D} = \frac{(M_D - M) + (W_D - W)}{M_D + W_D} = 1 - (\frac{M+W}{M_D + W_D}). \quad (22)$$

ϵ' is a Lagrangian strain, but Mirsky and Parmley (1974) have proposed that a natural strain (ϵ) is more characteristic of biological material properties. Consequently, in this section, all strains are defined as natural strain:

$$\epsilon = \ln(1 + \epsilon'). \quad (23)$$

When minor LV circumference is a minimum, $\pi(M_S + W_S)$, the maximum Lagrangian strain is

$$\epsilon'_{max} = \frac{(M_D - M_S) + (W_D - W_S)}{M_D + W_D} = 1 - (\frac{M_S + W_S}{M_D + W_D}). \quad (24)$$

The maximum natural strain is then

$$\epsilon_{max} = \ln(1 + \epsilon'_{max}) \tag{25}$$

Using eq. 22, strain is positive for an entire cardiac cycle (compressive strain). The quantity ϵ_{max} roughly approximates, though is not identical to, the %ΔM of Gould et al. (1974) and correlates with ejection fraction as noted by Lewis and Sandler (1971).

Specific strains examined are (a) the magnitude of systolic strain (ϵ_s') occuring ât peak V_{CF}:

$$\epsilon_S' = 1 - \frac{M(\text{Peak } V_{CF}) + W(\text{Peak } V_{CF})}{M_D + W_D} \tag{26}$$

which has a corresponding natural strain expression:

$$\epsilon_S = \ln(1 + \epsilon_S') \tag{27}$$

and (b) the magnitude of diastolic strain (ϵ_D') occurring at peak $-V_{CF}$:

$$\epsilon_D' = 1 - \frac{M(\text{Peak } -V_{CF}) + W(\text{Peak } -V_{CF})}{M_D + W_D} \tag{28}$$

which has a corresponding natural strain term:

$$\epsilon_D = \ln(1 + \epsilon_D') \tag{29}$$

Referring to the previous section entitled *Mathematical Approach*, we can define a systolic time constant (τ_S) by substituting eqs. 21 and 27 into eq. 18:

$$\tau_S = \frac{\epsilon_S}{V_{CF}} \tag{19}$$

Similarly, we can define a diastolic time constant (τ_D) by substituting eqs. 21 and 29 into eq. 18:

$$\tau_D = \frac{\epsilon_D}{|-V_{CF}|} \qquad (20)$$

Experimental Results

Table 9-III describes the clinical data for the two subject groups (a total of fourteen patients). The standard data includes age, sex, body surface area, and hemodynamic data is blood pressure, heart rate, end-diastolic volume, and ejection fraction.

Table 9-III

STATISTICAL AND HEMODYNAMIC DATA FOR THE NORMAL AND CAD POPULATIONS

Patient No.	Age	Sex	BSA[a]	BP[b]	HR[c]	EDV[d]	EF[e]
A	57	M	2.12	150/96	64	125	.736
B	49	M	1.86	140/78	62	111	.800
C	32	M	1.80	108/80	60	110	.675
D	19	F	1.60	116/86	68	104	.684
E	28	M	1.87	114/90	82	85	.683
F	18	M	1.78	114/72	61	60	.668
G	22	M	1.95	124/76	68	91	.730
H	29	M	1.73	124/76	52	80	.693

CAD

Patient No.	Age	Sex	BSA[a]	BP[b]	HR[c]	EDV[d]	EF[e]
I	53	F	1.75	126/82	88	111	.700
J	50	M	1.85	110/86	64	195	.590
K	50	M	1.85	114/90	80	195	.593
L	67	M	2.28	132/78	70	195	.672
M	67	M	2.17	124/76	78	141	.638
N	58	F	1.85	126/78	59	195	.760

[a]B.S.A.: Body Surface Area (M^2)
[b]B.P.: Blood Pressure (mm Hg)
[c]H.R.: Heart Rate (beats/minute)
[d]E.D.V.: End-Diastolic Volume (mls.)
[e]E.F.: Ejection Fraction

Table 9-IV summarizes the calculated values for the eight normal subjects. Individual values are shown for V_{CF}, ϵ_S, τ_S, ϵ_{max}, the absolute value of $-V_{CF}$, i.e. $|-V_{CF}|$, and ϵ_D, τ_D.

Table 9-IV

CALCULATED PARAMETERS
FOR THE EIGHT NORMAL PATIENTS

Patient No.	$+V_{CF}$ [a]	ϵ_s [b]	τ_s [c]	ϵ_{max} [d]	$\|-V_{CF}\|$ [e]	ϵ_D [f]	τ_D [g]
A	1.096	.051	.047	.181	.658	.124	.188
B	1.521	.125	.082	.271	2.091	.221	.106
C	1.365	.100	.073	.191	1.470	.122	.083
D	1.948	.046	.024	.174	1.407	.148	.105
E	2.053	.092	.045	.185	1.071	.161	.150
F	2.027	.157	.077	.287	2.413	.153	.064
G	2.261	.044	.019	.175	1.759	.100	.057
H	1.068	.029	.027	.245	1.709	.074	.043
	1.667 ±.464	.080 ±.045	.049 ±.025	.214 ±.046	1.572 ± .555	.138 ±.044	.100 ±.049

[a] $+V_{cf}$: Systolic Circumferential Fiber Velocity (circumferences per sec.)
[b] ϵ_s: Systolic Circumferential Fiber Strain
[c] τ_s: Systolic Time Constant (seconds)
[d] ϵ_{max}: Maximum Circumferential Fiber Strain
[e] $|-V_{cf}|$: Absolute Value of the Diastolic Circumferential Fiber Velocity (circumferences per second)
[f] ϵ_D: Diastolic Circumferential Fiber Strain
[g] τ_D: Diastolic System Time Constant (seconds)

Table 9-V illustrates the seven calculated values for the six subjects with coronary artery disease. There is no statistically significant difference between the mean V_{CF} or mean $|V_{CF}|$ for the two clinical populations due to large scatter of the values. The range of V_{CF} values is 0.648 to 2.261 circumferences per second, and the range of $|-V_{CF}|$ values is from 0.658 to 2.413 circumferences per second. Circumferential fiber strain at peak positive velocity (ϵ_S) is significantly lower for the eight normal subjects than it is for six CAD subjects ($p. < 0.025$). Circumferential fiber strain at peak negative velocity (ϵ_D) is significantly higher for the eight normal subjects compared to the 6 CAD subjects ($p. < 0.05$).

The systolic time constant (τ_S) is significantly lower for the eight normal subjects as compared to the six CAD subjects ($p. < 0.025$). This is consistent since τ_S is proportional to ϵ_S (from eq. 19). There is no significant difference between the mean maximum circumferential fiber strain (ϵ_{max}) or mean τ_D between the eight normal subjects and six CAD subjects.

Table 9-V

CALCULATED PARAMETERS FOR THE SIX CORONARY ARTERY DISEASE PATIENTS

Patient No.	$+V_{CF}$ [a]	ϵ_s	τ_s	ϵ_{max}	$\lvert -V_{CF} \rvert$	ϵ_D	τ_D
I	1.050	.133	.127	.191	1.050	.144	.137
J	.648	.112	.173	.144	.741	.043	.059
K	.926	.102	.110	.131	1.440	.016	.011
L	1.610	.185	.115	.226	2.034	.121	.059
M	2.228	.090	.040	.205	1.238	.099	.080
N	2.027	.173	.085	.223	2.196	.078	.035
	1.415 ±.638	.132 ±.039	.108 ±.044	.187 ±.040	1.500 ±.567	.084 ±.048	.064 ±.043
	N.S.[b]	p < .025	p < .025	N.S.	N.S.	p < .05	N.S.

[a]For nomenclature of the various symbols, see Table 9-IV.
[b]N.S.: No significance (p > 0.05).

Table 9-VI examines the three inequalities between systolic and diastolic velocity, strain and time constant. The systolic/diastolic velocity inequality is satisfied in half the normal subjects (four of the eight). However, the systolic/diastolic strain and time constant inequalities are satisfied in only 25 per cent of the subjects (two of the eight).

Table 9-VII presents the same three inequalities for all six CAD subjects. Each inequality is satisfied in five out of six for all the subjects. The systolic/diastolic strain inequality and the systolic/diastolic time constant inequality are satisfied concurrently in 83 per cent of the CAD subjects.

Table 9-VI

CALCULATED PARAMETER INEQUALITIES FOR THE EIGHT NORMAL PATIENTS

Patient No.	$+V_{CF} \leq \lvert -V_{CF} \rvert$	$\epsilon_D \leq \epsilon_S$	$\tau_D \leq \tau_S$
A	−[a]	−	−
B	+[b]	−	−
C	+	−	−
D	−	−	−
E	−	−	−
F	+	+	+
G	−	−	−
H	+	+	+

[a]Minus sign indicates that the inequality is not satisfied.
[b]Plus sign indicates that the inequality is satisfied.

Table 9-VII

CALCULATED PARAMETER INEQUALITIES FOR THE SIX CORONARY ARTERY DISEASE PATIENTS

Patient No.	$+V_{CF} \leq \lvert -V_{CF} \rvert$	$\epsilon_D \leq \epsilon_S$	$\tau_D \leq \tau_S$
I	+[a]	−	−
J	+	+	+
K	+	+	+
L	+	+	+
M	−	+	+
N	+	+	+

[a]Plus sign indicates that the inequality is satisfied.
[b]Minus sign indicates that the inequality is not satisfied.

Concluding Remarks

In a previous study with cineangiographic data (Phillips, 1977), the $+V_{CF}$ and $|-V_{CF}|$ are somewhat lower ($p. < 0.025$ and $p. < 0.05$, respectively) and the ϵ_S is somewhat higher ($p. < 0.05$) as compared to this present study. However, data processing techniques were quite different between that previous and current study. A filtered fifth order Fourier curve fit (Phillips et al., 1978) was utilized in the previous study, whereas the cardiologist used an empirical eyeball curve fit in our present study. The Fourier technique tends to emphasize the periodicity of a wave, whereas our eyeball technique is more sensitive to instantaneous variations (Phillips, personal observation). Moreover, it is not entirely appropriate to directly compare circumferential fiber strains (either ϵ_S or ϵ_D) between the previous study (Phillips, 1977) and our present one since the previous study utilizes Lagrangian strain and our current study utilizes natural strain. Mirsky (1974) has stated that natural strain is consistently lower than a corresponding Lagrangian strain.

By examining the group means of the two patient populations, it is interesting to note that the systolic time constant (and systolic strain) is significantly lower for the normal group compared to the CAD group ($p. < 0.025$ for both groups). In essense, the normal population has the less viscoelastic compliant, i.e. stiffer, left ventricle than the CAD group. This is precisely what we would expect from activated cardiac muscle in which there is significant cross-bridge coupling between actin and myosin filaments, i.e. normally activated cardiac muscle is stiff. This is reflected by the high modulus of elasticity calculated for activated cardiac muscle by Ghista et al. (1975) and Phillips et al. (1981).

However, we certainly should expect that the CAD group has fibrous (noncontractile) tissue in series and parallel with contractile tissue. It is possible that this slack elastic element in series with the actively contracting portion of the left ventricle could increase cardiac work and myocardial oxygen consumption (Ross, 1974). Certainly it would contribute to the generalized increase in left ventricular systolic viscoelastic compliance (less stiffness).

Admittedly, M-mode echocardiographic studies examine a

relatively small portion of the left ventricle, i.e. provides an "ice pick" view of the heart. Since coronary artery disease and myocardial infarction tend to produce segmented or regional defects in the left ventricle, it might be argued that calculations made on the basis of M-mode echo studies may not necessarily be representative of conditions existing in the left ventricle.

However, we would propose that the regional alterations in viscoelastic compliance of the postinfarction akinetic or dyskinetic segment could affect "global" viscoelasticity since it is *in series* or *in parallel* with the normally functioning contractile elements. Consequently, measurements (as in our study) which represent indices of viscoelastic compliance may be reasonably representative although other global indices, such as blood pressure, stroke volume, or ejection fraction, may not be. As an example, consider a ventricular myocardium as modeled by a contractile element (CE) in series with a series elastic element (SE), i.e. the two-element Hill model (1938). Now, if a part of the SE is made less stiff (the rest of the SE stays the same), the global result when force and length changes are observed at the muscle ends will be that the same forces will be generated and the overall length of shortening will stay the same; however, the *time* to develop those forces and length changes will increase, i.e. it will take longer to transmit CE force and velocity changes through a less stiff SE. That is precisely what we have observed in left ventricles with CAD. The systolic time constant (τ_S) is longer for CAD patients and shorter for normal patients (Tables 9-IV and 9-V) even though blood pressures (force) and ejection fractions (length change) are normal for CAD patients (Table 9-III).

None of the patients in this study had signs or symptoms consistent with ventricular dysfunction (clinical congestive heart failure). However, this does not rule out that a proportion of the surface area of some of the ventricles may be akinetic or dyskinetic (aneurysmic). However, the evidence suggests that when the dyskinetic or akinetic area is *sizeable*, the remainder of the normally contracting myocardium might not be able to sustain normal cardiac output (Ross, 1974). This allows us to at least infer that if aneurysms are present in some patients, they are not of significant size (to encroach on the hemodynamic function by depleting cardiac reserve).

Directing our attention to the system inequalities, five out of six CAD patients have $\epsilon_D \leqslant \epsilon_S$ and/or $\tau_D \leqslant \tau_S$, a condition present in only two of the eight normals. Inequalities compare diastolic viscoelastic compliance (τ_D) occurring during passive filling of the left ventricle) to systolic viscoelastic compliance (τ_S)(occurring during active ejection of blood). Systolic viscoelastic compliance reflects the stiffness of the SE (previously discussed for the two-element Hill model). Diastolic viscoelastic compliance reflects the stiffness of the parallel elastic element (PE) and requires the three-element Maxwell model, which is essentially the Hill 2-element model with a third element (the PE) in parallel with the CE and SE. During passive diastolic filling, the actin-myosin cross-bridges are assumed to uncouple so that the CE is freely distensible. The resting tension of the muscle is born entirely by the PE, which progressively lengthens in response to increasing diastolic filling pressure. However, the PE is many times less stiff (much more highly compliant) than the SE (Sonnenblick, 1964). Thus, when an akinetic or dyskinetic segment, i.e. and area of fibrosis, is part of the PE (the remainder of the PE remains unaltered), the fibrotic segment is now relatively more stiff and actually decreases the overall viscoelastic compliance of the PE. This is consistent with the higher filling pressures needed to attain the same end-diastolic volume seen in ischemic myocardium (Mann et al., 1976).

A stiffer system, i.e. one with lower viscoelastic compliance, is characterized by a shorter system time constant (Doeblin, 1962). An inequality utilizes τ_D, an index of diastolic viscoelastic compliance, which tends to be shorter in CAD patients and compares it with τ_S, an index of systolic viscoelastic compliance, which tends to be longer in CAD patients (Tables 9-IV and 9-V). It is not surprising, therefore, that the inequality $\tau_D \leqslant \tau_S$ is so often satisfied in CAD patients, and also the inequality $\epsilon_D \leqslant \epsilon_S$ (since τ_D and τ_S are linearly proportional to ϵ_D and ϵ_S respectively, *see* equations 19 and 20).

Inequalities appear to be a very good way to characterize patients on an individual basis since there is always some degree of overlap in individual values of ϵ_S and τ_S between the two clinical populations. However, if one takes a value of $\tau_s = 0.080$ seconds, only one normal patient falls above this value, and only one CAD patient falls below this value. Therefore, an absolute cut-off value of τ_s may in reality be the best way to characterize individual patients. Both approaches have been taken in this section, and the final selection of a cut-off τ_s value or an inequality will have to await the analysis of larger statistical patient populations.

In concluding, the velocity-strain relationship apparently characterizes coronary artery disease on an individual basis. Such an ability to characterize individuals is not present with standard hemodynamic parameters (BP, HR, EDV, and EF) or with single indices (such as $\%\Delta M$, V_{CF}, V'_{CF}, or V''_{CF}) due to a large degree of individual overlap between the two groups. The velocity-strain relationship requires measurement of ventricular geometry over an entire cardiac cycle; however, once obtained, the equations are relatively simple and can be processed on most small calculators. Significantly, this preliminary study shows that the velocity-strain relationship can be obtained noninvasively using echocardiography, making it possible to acquire this information during the routine screening of large patient populations.

REFERENCES

Badeer, H.S. (1964) The stimulus to hypertrophy of the myocardium. *Circulation*, 30:128-136

Burton, A.C. (1957) The importance of the shape and size of the heart. *Am. Heart J.*, 54:801-810

Dodge, H.T., and Baxley, W.A. (1969). Left ventricular volume and mass and their significance in heart disease. *Am. J. Cardiology*, 23:528-537

Doeblin, E.O. (1962) *Dynamic Analysis and Feedback Control.* New York, McGraw-Hill Co.

Falsetti, H.L., Mates, R.E., Grant, C., Greene, D.G., and Bunnel, I.L. (1970) Left ventricular wall stress calculated from one-plane cineangiography. *Circulation Research*, 26:71-83

Feigenbaum, J. (1976) *Echocardiography*, 2nd edition. Philadelphia, Lea and Febiger

Gault, J.H., Ross, Jr., J., and Braunwald, E. (1968) Contractile state of the left ventricle in man. *Circulation Research*, 22: 451-463

Ghista, D.N., Vayo, W.H., and Sandler, H. (1975). Elastic modulus of the human intact left ventricle. Determination and physiological interpretation. *Medical and Biological Engineering*, 2: 151-161

Gould, K.L., Lipscomb, K., Hamilton, G.W., and Kennedy, J.W. (1974) The rate of change of left ventricular volume in man. *Am. J. Cardiology*, 34:627-634

Hammermeister, K.E., Brooks, R.C., and Warbasse, J.R. (1974) The rate of change of left ventricular volume in man. *Circulation*, 49:729-738

Hill, A.V. (1938) Heat of shortening and dynamic constants of muscle. *Proc. Roy. Soc. Ser. B.*, 126:136-152

Karliner, J.S., Gault, J.H., Eckberg, D., Mullins, C.B., and Ross, Jr., J. (1971) Mean velocity of fiber shortening. *Circulation*, 44:323-333

Lewis, R.P. and Sandler, H. (1971) Relationship between changes in the left ventricular dimensions and the ejection fraction in man. *Circulation*, 44:548-557

Mack, C. (1967) *Essentials of Statistics for Scientists and Technologists*. New York, Plenum Press

Mann, T., Brodie, B., and McLaurin, L. (1976) The effect of Ischemia on the left ventricular diastolic pressure-volume relationship. *Circulation* (Abst.), 54:II-65

Mirsky, I. (1974) Basic terminology and formulae for left ventricular wall stress. In *Cardiac Mechanics: Physiological, Clinical and Mathematical Correlations*. (ed.) Mirsky, I., Ghista, D., and Sandler, H. New York, Wiley

Mirsky, I. and Parmley, W.W. (1974) Evaluation of the passive elastic stiffness for the left ventricle and isolated heart muscle. In *Cardiac Mechanics: Physiological, Clinical and Mathematical Correlations*. (ed.) Mirsky, I., Ghista, D., and Sandler, H. New York, Wiley

Mirsky, I. (1976) Assessment of passive elastic stiffness of cardiac muscle. *Prog. Cardiovasc. Dis.*, 18:277-308

Phillips, C.A. (1977) The velocity-strain relationship: Application in normal and abnormal left ventricular function. *Annals of Biomedical Engineering*, 5:329-342

Phillips, C.A., Cox, T.L., and Petrofsky, J.S. (1981) Active Material Properties of the Myocardium: Correlation with Left Ventricular Function in Man. *Ohio J. Science*, 81:153-160

Phillips, C.A., Grood, E.S., Mates, R.E., and Falsetti, H.L. (1978) Left ventricular function: Correlation with deformation of the myocardium. *J. Biomechanical Engineering*, 100:99-104

Phillips, C.A. and Petrofsky, J.S. (1981) An echocardiographic study of myocardial wall deformation in left ventricular pressure overload. *J. Clinical Engineering*, 6:213-218

Phillips, C.A. and Petrofsky, J.S. (1982) An echocardiographic study of the velocity-strain relationship in coronary artery disease. *J. Clinical Engineering*, 7:123-130

Ross, J., Jr. (1974) Hemodynamic changes in acute myocardial infarction. In *The Myocardium: Failure and Infarction*. (ed.) Braunwald, E. New York, HP Publishing Co.

Sandler, H., and Dodge, H.T. (1963) Left ventricular tension and stress in man. *Circulation Research*, 13:91-104

Sonnenblick, E.H. (1964) Series elastic and contractile elements in heart muscle: Changes in muscle length. *Am. J. Physiol.*, 207:1330-1338

Walburn, F.J. and Phillips, C.A. (1978) Discussion. *J. Biomechanical Engineering*, 100:104

Wood, R.H. (1892) A few applications of physical theorem to membranes the human body in a state of tension. *J. Anatomical Physiology*, 26:362-370

INDEX

QP
321
.M339

1983